ABEL'S THEOREM IN PROBLEMS AND SOLUTIONS

Abel's Theorem in Problems and Solutions

Based on the lectures of Professor V.I. Arnold

by

V.B. Alekseev
Moscow State University,
Moscow, Russia

KLUWER ACADEMIC PUBLISHERS
DORDRECHT / BOSTON / LONDON

A C.I.P. Catalogue record for this book is available from the Library of Congress.

ISBN 1-4020-2186-0 (HB)
ISBN 1-4020-2187-9 (e-book)

Published by Kluwer Academic Publishers,
P.O. Box 17, 3300 AA Dordrecht, The Netherlands.

Sold and distributed in North, Central and South America
by Kluwer Academic Publishers,
101 Philip Drive, Norwell, MA 02061, U.S.A.

In all other countries, sold and distributed
by Kluwer Academic Publishers,
P.O. Box 322, 3300 AH Dordrecht, The Netherlands.

Printed on acid-free paper

Printed in the Netherlands.

Contents

Preface to the English edition by V.I. Arnold

Abel's Theorem, claiming that there exists no finite combinations of radicals and rational functions solving the generic algebraic equation of degree 5 (or higher than 5), is one of the first and the most important impossibility results in mathematics.

I had given to Moscow High School children in 1963–1964 a (half year long) course of lectures, containing the topological proof of the Abel theorem.

Starting from the definition of complex numbers and from geometry, the students were led to Riemannian surfaces in a sequence of elementary problems. Next came the basic topological notions, such as the fundamental group, coverings, ramified coverings, their monodromies, braids, etc..

These geometrical and topological studies implied such elementary general notions as the transformations groups and group homomorphisms, kernels, exact sequences, and relativistic ideas. The normal subgroups appeared as those subgroups which are relativistically invariant, that is, do not depend on the choice of the coordinate frame, represented in this case as a numbering or labelling of the group elements.

The regular polyhedra symmetry groups, seen from this point of view, had led the pupils to the five Kepler's cubes, inscribed into the dodecahedron. The 12 edges of each of these cubes are the diagonals of the 12 faces of the dodecahedron.

Kepler had invented these cubes in his *Harmonia Mundi* to describe the distances of the planets from the Sun. I had used them to obtain the natural isomorphism between the dodecahedron rotations group and the group of the 60 even permutations of 5 elements (being the Kepler cubes). This elementary theory of regular polyhedra provides the non-solubility proof of the 5 elements permutation group: it can not be constructed from the commutative groups by a finite sequence of the extensions with commutative kernels.

The situation is quite different for the permutation groups of less than five elements, which are soluble (and responsible for the solvability of the equations of degree smaller than 5). This solubility depends on the inscription of two tetrahedra inside the cube (similar to the inscription of the 5 Kepler cubes inside the dodecahedron and mentioned also by Kepler).

The absence of the non-trivial relativistically invariant symmetry subgroups of the group of rotations of the dodecahedron is an easy result of elementary geometry. Combining these High School geometry arguments with the preceding topological study of the monodromies of the ramified coverings, one immediately obtains the Abel Theorem topological proof, the monodromy group of any finite combination of the radicals being soluble, since the radical monodromy is a cyclical commutative group, whilst the monodromy of the algebraic function $x(a)$ defined by the quintic equation $x^5 + ax + 1 = 0$ is the non-soluble group of the 120 permutations of the 5 roots.

This theory provides more than the Abel Theorem. It shows that the insolvability argument is topological. Namely, no function having the same topological branching type as $x(a)$ is representable as a finite combination of the rational functions and of the radicals.

I hope that my topological proof of this generalized Abel Theorem opens the way to many topological insolvability results. For instance, one should prove the impossibility of representing the generic abelian integrals of genus higher than zero as functions topologically equivalent to the elementary functions.

I attributed to Abel the statements that neither the generic elliptic integrals nor the generic elliptic functions (which are inverse functions of these integrals) are topologically equivalent to any elementary function.

I thought that Abel was already aware of these topological results and that their absence in the published papers was, rather, owed to the underestimation of his great works by the Paris Academy of Sciences (where his manuscript had been either lost or hidden by Cauchy).

My 1964 lectures had been published in 1976 by one of the pupils of High School audience, V.B. Alekseev. He has somewhere algebraized my geometrical lectures.

Some of the topological ideas of my course had been developed by A.G. Khovanskii, who had thus proved some new results on the insolvability of the differential equations. Unfortunately, the topological insolvability proofs are still missing in his theory (as well as in the Poincaré theory of the absence of the holomorphic first integral and in many other insolvability problems of differential equations theory).

I hope that the description of these ideas in the present translation of Alekseev's book will help the English reading audience to participate in the development of this new topological insolvability theory, started with the topological proof of the Abel Theorem and involving, say, the topologically non elementary nature of the abelian integrals as well as the topological non-equivalence to the integrals combinations of the complicated differential equations solutions.

The combinatory study of the Kepler cubes, used in the Abel theorem's proof, is also the starting point of the development of the theory of finite groups. For instance, the five Kepler cubes depend on the 5 Hamilton subgroups of the projective version $PSL_2(\mathbb{Z}_5)$ of the group of matrices of order 2 whose elements are residues modulo 5.

A Hamilton subgroup consists of 8 elements and is isomorphic to the group $\{\pm 1, \pm i, \pm j, \pm k\}$ of the quaternionics units.

The peculiar geometry of the finite groups includes their squaring monads, which are the oriented graphs whose vertices are the group elements and whose edges connect every element directly to its square.

The $PSL_2(\mathbb{Z}_5)$ monads theory leads to the unexpected Riemannian surfaces (including the monads as subgraphs), relating Kepler's cubes to the peculiarities of the geometry of elliptic curves.

The $PSL_2(\mathbb{Z}_7)$ extension of the Hamilton subgroups and of Kepler's cubes leads to the extended four colour problem (for the genus one toroidal surface of an elliptic curve), the 14 Hamilton subgroups providing the proof of the 7 colours necessity for the regular colouring of maps of a toroidal surface).

I hope that these recent theories will be developed further by the readers of this book.

V. Arnold

Preface

In high school algebraic equations in one unknown of first and second degree are studied in detail. One learns that for solving these equations there exist general formulae expressing their roots in terms of the coefficients by means of arithmetic operations and of radicals. But very few students know whether similar formulae do exist for solving algebraic equations of higher order. In fact, such formulae also exist for equations of the third and fourth degree. We shall illustrate the methods for solving these equations in the introduction. Nevertheless, if one considers the generic equation in one unknown of degree higher than four one finds that it is not solvable by radicals: there exist no formulae expressing the roots of these equations in terms of their coefficients by means of arithmetic operations and of radicals. This is exactly the statement of the Abel theorem.

One of the aims of this book is to make known this theorem. Here we will not consider in detail the results obtained a bit later by the French mathematician Évariste Galois. He considered some special algebraic equation, i.e., having particular numbers as coefficients, and for these equations found the conditions under which the roots are representable in terms of the coefficients by means of algebraic equations and radicals[1].

From the general Galois results one can, in particular, also deduce the Abel theorem. But in this book we proceed in the opposite direction: this will allow the reader to learn two very important branches of modern mathematics: group theory and the theory of functions of one complex variable. The reader will be told what is a group (in mathematics), a field, and which properties they possess. He will also learn what the complex numbers are and why they are defined in such a manner and not

[1] To those who wish to learn the Galois results we recommend the books: Postnikov M.M., Boron L.F., Galois E., *Fundamentals of Galois Theory*, (Nordhoff: Groningen), (1962); Van der Waerden B.L., Artin E., Noether E., Algebra, (Ungar: New York, N.Y.) (1970).

otherwise. He will learn what a Riemann surface is and of what the 'basic theorem of the complex numbers algebra' consists.

The author will accompany the reader along this path, but he will also give him the possibility of testing his own forces. For this purpose he will propose to the reader a large number of problems. The problems are posed directly within the text, so representing an essential part of it. The problems are labelled by increasing numbers in bold figures. Whenever the problem might be too difficult for the reader, the chapter 'Hint, Solutions, and Answers' will help him.

The book contains many notions which may be new to the reader. To help him in orienting himself amongst these new notions we put at the end of the book an alphabetic index of notions, indicating the pages where their definitions are to be found.

The proof of the Abel theorem presented in this book was presented by professor Vladimir Igorevich Arnold during his lectures to the students of the la 11th course of the physics-mathematics school of the State University of Moscow in the years 1963–64. The author of this book, who at that time was one of the pupils of that class, during the years 1970–71 organized for the pupils of that school a special seminar dedicated to the proof of the Abel theorem. This book consists of the material collected during these activities. The author is very grateful to V.I. Arnold for having made a series of important remarks during the editing of the manuscript.

V.B. Alekseev

Introduction

We begin this book by examining the problem of solving algebraic equations in one variable from the first to the fourth degree. Methods for solving equations of first and second degree were already known by the ancient mathematicians, whereas the methods of solution of algebraic equations of third and fourth degree were invented only in the XVI century.

An equation of the type:

$$a_0x^n + a_1x^{n-1} + \dots a_{n-1}x + a_n = 0,$$

in which $a_0 \neq 0$ [2], is called the *generic algebraic equation of degree* n *in one variable.*

For $n = 1$ we obtain the linear equation

$$a_0x + a_1 = 0.$$

This equation has the unique solution

$$x = -\frac{a_1}{a_0}$$

for any value of the coefficients.

For $n = 2$ we obtain the quadratic equation

$$ax^2 + bx + c = 0, \quad a \neq 0$$

(in place of a_0, a_1, a_2 we write a, b, c, as learnt in school). Dividing both members of this equation by a and putting $p = b/a$ and $q = c/a$ we obtain the reduced equation

$$x^2 + px + q = 0. \qquad (1)$$

[2] For the time being the coefficients $a_0, a_1, \dots, a_n$ may be considered to be arbitrary real numbers.

After some transformations we obtain

$$x^2 + px + \frac{p^2}{4} = \frac{p^2}{4} - q \quad \text{and} \quad \left(x + \frac{p}{2}\right)^2 = \frac{p^2}{4} - q. \qquad (2)$$

In high school one considers only the case $\frac{1}{4}p^2 - q \geq 0$. Indeed, if $\frac{1}{4}p^2 - q < 0$ then one says that Eq. (1) cannot be satisfied and that Eq. (2) has no real roots. In order to avoid these exclusions, in what follows we shall not restrict ourselves to algebraic equations over the field of the real numbers, but we will consider them over the wider field of complex numbers.

We shall examine complex numbers in greater detail (together with their definition) in Chapter 2. In the meantime it is sufficient for the reader to know, or to accept as true, the following propositions about the complex numbers:

1. the set of complex numbers is an extension of the set of real numbers, i.e., the real numbers are contained in the complex numbers, just as, for example, the integer numbers are contained in the real numbers;

2. the complex numbers may be added, subtracted, multiplied, divided, raised to a natural power; moreover, all these operations possess all the basic properties of the corresponding operations on the real numbers;

3. if z is a complex number different from zero, and n is a natural number, then there exist exactly n roots of nth degree of z, i.e., n complex numbers w such that $w^n = z$. For $z = 0$ we have $\sqrt[n]{z} = 0$. If w_1 and w_2 are square roots of the number z then $w_1 = -w_2$.

In the following we shall be interested not only in complex roots of equations as well as in the real ones, but also we will consider arbitrary complex numbers as coefficients of these equations. Hence, the arguments previously expounded about linear and quadratic equations remain true by virtue of what results from property 2 of complex numbers.

Let us continue to study the quadratic equation. In the field of complex numbers for any value of p and q Eq. (2) is equivalent to

$$x + \frac{p}{2} = \pm\sqrt{\frac{p^2}{4} - q}\,,$$

where by $\sqrt{\frac{1}{4}p^2 - q}$ is indicated whichever of the defined values of the square root. In so doing:

$$x_{1,2} = -\frac{p}{2} \pm \sqrt{\frac{p^2}{4} - q}\,. \tag{3}$$

Going back to the coefficients a, b, c we obtain

$$x_{1,2} = \frac{-b \pm \sqrt{b^2 - 4ac}}{2a}\,. \tag{4}$$

For what follows we need to recall two properties related to the equations of second degree.

1. *Viète's Theorem*[3]: The complex numbers x_1 and x_2 are the roots of the equation $x^2+px+q=0$ if and only if $x_1+x_2=-p$ and $x_1 \cdot x_2 = q$. Indeed, if x_1 and x_2 are roots of the equation $x^2+px+q=0$ then Eq. (3) is satisfied, from which $x_1+x_2=-p$ and $x_1 \cdot x_2 = q$. Conversely, if $x_1+x_2=-p$ and $x_1 \cdot x_2 = q$, then, substituting p and q in the equation $x^2+px+q=0$ by their expressions in terms of x_1 and x_2, we obtain $x^2-(x_1+x_2)x+x_1x_2=(x-x_1)(x-x_2)=0$, and therefore x_1 and x_2 are roots of the equation $x^2+px+q=0$;

2. The quadratic trinomial ax^2+bx+c is a perfect square, i.e.,

   $$ax^2+bx+c=[\sqrt{a}(x-x_0)]^2$$

   for some complex number x_0 if and only if the roots of the equation ax^2+bx+c coincide (they must be both equal to x_0). This happens if and only if $b^2-4ac=0$ (see formula (4)). The expression b^2-4ac is called the *discriminant* of the quadratic trinomial.

We consider now the reduced cubic equation

$$x^3+ax^2+bx+c=0. \tag{5}$$

The generic equation of third degree is reduced to Eq. (5) by dividing by a_0. After the substitution $x=y+d$ (where d will be chosen later) we obtain

$$(y+d)^3+a(y+d)^2+b(y+d)+c=0.$$

[3]François Viète (1540-1603) was a French mathematician.

Removing the brackets and collecting the terms of the same degree in y we obtain the equation

$$y^3 + (3d + a)y^2 + (3d^2 + 2ad + b)y + (d^3 + ad^2 + bd + c) = 0.$$

The coefficient of y^2 in this equation is equal to $3d + a$. Therefore if we put $d = -a/3$, after substituting $x = y - a/3$ we transform the equation into:

$$y^3 + py + q = 0\ , \tag{6}$$

where p and q are some polynomials in a, b and c.

Let y_0 be a root of Eq. (6). Representing it in the form $y_0 = \alpha + \beta$ (where α and β are temporarily unknown) we obtain

$$\alpha^3 + 3\alpha\beta(\alpha + \beta) + \beta^3 + p(\alpha + \beta) + q = 0$$

and

$$\alpha^3 + \beta^3 + (\alpha + \beta)(3\alpha\beta + p) + q = 0. \tag{7}$$

We check whether it is possible to impose that α and β satisfy

$$\alpha\beta = -\frac{p}{3}.$$

In this case we obtain two equations for α and β:

$$\begin{cases} \alpha + \beta = y_0, \\ \alpha\beta = -p/3. \end{cases}$$

By Viète's theorem, for any y_0 such α and β (which may be complex) indeed exist, and they are the roots of the equation

$$w^2 - y_0 w - \frac{p}{3} = 0.$$

If we take such (still unknown) α and β, then Eq. (7) is transformed into

$$\alpha^3 + \beta^3 + q = 0. \tag{8}$$

Raising either terms of the equation $\alpha\beta = -p/3$ to the third power, and comparing the obtained equation with Eq. (8), we have

$$\begin{cases} \alpha^3 + \beta^3 = -q, \\ \alpha^3 \cdot \beta^3 = -p^3/27. \end{cases}$$

By Viète's theorem α^3 and β^3 are the roots of the equation

$$w^2 + qw - \frac{p^3}{27} = 0.$$

In this way

$$\alpha^3 = -\frac{q}{2} + \sqrt{\frac{q^2}{4} + \frac{p^3}{27}} \quad \text{and} \quad \beta^3 = -\frac{q}{2} - \sqrt{\frac{q^2}{4} + \frac{p^3}{27}},$$

where again $\sqrt{(q^2/4) + (p^3/27)}$ indicates one defined value of the square root. Hence the roots of Eq. (6) are expressed by the formula

$$y_{1,2,3} = \sqrt[3]{-\frac{q}{2} + \sqrt{\frac{q^2}{4} + \frac{p^3}{27}}} + \sqrt[3]{-\frac{q}{2} - \sqrt{\frac{q^2}{4} + \frac{p^3}{27}}},$$

in which for each of the three values of the first cubic root[4] one must take the corresponding value of the second, in such a way that condition $\alpha\beta = -p/3$ be satisfied.

The obtained formula is named *Cardano's formula*[5]. Substituting in this formula p and q by their expressions in terms of a, b, c and subtracting $a/3$, we obtain the formula for Eq. (5). After the transformations $a = a_1/a_0$, $b = a_2/a_0$, $c = a_3/a_0$, we obtain the formula for the roots of the generic equation of third degree.

Now we examine the reduced equation of fourth degree

$$x^4 + ax^3 + bx^2 + cx + d = 0. \tag{9}$$

(the generic equation is reduced to the previous one by dividing by a_0). By making the change of variable $x = y - a/4$, similarly to the change made in the case of the equation of third degree, we transform Eq. (9) into

$$y^4 + py^2 + qy + r = 0, \tag{10}$$

where p, q and r are some polynomials in a, b, c, d.

We shall solve Eq. (10) by a method called *Ferrari's method*[6]. We transform the left term of Eq. (10) in the following way:

$$\left(y^2 + \frac{p}{2}\right)^2 + qy + \left(r - \frac{p^2}{4}\right) = 0,$$

[4]See the aforementioned Property 3 of complex numbers.

[5]G. Cardano (1501-1576) was an Italian mathematician.

[6]L. Ferrari (1522–1565) was an Italian mathematician, a pupil of Cardano.

and

$$\left(y^2 + \frac{p}{2} + \alpha\right)^2 - \left[2\alpha\left(y^2 + \frac{p}{2}\right) + \alpha^2 - qy + \frac{p^2}{4} - r\right] = 0, \qquad (11)$$

where α is an arbitrary number. We try now to determine α such that the polynomial of second degree in y

$$2\alpha y^2 - qy + \left(\alpha p + \alpha^2 + \frac{p^2}{4} - r\right),$$

within square brackets becomes a perfect square. As was noticed above, in order for it to be a perfect square it is necessary and sufficient that the discriminant of this polynomial vanish, i.e.,

$$q^2 - 8\alpha\left(\alpha p + \alpha^2 + \frac{p^2}{4} - r\right) = 0. \qquad (12)$$

Eliminating the brackets, to find α we obtain an equation of third degree which we are able to solve. If in place of α we put one of the roots of Eq. (12) then the expression in the square brackets of (11) will be a perfect square. In this case the left member of Eq. (11) is a difference of squares and therefore it can be written as the product of two polynomials of second degree in y. After that it remains to solve the two equations of second degree obtained.

Hence the equation of fourth degree can always be solved. Moreover, as in the case of the third order, it is possible to obtain a formula expressing the roots of the generic equation of fourth order in terms of the coefficients of the equation by means of the operations of addition, subtraction, multiplication, division, raising to a natural power, and extracting a root of natural degree.

For a long time mathematicians tried to find a method of solution by radicals of the generic equation of fifth order. But in 1824 the Norwegian mathematician Niels Henrik Abel (1802–1829) proved the following theorem.

Abel's Theorem. *The generic algebraic equation of degree higher than four is not solvable by radicals, i.e., formulæ do not exist for expressing roots of a generic equation of degree higher than four in terms of its coefficients by means of operations of addition, subtraction, multiplication, division, raising to a natural power, and extraction of a root of natural degree.*

We will be able to prove this theorem at the end of this book. But for this we need some mathematical notions, such as those of group, so-luble group, function of a complex variable, Riemann surface, etc.. The reader will become familiar with all these and other mathematical instruments after reading what follows in the forthcoming pages of this book. We start by examining the notion of group, a very important notion in mathematics.

Chapter 1

Groups

1.1 Examples

In arithmetic we have already met operations which put two given numbers in correspondence with a third. Thus, the addition puts the pair (3,5) in correspondence with the number 8 and the pair (2,2) with the number 4. Also the operation of subtraction if considered inside the set of all integer numbers, puts every pair of integers in correspondence with an integer: in this case, however, the order of numbers in the pair is important. Indeed, subtraction puts the pair (5,3) in correspondence with the number 2, whereas the pair (3,5) with the number -2. The pairs (5,3) and (3,5) thus have to be considered as different.

When the order of elements is specified a pair is said to be *ordered.*

DEFINITION. Let M be a set of elements of arbitrary nature. If every ordered pair of elements of M is put into correspondence with an element of M we say that in M a *binary operation* is defined.

For example, the addition in the set of natural numbers and the subtraction in the set of integer numbers are binary operations. The subtraction is not a binary operation in the set of natural numbers because, for example, it cannot put the pair (3,5) in correspondence with any natural number.

1. Consider the following operations: a) addition; b) subtraction; c) multiplication; in the following sets: 1) of all even natural numbers; 2) of all odd natural numbers; 3) of all negative integer numbers. In which cases does one obtain a binary operation[1]?

[1] Part of the problems proposed in the sequel has a practical character and is aimed

Let us still consider some examples of binary operations. We shall often return to these examples in future.

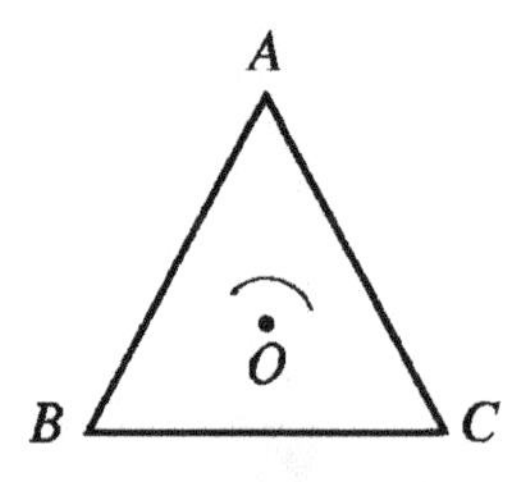

FIGURE 1

EXAMPLE 1. Let A, B, and C be the vertices of an equilateral triangle (Figure 1). We rotate the triangle by an angle of 120° around its centre O in the direction shown by the arrow. Then vertex A goes over vertex B, B over C, and C over A. In this way the final triangle coincides with the initial triangle (if we neglect the labels of the vertices). We say that the rotation by 120° around the point O is a transformation which sends the triangle into itself. We denote this transformation by a. We can write it in the form $a = \begin{pmatrix} A & B & C \\ B & C & A \end{pmatrix}$, where the first row contains all vertices of the triangle, and the second row indicates where each vertex is sent. A rotation by 240° in the same direction around the point O is also a transformation sending the triangle into itself. Denote this transformation by b, $b = \begin{pmatrix} A & B & C \\ C & A & B \end{pmatrix}$. There still exists one transformation sending the triangle into itself, and which is different from a and b: it is the rotation by 0°. We denote it by e: thus $e = \begin{pmatrix} A & B & C \\ A & B & C \end{pmatrix}$. It is easy to see that there are only three different rotations of the plane[2] transforming an equilateral triangle ABC into itself, namely e, a and b.

Let g_1 and g_2 be two arbitrary transformations of the triangle. Then we denote by $g_1 \cdot g_2$ (or simply $g_1 g_2$) the transformation g_3 obtained by carrying out first the transformation g_2 and later the transformation g_1; g_3 is called the *product* or *composition* of the transformations g_2 and g_1.

It is possible to make the *multiplication table* (Table 1) where every row, as well as every column, corresponds to some rotation transforming

at a better comprehension of notions by means of examples. The other problems are theoretical, and their results will be used later on. Therefore if the reader is unable to solve some problems, he must read their solutions in the Section Hints, Solutions, and Answers.

[2]We mean rotation of the plane around one axis perpendicular to the plane.

the triangle ABC into itself. We put the transformation corresponding to $g_1 \cdot g_2$ in the intersection of the row corresponding to the transformation g_2 with the column corresponding to the transformation g_2. So, for example, in the selected cell of Table 1 we have to put the transformation $a \cdot b$, which is obtained by first rotating the triangle by 240° and later by 120° more. Hence $a \cdot b$ is a rotation by 360°, i.e., it coincides with e. We obtain the same result by the following reasoning: transformation b sends vertex A onto vertex C, and a later sends C onto A. In this way the transformation $a \cdot b$ sends A onto A. In exactly the same way we obtain that B is sent onto B, and C onto C. Hence $ab = \binom{A\ B\ C}{A\ B\ C}$, i.e., $ab = e$.

Table 1

	e	a	b
e			
a			e
b			

2. Complete Table 1.

Any transformation of some geometrical figure into itself which maintains the distances between all its points is called a *symmetry* of the given figure. So the rotations of the equilateral triangle, considered in Example 1, are symmetries of it.

EXAMPLE 2. Besides rotations, the equilateral triangle still possesses 3 symmetries, namely, the reflections with respect to the axes l_1, l_2 and l_3 (Figure 2). We denote these transformations by c, d, and f, so that $c = \binom{A\ B\ C}{A\ C\ B}$, $d = \binom{A\ B\ C}{C\ B\ A}$, $f = \binom{A\ B\ C}{B\ A\ C}$. Here it is possible to imagine the composition of two transformations in two different ways. Consider,

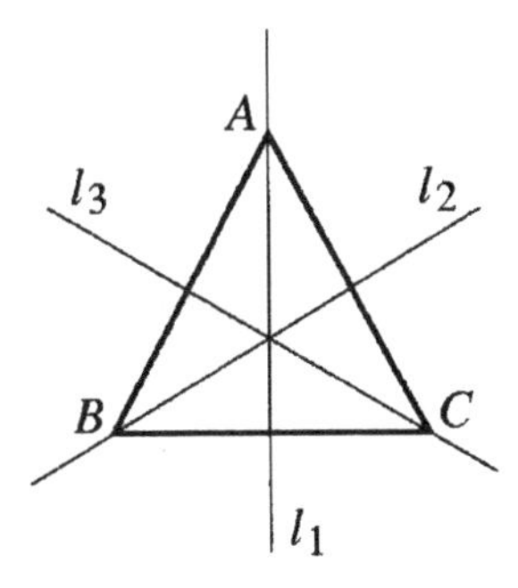

FIGURE 2

for example, the composition $c \cdot d$. We can imagine that the axis l_1 is sent by the transformation d into a new position (i.e., in the original position of the axis l_3), and after this, consider the transformation c as the reflection with respect to the new position of the axis l_1 (i.e., with respect to the original axis l_3). On the other hand, it is also possible to consider that the axes are not rigidly fixed to the figure, and that they do not move with it; therefore in the example which we examine, after the transformation d the transformation c is done as the reflection with respect to the original axis l_1. We will consider the compositions of two transformations in exactly this way. With this choice the reasoning about the vertices of the figure, analogously to the arguments presented immediately before Problem 2, is correct. It is convenient to utilize such arguments to calculate the multiplication table.

3. Write the multiplication table for all symmetries of the equilateral triangle.

EXAMPLE 3. Let e, a, b and c denote the rotations of a square by 0°, 180°, 90° and 270° in the direction shown by the arrow (Figure 3).

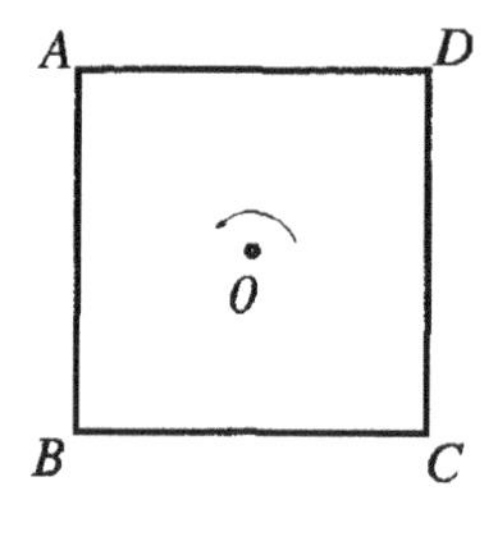

FIGURE 3

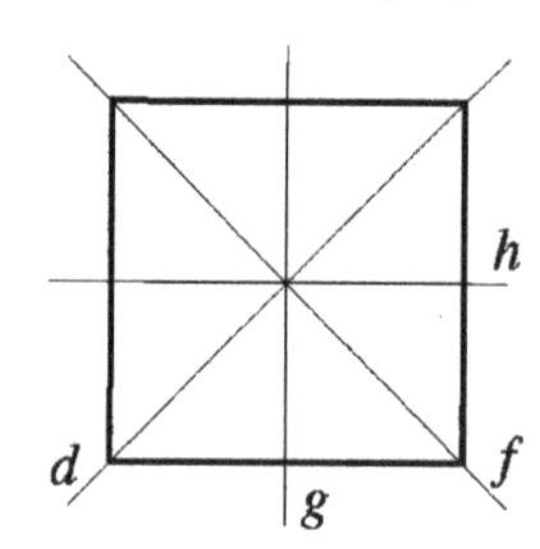

FIGURE 4

4. Write the multiplication table for the rotations of the square.
EXAMPLE 4. Let d, f, g and h denote the reflections of the square with respect to the axes shown in Figure 4.

5. Write the multiplication table for all symmetries of the square.

EXAMPLE 5. Let $ABCD$ be a rhombus, which is not a square.

6. Find all symmetries of the rhombus and write their multiplication table.

EXAMPLE 6. Let $ABCD$ be a rectangle, which is not a square.

7. Find all symmetries of the rectangle and write their multiplication table.

1.2 Groups of transformations

Let X and Y be two sets of elements of arbitrary nature, and suppose that every element x of X is put into correspondence with a defined element y of Y. Thus one says that there exists a *mapping* ϕ of the set X into the set Y: $\phi : X \to Y$. The element y is called the *image* of the element x, and x the *pre-image* of element y. One writes: $\phi(x) = y$.

DEFINITION. The mapping $\phi : X \to Y$ is called *surjective* (or, equivalently, a mapping of set X *onto* set Y) if for every element y of Y there exists an element x of X such that $\phi(x) = y$, i.e., every y of Y has a pre-image in X.

8. Let the mapping ϕ put every capital city in the world in correspondence with the first letter of its name in English (for example, ϕ(London) = L). Is ϕ a mapping of the set of capitals **onto** the entire English alphabet?

DEFINITION. The mapping $\phi : X \to Y$ is called a *one to one* (or *bijective*) mapping of the set X into the set Y if for every y in Y there exists a pre-image in X and this pre-image is unique.

9. Consider the following mappings of the set of all integer numbers into the set of the non-negative integer numbers:

$$\text{a) } \phi(n) = n^2; \quad \text{b) } \phi(n) = |n|; \quad \text{c) } \phi(n) = \begin{cases} 2n & \text{if } n \geq 0, \\ 2|n| - 1 & \text{if } n < 0. \end{cases}$$

Which amongst these mappings are surjective, which are bijective?

Let M be an arbitrary set. For brevity we shall call any bijective mapping of M into itself a *transformation* of set M.

Two transformations g_1 and g_2 will be considered equal if $g_1(A) = g_2(A)$ for every element A of M. Instead of term 'transformation' the term *permutation* is often used. We shall use this term only when the transformation is defined on a finite set. A permutation can thus be written in the form

$$\begin{pmatrix} A_1 & A_2 & \dots & A_n \\ A_{i_1} & A_{i_2} & \dots & A_{i_n} \end{pmatrix},$$

where the first row contains all the elements of the given set, and the second row indicates all the corresponding images under the permutation.

Since the transformation is one to one, for every transformation there exists the *inverse transformation* g^{-1}, which is defined in the following way: if $g(A) = B$ then $g^{-1}(B) = A$. So in Example 1 $a = \begin{pmatrix} A & B & C \\ B & C & A \end{pmatrix}$. Therefore $a^{-1} = \begin{pmatrix} A & B & C \\ C & A & B \end{pmatrix}$, i.e., $a^{-1} = b$.

10. Find the inverse transformations of all symmetries of the equilateral triangle (see Examples 1 and 2).

11. Consider the transformation of all real numbers given by $g(x) = 2x$. Find the inverse transformation.

The *multiplication of the transformations* g_1 and g_2 is defined as $g_1g_2(A) = g_1(g_2(A)) =$ (the transformation g_2 is done first, g_1 afterwards). If g_1 and g_2 are transformations of the set M then g_1g_2 is also a transformation of set M.

DEFINITION. Suppose that a set G of transformations possesses the following properties: 1) if two transformations g_1 and g_2 belong to G, then their product g_1g_2 also belongs to G; 2) if a transformation g belongs to G then its inverse transformation g^{-1} belongs to G. In this case we call such a set of transformations a *group of transformations.*

It is not difficult to verify that the sets of transformations considered in Examples 1–6 are, in fact, groups of transformations.

12. Prove that any group of transformations contains the identical transformation e such that $e(A) = A$ for every element A of the set M.

13. Prove that $eg = ge = g$ for any transformation g.

14. Prove that for any three transformations g_1, g_2, and g_3 the following equality holds[3]:

$$(g_1g_2)g_3 = g_1(g_2g_3).$$

1.3 Groups

To solve Problems **6** and **7** we wrote the multiplication tables for the symmetries of the rhombus and of the rectangle. It has turned out that in our

[3]This equality is true not only for transformations but also for any three mappings g_1, g_2, and g_3 such that $g_3 : M_1 \to M_2$, $g_2 : M_2 \to M_3$, $g_1 : M_3 \to M_4$

notations (see the solutions) these tables coincide. For many purposes it is natural to consider such groups of transformations as coinciding. Therefore we shall consider abstract objects rather than sets of real elements (in our case of transformations). Furthermore, we shall consider those binary operations on arbitrary sets which possess the basic properties of the binary operation in a group of transformations. Thus any binary operation will be called a multiplication; if to the pair (a, b) there corresponds the element c, we call c the product of a and b, and we write $ab = c$. In some special cases the binary operation will be called differently, for example, composition, addition, etc..

DEFINITION. A set G of elements of an arbitrary nature, on which one can define a binary operation such that the following conditions are satisfied, is called a *group*:

1) *associativity* : $(ab)c = a(bc)$ for any elements a, b and c of G;

2) in G there is an element e such that $ae = ea = a$ for every element a of G; such element e is called the *unit* (or *neutral element*) of group G;

3) for every element a of G there is in G an element a^{-1} such that $aa^{-1} = a^{-1}a = e$: such an element is called the *inverse of element* a.

From the results of Problems **12**–**14** we see that any group of transformations is a group (in some sense the converse statement is also true (see **55**)). In this way we have already seen a lot of examples of groups. All these groups contain a finite number of elements: such groups are called *finite* groups. The number of elements of a finite group is called the *order of the group*. Groups containing an infinite number of elements are called *infinite* groups.

Let us give some examples of infinite groups.

EXAMPLE 7. Consider the set of all integer numbers. In this set we shall take as binary operation the usual addition. We thus obtain a group. Indeed, the role of the unit element is played by 0, because $0 + n = n + 0 = n$ for every integer n. Moreover, for every n there exists the inverse element $-n$ (which is called in this case the *opposite element*), because $n + (-n) = (-n) + n = 0$. The associativity follows from the rules of arithmetic. The obtained group is called the *group of integers under addition.*

15. Consider the following sets: 1) all the real numbers; 2) all the real numbers without zero. Say whether the sets 1 and 2 form a group under multiplication.

16. Say whether all real positive numbers form a group under multiplication.

17. Say whether all natural numbers form a group: a) under addition; b) under multiplication.

18. Prove that in every group there exists one unique unit element.

19. Prove that for every element a of a group there exists one unique inverse element a^{-1}.

20. Prove that: 1) $e^{-1} = 1$; 2) $(a^{-1})^{-1} = a$.

If a and b are elements of a group then by the definition of binary operation the expression $a \cdot b$ gives some defined element of the group. Hence also expressions like $(a \cdot b) \cdot c$, $a \cdot (b \cdot c)$, $(a \cdot b) \cdot (c \cdot d)$ give some defined elements of the group. Any two of the obtained elements can be multiplied again, obtaining again an element of the group, and so on. Therefore, in order to set up uniquely at every step which operation will be performed at the next step we shall put into brackets the two expressions which have to be multiplied (we may not enclose in brackets the expressions containing only one letter). We call all expressions that we can write in this way *well arranged expressions.* For example $(a \cdot b) \cdot (c \cdot (a \cdot c))$ is a well arranged expression, whereas $(a \cdot b) \cdot c \cdot (c \cdot d)$ is not well arranged, because it is not clear in which order one has to carry out the operations. When we consider the product $a_1 \cdot a_2 \cdot \ldots \cdot a_n$ of the real numbers $a_1, a_2, \ldots, a_n$, we do not put any bracket, because the result does not depend on the order in which the operations are carried out — i.e., for every arrangement of the brackets giving a well arranged expression the result corresponding to this product is the same. It turns out that this property is satisfied by any group, as follows from the result of the next question.

21. Suppose that a binary operation $a \cdot b$ possesses the associativity property, i.e., $(a \cdot b) \cdot c = a \cdot (b \cdot c)$ for any elements a, b, c. Prove that every well arranged expression in which the elements from left to right are $a_1, a_2, \ldots, a_n$ gives the same element as the multiplication

$$(\ldots((a_1 \cdot a_2) \cdot a_3) \cdot \ldots \cdot a_{n-1}) \cdot a_n).$$

In this way if the elements $a_1, a_2, \ldots, a_n$ are elements of a group then all the well arranged expressions containing elements $a_1, a_2, \ldots, a_n$ in this order and distinguished only by the disposition of brackets give the same element, which we will denote by $a_1 \cdot a_2 \cdot \ldots \cdot a_{n-1} \cdot a_n$ (eliminating all brackets).

The multiplication of real numbers possesses yet another important property: the product $a_1 \cdot a_2 \cdot \ldots \cdot a_{n-1} \cdot a_n$ does not change if the factors are permuted arbitrarily. However, not all groups possess this property.

DEFINITION. Two elements a and b of a group are called *commuting* if $ab = ba$. (One says also that a and b *commute.*) If in a group any two elements commute, the group is said to be *commutative* or *abelian.*

There exist non-commutative groups. Such a group is, for example, the group of symmetries of the triangle (see Example 2, where $ac = f$, $ca = d$ i.e., $ac \neq ca$).

22. Say whether the following groups are commutative (see **2**, **4**–**7**): 1) the group of rotations of the triangle; 2) the group of rotations of the square; 3) the group of symmetries of the square; 4) the group of symmetries of a rhombus; 5) the group of symmetries of a rectangle.

23. Prove that in any group:
1)$(ab)^{-1} = b^{-1}a^{-1}$; 2) $(a_1 \cdot \ldots \cdot a_n)^{-1} = a_n^{-1} \cdot \ldots \cdot a_1^{-1}$.

REMARK. The jacket is put on after the shirt, but is taken off before it.

If a certain identity $a = b$ holds in a group G (a and b being two expressions giving the same element of G) then one obtains a new identity by multiplying the two members of the initial identity by an arbitrary element c of the group G. However, since in a group the product may depend on the order of its factors, one can multiply the two members of the identity by c either on the left (obtaining $ca = cb$) or on the right (obtaining $ac = bc$).

24. Let a, b be two arbitrary elements of a group G. Prove that each one of the equations $ax = b$ and $ya = b$ has one and only one solution in G.

The uniqueness of the solution in Problem **24** can be also enunciated in this way: if $ab_1 = ab_2$ or $b_1a = b_2a$ then $b_1 = b_2$.

25. Let us suppose that $a \cdot a = e$ for every element a of a group G. Prove that G is commutative.

Let a be an arbitrary element of a group G. We will denote by a^m the product $a \cdot a \cdot \ldots \cdot a$, where m is the number of factors, all equal to a.

26. Prove that $(a^m)^{-1} = (a^{-1})^m$, where m is an integer.

In this way $(a^m)^{-1}$ and $(a^{-1})^m$ for every integer m indicate the same element, which we will denote by a^{-m}. Moreover, $a^0 = e$ for every element a.

27. Prove that $a^m \cdot a^n = a^{m+n}$ for any integers m and n.

28. Prove that $(a^m)^n = a^{mn}$ for any integers m and n.

1.4 Cyclic groups

The simplest groups are the cyclic groups. They are, however, very important.

DEFINITION. Let a be an element of a group G. The smallest integer n such that the element $a^n = e$ is called the *order of the element a*. If such an integer does not exist one says that a is an *element of infinite order*.

29. Find the order of all elements of the groups of symmetries of the equilateral triangle, of the square and of the rhombus (see **3**,**5**,**6**).

30. Let the order of an element a be equal to n. Prove that: 1) elements $e, a, a^2, \ldots, a^{n-1}$ are all distinct; 2) for every integer m the element a^m coincides with one of the elements listed above.

DEFINITION. If an element a has order n and in a group G there are no other elements but $e, a, a^2, \ldots, a^{n-1}$, the group G is called the *cyclic group of order n, generated by the element a*, and the element a is called a *generator* of the group.

EXAMPLE 8. Consider a regular n-gon (polygon with n sides) and all rotations of the plane that transform the n-gon into itself.

31. Prove that these rotations form a cyclic group of order n.

32. Find all generators in the group of rotations of the equilateral triangle and in the group of rotations of the square (see Examples 1 and 3 in §1.1).

33. Let the order of an element a be equal to n. Prove that $a^m = e$ if and only if $m = nd$, where d is any integer.

34. Suppose that the order of an element a is equal to a prime number p and that m is an arbitrary integer. Prove that either $a^m = e$ or a^m has order p.

35. Suppose that d is the maximal common divisor of the integers m and n and that a has order n. Prove that the element a^m has order n/d.

36. Find all generators of the group of rotations of the regular dodecagon.

37. Let a be an element of infinite order. Prove that the elements $\dots, a^{-2}, a^{-1}, e, a, a^2, \dots$ are all distinct.

DEFINITION. If a is an element of infinite order and group G has no other elements but $\dots, a^{-2}, a^{-1}, e, a, a^2, \dots$, then G is called an *infinite cyclic group* and a its *generator*.

38. Prove that the group of the integers is a cyclic group under addition (see Example 7, §1.3). Find all generators.

EXAMPLE 9. Let n be an integer different from zero. Consider all the possible remainders of the division of integers by n, i.e., the numbers $0, 1, 2, \dots, n-1$. Let us introduce in this set of remainders the following binary operation. After adding two remainders as usually, we keep the remainder of the division by n of the obtained sum. This operation is called the *addition modulo n*. So we have, summing modulo 4, $1+2=3$, but $3+3=2$.

39. Write the multiplication table for the addition modulo: a) 2; b) 3; c) 4.

40. Prove that the set of remainders with the addition modulo n form a group, and that this group is a cyclic group of order n.

Consider again an arbitrary cyclic group of order n: $e, a, a^2, \dots, a^n$.

41. Prove that $a^m \cdot a^r = a^k$, where $0 \le m < n$, $0 \le r < n$ and $0 \le k < n$ if and only if modulo n one has $m + r = k$.

From the result of the preceding problem it follows that to the multiplication of the elements in an arbitrary cyclic group there corresponds the addition of the remainders modulo n. Similarly to the multiplication of two elements in an infinite cyclic group there corresponds the addition of integers (see **7**). We come in this way to an important notion in the theory of groups: the notion of isomorphism.

1.5 Isomorphisms

DEFINITION. Let two groups G_1 and G_2 be given with a bijective mapping

ϕ from G_1 into G_2 (see §1.2) with the following property: if $\phi(a) = a'$, $\phi(b) = b'$, $\phi(c) = c'$ and $ab = c$ in group G_1, then $a'b' = c'$ in group G_2. In other words, to the multiplication in G_1 there corresponds under ϕ the multiplication in G_2. The mapping ϕ is thus called an *isomorphism* between groups G_1 and G_2, and the groups G_1 and G_2 are said to be *isomorphic*. The condition for a bijective mapping ϕ to be an isomorphism can also be expressed by the following condition: $\phi(ab) = \phi(a) \cdot \phi(b)$ for all elements a and b of group G_1; here the product ab is taken in the group G_1 and the product $\phi(a) \cdot \phi(b)$ in the group G_2.

42. Which amongst the following groups are isomorphic: 1) the group of rotations of the square; 2) the group of symmetries of the rhombus; 3) the group of symmetries of the rectangle; 4) the group of remainders under addition modulo 4?

43. Let $\phi : G_1 \to G_2$ be an isomorphism. Prove that the inverse mapping $\phi^{-1} : G_2 \to G_1$ is an isomorphism.

44. Let $\phi_1 : G_1 \to G_2$ and $\phi_2 : G_2 \to G_3$ be two isomorphisms. Prove that the compound mapping $\phi_2\phi_1 : G_1 \to G_3$ is an isomorphism.

From the two last problems it follows that two groups which are isomorphic to a third group are isomorphic to each other.

45. Prove that every cyclic group of order n is isomorphic to the group of the remainders of the division by n under addition modulo n.

46. Prove that every infinite cyclic group is isomorphic to the group of integers under addition.

47. Let $\phi : G \to F$ be an isomorphism. Prove that $\phi(e_G) = e_F$, where e_G and e_F are the unit elements in groups G and F.

48. Let $\phi : G \to F$ be an isomorphism. Prove that $\phi(g^{-1}) = [\phi(g)]^{-1}$, for every element g of group G.

49. Let $\phi : G \to F$ be an isomorphism and let $\phi(g) = h$. Prove that g and h have the same order.

If we are interested in the group operation and not in the nature of the elements of the groups (which, in fact, does not play any role), then we can identify all groups which are isomorphic. So, for example, we shall say that there exists, up to isomorphism, only one cyclic group of order n (see **45**), which we denote by $\mathbb{Z}_n$, and only one infinite cyclic group (see **46**), which we indicate by $\mathbb{Z}$.

If a group G_1 is isomorphic to a group G_2 then we write $G_1 \cong G_2$.

50. Find (up to isomorphism) all groups containing: a) 2 elements; b) 3 elements.

51. Give an example of two non-isomorphic groups with the same number of elements.

52. Prove that the group of all real numbers under addition is isomorphic to the group of the real positive numbers under multiplication.

53. Let a be an arbitrary element of a group G. Consider the mapping ϕ_a of a group G into itself defined in the following way: $\phi_a(x) = ax$ for every x of G. Prove that ϕ_a is a permutation of the set of the elements of group G (i.e., a bijective mapping of the set of the elements of G into itself).

54. For every element a of a group G let ϕ_a be the permutation defined in Problem **53**. Prove that the set of all permutations ϕ_a forms a group under the usual law of composition of mappings.

55. Prove that group G is isomorphic to the group of permutations defined in the preceding problem.

1.6 Subgroups

In the set of the elements of a group G consider a subset H. It may occur that H is itself a group under the same binary operation defined in G.

In this case H is called a *subgroup* of the group G. For example, the group of rotations of the regular n-gon is a subgroup of the group of all symmetries of the n-gon.

If a is an element of a group G, then the set of all elements of type a^m is a subgroup of G (this subgroup is cyclic, as we have seen in §1.4).

56. Let H be a subgroup of a group G. Prove that: a) the unit elements in G and in H coincide; b) if a is an element of subgroup H, then the inverse elements of a in G and in H coincide.

57. Prove that in order for H to be a subgroup of a group G (under the same binary operation) the following conditions are necessary and sufficient: 1) if a and b belong to H then the element ab (product in group G) belongs to H; 2) e (the unit element of group G) belongs to H; 3) if a belongs to H then also a^{-1} (taken in group G) belongs to H.

Remark. Condition 2 follows from conditions 1 and 3.

58. Find all subgroups of the following groups: 1) of symmetries of the equilateral triangle, 2) of symmetries of the square.

59. Find all subgroups of the following cyclic groups: a) $\mathbb{Z}_5$; b) $\mathbb{Z}_8$; c) $\mathbb{Z}_{15}$.

60. Prove that all subgroups of $\mathbb{Z}_n$ have the form $\{e, a^d, a^{2d}, \ldots, a^{n-d}\}$, where d divides n and a is a generator of the group $\mathbb{Z}_n$.

61. Prove that all subgroups of an infinite cyclic group are of the type $\{\ldots, a^{-2r}, a^{-r}, e, a^r, a^{2r}, \ldots\}$, where a is a generator and r is an arbitrary non zero integer number.

62. Prove that an infinite cyclic group has an infinite number of subgroups.

63. Prove that the intersection of an arbitrary number of subgroups[4] of a group G is itself a subgroup of group G.

EXAMPLE 10. Consider a regular tetrahedron, with vertices marked with the letters A, B, C, and D. If we look at the triangle ABC from point the D, then the rotation defined by the cyclic order of points A, B, C may be a clockwise or counterclockwise rotation (see Figure 5). We shall distinguish these two different orientations of the tetrahedron.

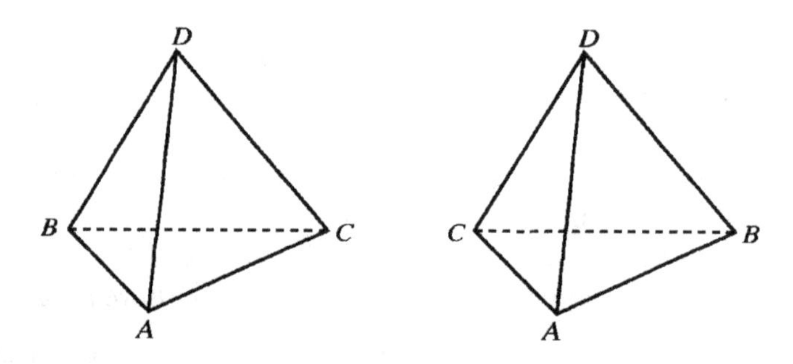

FIGURE 5

64. Is the orientation of the tetrahedron preserved by the following permutations: $a = \begin{pmatrix} A & B & C & D \\ B & C & A & D \end{pmatrix}$ (rotation by 120° around the alti-

[4]The intersection of many sets is the set of all elements belonging at the same time to all the sets.

tude); $b = \begin{pmatrix} A & B & C & D \\ D & C & B & A \end{pmatrix}$ (rotation by 180° around the axis through the middle points of the edges AD and BC); $c = \begin{pmatrix} A & B & C & D \\ A & C & B & D \end{pmatrix}$ (reflection with respect to the plane containing edge AD and the middle point of edge BC); $d = \begin{pmatrix} A & B & C & D \\ B & C & D & A \end{pmatrix}$ (cyclic permutation of the vertices)?

All symmetries of the regular tetrahedron obviously form a group, which is called the *group of symmetries of the tetrahedron.*

65. How many elements does the group of symmetries of tetrahedron contain?

66. In the group of symmetries of the tetrahedron find the subgroups isomorphic to: a) the group of symmetries of the equilateral triangle; b) the cyclic group $\mathbb{Z}_4$.

67. Prove that all symmetries of the tetrahedron preserving its orientation form a subgroup. How many elements does it contain?

The group of symmetries of the tetrahedron preserving its orientation is called the *group of rotations of the tetrahedron.*

68. Find in the group of rotations of the tetrahedron the subgroups isomorphic to: a) $\mathbb{Z}_2$; b) $\mathbb{Z}_3$.

1.7 Direct product

Starting from two groups one may define a third group.

DEFINITION. The *direct product* $G \times H$ of groups G and H is the set of all the ordered pairs (g, h), where g is any element of G and h any element of H, with the following binary operation: $(g_1, h_1) \cdot (g_2, h_2) = (g_1 g_2, h_1 h_2)$, where the product $g_1 g_2$ is taken in the group G, and $h_1 h_2$ in the group H.

69. Prove that $G \times H$ is a group.

70. Suppose that a group G has n elements, and that a group H has k elements. How many elements does the group $G \times H$ contain?

71. Prove that the groups $G \times H$ and $H \times G$ are isomorphic.

72. Find the subgroups of $G \times H$ isomorphic to the groups G and H.

73. Let G and H be two commutative groups. Prove that the group $G \times H$ is also commutative.

74. Let G_1 be a subgroup of a group G and H_1 a subgroup of a group H. Prove that $G_1 \times H_1$ is a subgroup of the group $G \times H$.

75. Let G and H be two arbitrary groups. Is it true that every subgroup of the group $G \times H$ can be represented in the form $G_1 \times H_1$, where G_1 is a subgroup of the group G and H_1 a subgroup of the group H?

76. Prove that the group of symmetries of the rhombus is isomorphic to the group $\mathbb{Z}_2 \times \mathbb{Z}_2$.

77. Is it true that: 1) $\mathbb{Z}_2 \times \mathbb{Z}_3 \cong \mathbb{Z}_6$; 2) $\mathbb{Z}_2 \times \mathbb{Z}_4 \cong \mathbb{Z}_8$?

78. Prove that $\mathbb{Z}_m \times \mathbb{Z}_n \cong \mathbb{Z}_{mn}$ if and only if the numbers m and n are relatively prime.

1.8 Cosets. Lagrange's theorem

For every subgroup H of a group G there exists a partition of the set of the elements of G into subsets. For each element x of G consider the set of all elements of the form xh, where h runs over all elements of a subgroup H. The set so obtained, denoted by xH, is called the *left coset of* H (or *left lateral class of* H) *in* G, *generated by the element* x.

79. Find all left cosets of the following subgroups of the group of symmetries of the equilateral triangle: a) the subgroup of rotations of the triangle; b) the group generated by the reflection with respect to a single axis $\{e, c\}$ (see Examples 1 and 2, §1.1).

80. Prove that given a subgroup H of a group G each element of G belongs to one left coset of H in G.

81. Suppose that an element y belongs to the left coset of H generated by an element x. Prove that the left cosets of H generated by elements x and y coincide.

82. Suppose that the left cosets of H, generated by elements x and y, have a common element. Prove that these left cosets coincide.

Hence left cosets generated by two arbitrary elements either are disjoint or coincide. In this way we have obtained a partition of all elements of a group G into disjoint classes. Such a partition is called the *left partition of the group G by the subgroup H.*

The number of elements of a subgroup is called the *order of the subgroup.* Let m be the order of a subgroup H. If h_1 and h_2 are two different elements of H then $xh_1 \neq xh_2$. Every left coset thus contains m elements. Hence if n is the order of group G and r is the number of the left cosets of the partition of G by H, then $m \cdot r = n$ and we have proved the following theorem.

THEOREM 1. (Lagrange's theorem[5]) The order of a subgroup H of a group G divides the order of the group G.

83. Prove that the order of an arbitrary element (see definition in §1.4) divides the order of the group.

84. Prove that a group whose order is a prime number is cyclic and that every element of it different from the unit is its generator.

85. Suppose that a group G contains exactly 31 elements. How many subgroups does it contain?

86. Let p be a prime number. Prove that all groups having the same order p are isomorphic to each other.

87. Suppose that m divides n. Obtain a group of order n containing a subgroup isomorphic to a given group G of order m.

88. Suppose that m divides n. Is it possible that a group of order n does not contain any subgroup of order m?

One can obtain as well the *right cosets Hx* and the *right partition* of a group G by a subgroup H. If the order of a subgroup H is equal to m, then each right coset contains m elements and the number of cosets is equal to the integer n/m, where n is the order of the group. Hence the number of right cosets coincides with the number of the left cosets.

89. Find the left and the right partitions of the group of symmetries of the equilateral triangle by the following subgroups: a) the subgroup of rotations $\{e, a, b\}$; b) the subgroup $\{e, c\}$, generated by the reflection with respect to one axis.

[5]Joseph Louis Lagrange (1736–1813), French mathematician.

90. Find the left and right partitions of the group of symmetries of the square by the following subgroups: a) the subgroup $\{e, a\}$, generated by the central symmetry; b) the subgroup $\{e, d\}$, generated by the reflection with respect to one diagonal.

91. Find the partition of the group of all integers (under addition)[6] by the subgroup of the numbers divisible by 3.

92. Find all groups (up to isomorphism) of order: a) 4; b) 6; c) 8.

1.9 Internal automorphisms

Let us start with an example. Consider the group of symmetries of the equilateral triangle. If the letters A, B, and C denote the vertices of the triangle, then each element of the group defines a permutation of the three letters. For example, the reflection of the triangle with respect to the altitude drawn from the vertex A to the base BC will be written in the form $\begin{pmatrix} A & B & C \\ A & C & B \end{pmatrix}$. To multiply two elements of the group of symmetries of the triangle it suffices to carry out the corresponding permutations one after the other. In this way we obtain an isomorphism between the group of symmetries of the triangle and the group of permutations of letters A, B, and C.

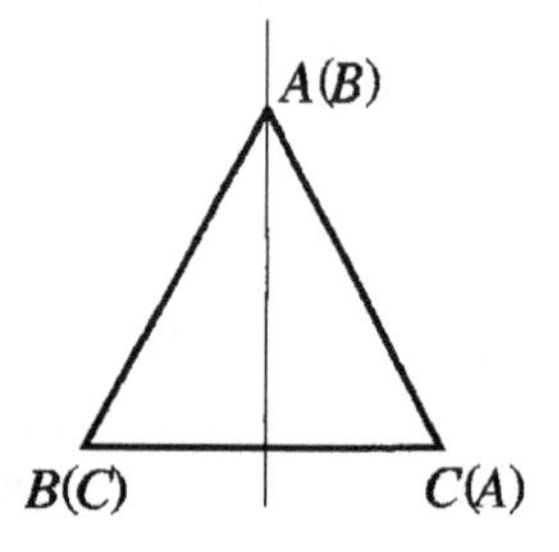

FIGURE 6

Now we observe that this isomorphism is not uniquely defined: it depends on which vertex is named A, which B, and which C. The change of notations of the three vertices of the triangle may be also considered

[6]Here we do not mention the type of the partition (left or right) because in commutative groups the two partitions obviously coincide.

as a permutation of the three letters A, B and C. For example $g = \begin{pmatrix} A & B & C \\ B & C & A \end{pmatrix}$ corresponds to the following change of notations:

old notation	A	B	C
new notation	B	C	A

By the new notations of vertices each element of the group of symmetries of the triangle obtains a new notation in terms of permutation of letters A, B, C. For example, the reflection of the triangle with respect to its vertical altitude (Figure 6) is written in the following way:

$$\text{old notation} \quad \begin{pmatrix} A & B & C \\ A & C & B \end{pmatrix}$$

$$\text{new notation} \quad \begin{pmatrix} A & B & C \\ C & B & A \end{pmatrix}$$

93. Consider an element of the group of symmetries of the triangle, to which, for some notation of vertices, there corresponds a permutation h. What permutation will correspond to the same element of the group of symmetries of the triangle after the change of notation g of the vertices?

We observe now that by the change of notation, g, an element h of a group of transformations is sent to ghg^{-1} not only in the example considered of the group of symmetries of the triangle, but also in the more general case. So the study of the changes of notations leads to the following definition.

DEFINITION. Let G be a group and g one of its elements. Define the mapping ϕ_g of the group G into itself by the formula $\phi_g(h) = ghg^{-1}$ (where h is any element of the group). This mapping is called the *internal automorphism* of group G generated by the element g.

94. Prove that an internal automorphism of a group is an isomorphism of the group into itself.

95. Which is the image of the reflection of the triangle with respect to its altitude under all possible internal automorphisms of the group of symmetries of the triangle?

96. Which is the image of the rotation by 120° of the triangle under all possible internal automorphisms of the group of symmetries of the triangle?

97. Which are the pairs of elements of the group of symmetries of the tetrahedron that can be sent one on the other by an internal automorphism? Which elements can not? The same question for the group of rotations of the tetrahedron.

98. Prove that in any group the orders of elements ab and ba are equal.

Note that the image of a subgroup under any internal automorphism of the whole group (as well as under any isomorphism) is, in general, different (for example, the reflection with respect to one altitude of the triangle is sent to the reflection with respect to another altitude). However, some subgroups, 'super-symmetric', are invariants under all internal automorphisms (for example, the subgroup of rotations of the triangle in the group of symmetries of the triangle). We will study these subgroups in the next section.

1.10 Normal subgroups

DEFINITION. A subgroup of a group is called a *normal subgroup* if it is mapped onto itself by all internal automorphisms of the group. In other words, a subgroup N of a group G is called a normal subgroup of G if for every element a of N and for every element g of G the element gag^{-1} belongs to N.

Hence in the group of symmetries of a triangle ABC the subgroup of rotations is normal, whereas the group generated by the reflection with respect to the altitude drawn from A to the base BC (containing two elements) is not normal.

99. Prove that in a commutative group every subgroup is normal.

100. Is it true that the subgroup of the group of symmetries of the square, consisting of elements $\{e, a\}$, where a is the central symmetry (see Example 3,4 §1.1), is normal?

THEOREM 2. *A subgroup N of a group G is normal if and only if the left and the right partitions of group G by the subgroup N coincide*[7].

101. Prove Theorem 2.

---

[7] In this case the partition obtained is called the partition by the normal subgroup.

102. Let n be the order of a group G, m the order of a subgroup H and $m = n/2$. Prove that H is a normal subgroup of the group G.

103. Prove that the intersection (see footnote to Problem **63**) of an arbitrary number of normal subgroups of a group G is a normal subgroup of the group G.

DEFINITION. The set of elements of a group G which commute with all elements of the group is called the *centre* of group G.

104. Prove that the centre of a group G is a subgroup and, moreover, a normal subgroup of the group G.

105. Let N_1 and N_2 be two normal subgroups of two groups G_1 and G_2, respectively. Prove that $N_1 \times N_2$ is a normal subgroup of the group $G_1 \times G_2$.

The following example shows that a normal subgroup of a subgroup of a group G can be a non-normal subgroup of the group G.

EXAMPLE 11. Consider the subgroup of the group of symmetries of the square, generated by the reflections with respect to the diagonals and to the centre (see Examples 3,4 §1.1, the subgroup $\{e, a, d, f\}$). This subgroup contains one half of the elements of the group of symmetries of the square, and it is therefore a normal subgroup (see **102**). The subgroup $\{e, d\}$, generated by the reflection with respect to one of the diagonals, contains one half of the elements of the subgroup $\{e, a, d, f\}$, and it is therefore a normal subgroup of this subgroup. But the subgroup $\{e, d\}$ is not a normal subgroup of the group of symmetries of the square, because d is sent by an internal automorphism to the reflection with respect to the other diagonal: $bdb^{-1} = f$.

1.11 Quotient groups

Let us start with an example. Consider the partition of the group of symmetries of the square by the normal subgroup e, a, generated by the central symmetry a (see Example 3,4 §1.1). It is easy to see that the partition of our group into four cosets has the form shown in Table 2.

Table 2

e	a	b	g
a	c	f	g
E	A	B	C

Denote each coset by a letter, for example E, A, B, and C. If one multiplies an arbitrary element of coset A by an arbitrary element of coset B, then the result belongs to one and only one coset C, independently of the particular elements chosen in cosets A and B. From the solution of the next problem it follows that this result is not casual.

106. Let there be given the partition of group G by the normal subgroup N. Suppose that the elements x_1 and x_2 belong to one coset and that the elements y_1 and y_2 belong to another coset. Prove that the elements x_1y_1 and x_2y_2 belong to the same coset.

In this way, multiplying in a given order two elements representant of two different cosets, one obtains an element of a coset which does not depend on the chosen representant elements. As a consequence, given a partition of a group by a normal subgroup, we can define a binary operation in the following way: whenever $A = xN$, $B = yN$, we write $A \cdot B = (xy)N$ ($A \cdot B$ is also denoted by AB). The result of Problem **106** shows that this operation is uniquely defined and does not depend on the elements x and y which generate the cosets A and B. So in the above example we have $A \cdot B = C$.

In problems **107**–**109** the subgroups have to be considered as normal.

107. Let T_1, T_2, T_3 be three cosets. Prove that $(T_1T_2)T_3 = T_1(T_2T_3)$.

108. Let E be a normal subgroup. Prove that $ET = TE = T$ for every coset T.

109. Prove that for every coset T there exists a coset T^{-1} such that $TT^{-1} = T^{-1}T = E$.

From problems **107**–**109** it follows that the set of all cosets with the binary operation just defined forms a group. This group is called the *quotient group of the group G by the normal subgroup N* and is denoted by G/N.

It is evident that $G/\{e\} \cong G$ and $G/G \cong \{e\}$. It is evident as well that the order of the quotient group is equal to the integer n/m, where n is the order of the group and m the order of the normal subgroup. For example, the quotient group of the group of symmetries of the square by the subgroup generated by the central symmetry contains 4 elements.

110. Calculate whether the quotient group of the group of symmetries of the square by the subgroup generated by the central symmetry is isomorphic to the group of rotations of the square or to the group of symmetries of the rhombus.

111. Find all normal subgroups and the corresponding quotient groups of the following groups[8]: a) the group of symmetries of the triangle; b) $\mathbb{Z}_2 \times \mathbb{Z}_2$; c) the group of symmetries of the square; d) the group of quaternions (see solution of Problem **92**).

112. Describe all normal subgroups and the corresponding quotient groups of the following groups: a) $\mathbb{Z}_n$; b) $\mathbb{Z}$.

113. Find all normal subgroups and the corresponding quotient groups of the group of rotations of the tetrahedron.

114. Consider the subgroup $G_1 \times \{e\}$ in the direct product $G_1 \times G_2$. Prove that it is a normal subgroup and that the corresponding quotient group is G_2.

1.12 Commutant

Recall that two elements a and b of a group G are said to be *commuting* if $ab = ba$. The degree of non-commutativity of two elements of a group can be measured by the product $aba^{-1}b^{-1}$, which is equal to the unit element if and only if a and b commute.

DEFINITION. The element $aba^{-1}b^{-1}$ is called the *commutator* of the elements a and b. The set of all possible products of a finite number of commutators of a group G is called the *commutant* of the group G and it is denoted by $K(G)$.

115. Prove that the commutant is a subgroup.

116. Prove that the commutant is a normal subgroup.

117. Prove that the commutant coincides with the unit element $\{e\}$ if and only if the group is commutative.

118. Find the commutant in the following groups: a) of symmetries of the triangle; b) of symmetries of the square; c) the group of quaternions (see solution of Problem **92**).

119. Prove that the commutant in the group of symmetries of the regular n-gon is isomorphic to the group $\mathbb{Z}_n$ if n is odd and to the group $\mathbb{Z}_{n/2}$ if n is even.

---

[8]In the sequel finding the quotient group will mean showing a group, among those already considered, to be isomorphic to the group requested.

120. Find the commutant in the group of symmetries of the tetrahedron.

121. Prove that if a normal subgroup of the group of rotations or of the group of symmetries of the tetrahedron contains just a sole rotation around one axis through a vertex, then it contains all rotations of the tetrahedron.

122. Find the commutant in the group of symmetries of the tetrahedron.

Consider two groups: the *group of rotations of the cube* and the *group of rotations of the regular octahedron.*

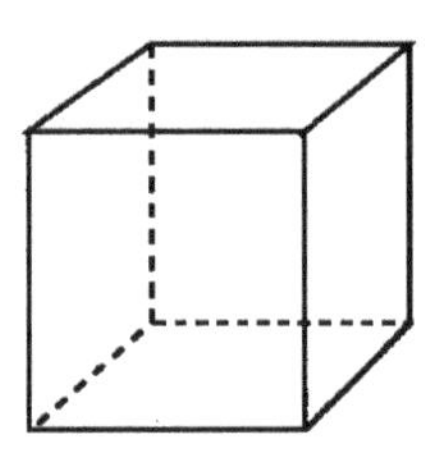

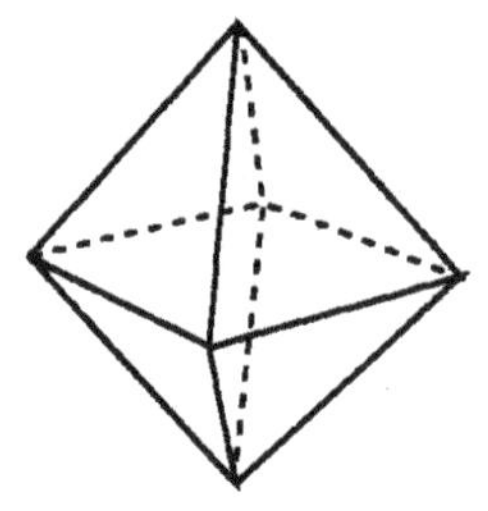

FIGURE 7

123. How many elements are contained in these groups? Calculate the elements of the group of rotations of the cube.

124. Prove that the groups of rotations of the cube and of the octahedron are isomorphic.

125. In how many different ways is it possible to colour the surface of a cube with 6 colours (a different colour for each face) if one considers two coloured cubes as different if they do not coincide even after some rotation? The same question for a box of matches.

126. Which group amongst those you know is isomorphic to the group of rotations of a box of matches?

To calculate the commutant of the group of rotations of the cube we inscribe in the cube a tetrahedron (see Figure 8).

Joining the remaining vertices B, D, A_1, and C_1 one obtains a second tetrahedron. Any rotation of the cube either sends each tetrahedron onto itself or exchanges the tetrahedra with each other.

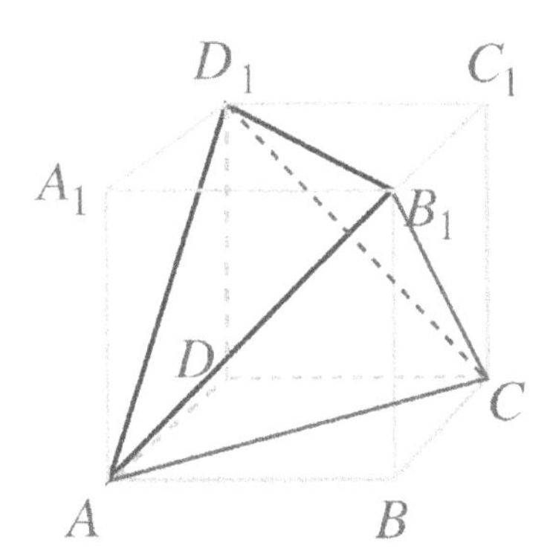

FIGURE 8

127. Prove that all rotations of the cube sending each tetrahedron onto itself form: a) a subgroup; b) a normal subgroup of the group of rotations of the cube.

128. Prove that the commutant of the group of rotations of the cube is isomorphic to the group of rotations of the tetrahedron.

We now prove three properties of the commutant which will be of use later on.

129. Prove that the quotient group of an arbitrary group G by its commutant is commutative.

130. Let N be a normal subgroup of a group G and let the quotient group G/N be commutative. Prove that N contains the commutant of the group G.

131. Let N be a normal subgroup of a group G and $K(N)$ the commutant of a subgroup N. Prove that $K(N)$ is a normal subgroup of G (compare with Example 11, §1.10).

1.13 Homomorphisms

Let G and F be two groups. A mapping $\phi : G \to F$ such that $\phi(ab) = \phi(a)\phi(b)$ for all elements a and b of the group G (here the product ab is taken in G and $\phi(a) \cdot \phi(b)$ in F) is called a *homomorphism* of G into F. Homomorphisms are distinguishable from isomorphisms because the homomorphisms are not necessarily bijective.

EXAMPLE 12. Let G be the group of rotations of the cube, and $\mathbb{Z}_2$ the group of permutations of the two tetrahedra, inscribed inside the cube (see §1.12). To each rotation of the cube there corresponds a well defined permutation of tetrahedra. When we carry out two rotations of the cube one after the other, the permutation of the tetrahedra so obtained is the product of the permutations of the tetrahedra corresponding to these rotations. Therefore the mapping of the group of rotations of the cube into the group of permutations of two tetrahedra is a homomorphism.

132. Let $\phi : G \to F$ be a surjective homomorphism of a group G onto a group F. Prove that if the group G is commutative then F is commutative. Is the converse proposition true?

133. Prove that a homomorphism of a group G into a group F sends the unit of the group G onto the unit of the group F.

134. Prove that $\phi(a^{-1}) = [\phi(a)]^{-1}$, where $\phi : G \to F$ is a homomorphism. Note that the inverse element appearing in the left member of the equation is taken in the group G, whereas in the right member it is taken in the group F.

135. Let $\phi_1 : G \to F$ and $\phi_2 : F \to H$ be two homomorphisms. Prove that $\phi_2 \circ \phi_1 : G \to H$ is a homomorphism.

Important examples of homomorphisms are obtained by means of the construction of the 'natural homomorphism'. Let N be a normal subgroup of a group G. Consider the mapping ϕ of the group G into the quotient group G/N which sends each element g of the group G to a coset T of N containing the element g.

136. Prove that $\phi : G \to G/N$ is a surjective homomorphism of G onto G/N.

DEFINITION. The surjective mapping ϕ is called the *natural homomorphism* of a group G into the quotient group G/N.

We have proved that to every normal subgroup there corresponds a homomorphism. We shall now prove that, inversely, every homomorphism surjective of a group G onto a group F can be seen as a natural homomorphism of G onto the quotient group G/N by a suitable normal subgroup N.

DEFINITION. Let $\phi : G \to F$ be a homomorphism. The set of elements g such that $\phi(g) = e_F$ is called the *kernel of the homomorphism* ϕ and is denoted by $\ker \phi$.

137. Prove that $\ker\phi$ is a subgroup of group G.

138. Prove that $\ker\phi$ is a normal subgroup of group G.

Consider the partition of G by the kernel $\ker\phi$.

139. Prove that g_1 and g_2 belong to the same coset if and only if $\phi(g_1) = \phi(g_2)$.

THEOREM 3. *Let $\phi : G \to F$ be a surjective homomorphism of a group G onto a group F. The mapping $\psi : G/\ker\phi \to F$ sending each coset to the image by ϕ of a certain element of the coset (and thus of an arbitrary element (see* **139***)), is an isomorphism.*

The proof of this theorem is contained in the solutions of the following problems.

140. Prove that ψ is surjective.

141. Prove that ψ is bijective.

142. Prove that ψ is an isomorphism.

We will consider some applications of this theorem.

EXAMPLE 13. Problem **110** asked whether the quotient group of the group of symmetries of the square by the normal subgroup generated by the central symmetry is isomorphic either to the group of rotations of the square or to the group of symmetries of the rhombus. To each element of the group of symmetries of the square there corresponds some permutation of the axes of symmetry l_1, l_2, l_3, l_4 (Figure 9). This permutation can just exchange between each other the diagonals l_1 and l_3, as well as the axes l_2 and l_4.

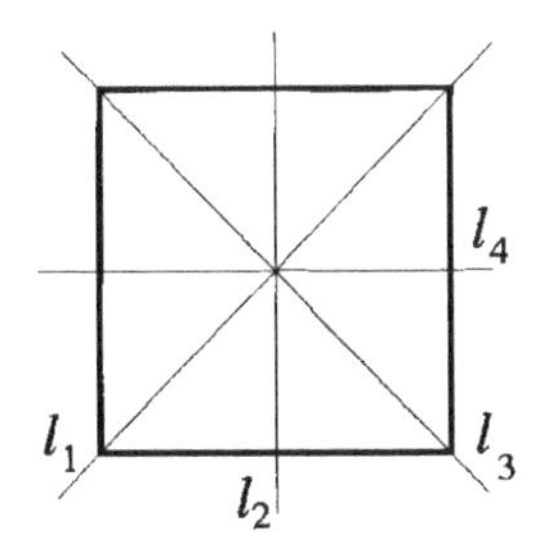

FIGURE 9

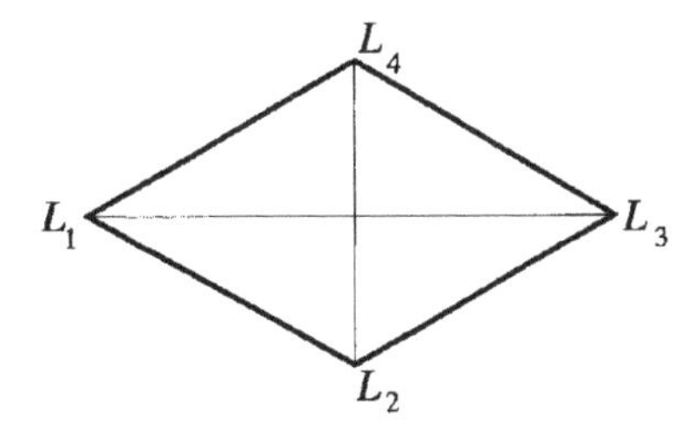

FIGURE 10

We thus obtain a mapping of the group of symmetries of the square into a group of permutations of four elements l_1, l_2, l_3 and l_4. This mapping is a homomorphism surjective onto the group of those permutations which send $\{l_1, l_3\}$ to $\{l_1, l_3\}$, and $\{l_2, l_4\}$ to $\{l_2, l_4\}$ (verify). This group consists of four permutations and is isomorphic to the group of symmetries of the rhombus $L_1L_2L_3L_4$ (Figure 10).

The kernel of the homomorphism so obtained contains all symmetries of the square sending each axis of symmetry onto itself. It is not difficult to verify that these transformations are just e and the central symmetry a. Therefore by Theorem 3 the subgroup $\{e, a\}$ is a normal subgroup of the group of symmetries of the square, and the corresponding quotient group is isomorphic to the group of symmetries of the rhombus.

The following problems may be solved in a similar way.

143. Prove that the rotations of the tetrahedron by 180° around the axes through the middle points of opposite edges form, together with the identity, a normal subgroup of the group of symmetries of the tetrahedron. Find the corresponding quotient group.

144. Prove that the rotations of the cube by 180° around the axes through the centres of opposite faces form, together with the identity, a normal subgroup of the group of rotations of the cube. Find the corresponding quotient group.

145. Let there be given on the plane a regular n-gon with centre O. Let R be the group of rotations of the plane around the point O. Consider the subgroup $\mathbb{Z}_n$ of the rotations of the plane sending the regular n-gon to itself. Prove that this subgroup is a normal subgroup of R and that $R/\mathbb{Z}_n$ is isomorphic to R.

146. Let N_1 and N_2 be two normal subgroups of groups G_1 and G_2, respectively. Prove that $N_1 \times N_2$ is a normal subgroup of $G_1 \times G_2$ and that $(G_1 \times G_2)/(N_1 \times N_2) \cong (G_1/N_1) \times (G_2/N_2)$.

147. Is it possible that two normal subgroups of two non-isomorphic groups are isomorphic to each other, and that the corresponding quotient groups are isomorphic?

148. Is it possible that two normal subgroups of the same group are isomorphic and that the corresponding quotient groups are not isomorphic?

149. Is it possible that two normal subgroups of the same group are not isomorphic and that the corresponding quotient groups are isomorphic?

We now observe what happens to subgroups, to normal subgroups, and to commutants under the action of a homomorphism. Let $\phi : G \to F$ be a homomorphism. Chose in G a subset M. The set of the elements of F having at least a pre-image in M is called the *image of the set* M by the homomorphism ϕ (denoted by $\phi(M)$). Conversely, let P be a subset of F; the set of all elements of G having an image in P is called the *pre-image* of P (denoted by $\phi^{-1}(P)$). Note that the symbol ϕ^{-1} has no meaning outside P: a homomorphism, in general, has no inverse mapping. Note also that if $\phi(M) = P$, then $\phi^{-1}(P)$ is contained in M, but it does not necessarily coincide with M (see Figure 11).

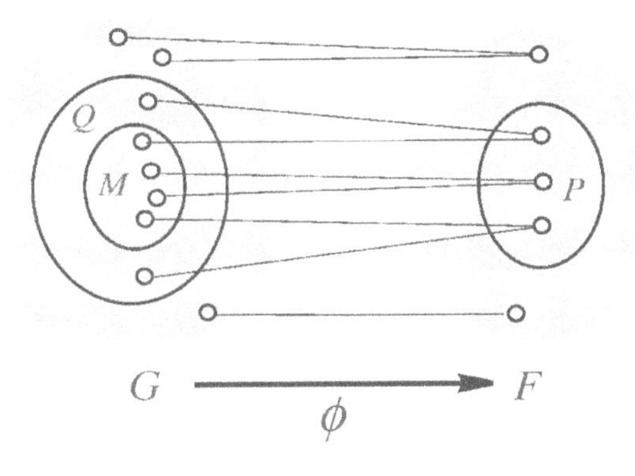

FIGURE 11

150. Prove that the image of a subgroup H of a group G under a homomorphism $\phi : G \to F$ is a subgroup of the group F.

151. Let H be a subgroup of F and $\phi : G \to F$ a homomorphism. Prove that $\phi^{-1}(H)$ is a subgroup of G.

152. Let N be a normal subgroup of a group F and $\phi : G \to F$ a homomorphism. Prove that $\phi^{-1}(N)$ is a normal subgroup of the group G.

153. Let $\phi : G \to F$ be a homomorphism, K_1 and K_2 the commutants of G and F. Prove that $\phi(K_1)$ is contained in K_2 and that K_1 is contained in $\phi^{-1}(K_2)$.

154. Let N be a normal subgroup of a group G and $\phi : G \to F$ a homomorphism surjective of group G onto a group F. Prove that $\phi(N)$ is a normal subgroup of F.

155. Let K_1 and K_2 be the commutants of groups G and F and $\phi : G \to F$ a surjective homomorphism of G onto F. Prove that $\phi(K_1) = K_2$. Is it true that $K_1 = \phi^{-1}(k_2)$?

1.14 Soluble groups

There exist an important class of groups which is similar to the commutative groups: that of soluble groups. This appellation comes from the possibility of solving algebraic equations by radicals depends on the solubility of some groups, as we will see in the next chapter.

Let G be a group and $K(G)$ its commutant. The commutant $K(G)$ is itself a group, and one can consider its commutant $K(K(G))$ as well. In the group obtained one can again consider the commutant, etc.. One obtains the group $\underbrace{K(K(\ldots(K}_{r}(G))\ldots))$, in short $K_r(G)$. So $K_{r+1}(G) = K(K_r(G))$.

DEFINITION. A group G is said to be *soluble* if the sequence of groups $G, K(G), K_2(G), K_3(G), \ldots$ ends, for a finite n, with the unit group, i.e., for some n one has $K_n(G) = \{e\}$.

For example, all commutative groups are soluble, because if G is commutative, then at the first step one already has $K(G) = \{e\}$. A group G is also soluble whenever its commutant is commutative, because in this case $K_2(G) = \{e\}$.

156. Say whether the following groups are soluble or not: a) the cyclic group $\mathbb{Z}_n$; b) the group of symmetries of the equilateral triangle; c) the group of symmetries of the square; d) the group of quaternions (see **92**); e) the group of rotations of the tetrahedron; f) the group of symmetries of the tetrahedron; g) the group of rotations of the cube.

All groups considered in Problem **156** are soluble. It is thus natural to ask whether there exist in general non-soluble groups. We will prove that the *group of rotations of the regular dodecahedron* (Figure 12) is not soluble.

157. How many elements are contained in the group of rotations of the dodecahedron?

All rotations of the dodecahedron can be divided into four classes: 1) the identity transformation; 2) rotations around the axes through the centres of opposite faces; 3) rotations around the axes through opposite

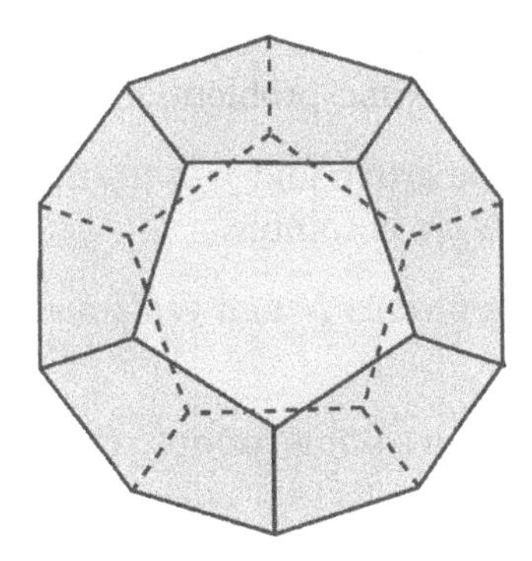

FIGURE 12

vertices; 4) rotations around the axes through the middle points of opposite edges.

158. How many elements are contained in each class (without counting the identity transformation in classes 2–4)?.

159. Let N be an arbitrary normal subgroup of the group of rotations of the dodecahedron and suppose that N contains at least one element of one among classes 1–4. Prove that N contains the entire class of this element.

As a consequence each one of classes 1–4 either belongs entirely to N or has no elements in common with N.

160. Prove that in the group of rotations of the dodecahedron there are no other normal subgroups except $\{e\}$ and the whole group.

161. Suppose that a group G is not commutative and that it has no normal subgroups other than $\{e\}$ and G. Prove that G is not soluble.

From problems **160** and **161** it follows that the group of rotations of the dodecahedron is not soluble.

We shall consider some more problems whose results will be of use later on.

162. Prove that every subgroup of a soluble group is soluble.

163. Let $\phi : G \to F$ a homomorphism surjective of a group G onto a group F and suppose that group G is soluble. Prove that the group F is also soluble.

164. Give an example in which the group F is soluble whereas the group G is not (see the preceding problem).

165. Let G be a soluble group and N a normal subgroup of G. Prove that the quotient group G/N is soluble.

166. Prove that if the groups N and G/N are soluble then the group G is soluble.

167. Let G and F be two soluble groups. Prove that the group $G \times F$ is soluble.

168. Let G be a soluble group. Prove that there exists a sequence of groups $G_0, G_1, \ldots, G_n$ such that: 1) $G_0 = G$; 2) each group G_i $(1 \leq i \leq n)$ is a normal subgroup of the group G_{i-1}, and all quotient groups G_{i-1}/G_i are commutative; 3) the group G_n is commutative.

169. Suppose that for a group G there exists a sequence of groups with the properties described in the preceding problem. Prove that the group G is soluble.

The results of Problems **168** and **169** show that for a group G the existence of a sequence of groups with the properties described in Problem **168** is equivalent to the condition of solubility and can as well be considered as a definition of solubility. One may obtain yet another definition of solubility using the results of the next two problems.

170. Let G be a soluble group. Prove that there exists a sequence of groups $G_0, G_1, \ldots, G_n$ such that: 1) $G_0 = G$; 2) every group G_i $(1 \leq i \leq n-1)$ contains a commutative normal subgroup N_i such that the quotient group $G_i/N_i \cong G_{i+1}$; 3) the group G_n is commutative.

171. Suppose that for a group G there exists a sequence of groups with the properties described in Problem 170. Prove that the group G is soluble.

1.15 Permutations

We consider now, more attentively, the *permutations* (i.e., the transformations) of the set of integers $1, 2, \ldots, n$; these permutations are called *permutations of degree n*. We observe that any permutation in an arbitrary set of n elements can be considered as a permutation of degree n: it suffices to enumerate the elements of the set by the integers

$1, 2, \ldots, n$. Every permutation of degree n can be written in the form $\begin{pmatrix} 1 & 2 & \ldots & n \\ i_1 & i_2 & \ldots & i_n \end{pmatrix}$, where i_m is the image of the element m under the permutation. Recall that a permutation is a bijective mapping; as a consequence the elements of the second row are all distinct.

172. Which is the number of all permutations of degree n?

DEFINITION. The group of all permutations of degree n with the usual operation of multiplication (i.e., composition) of permutations[9] is called the *symmetric group of degree n* and is denoted by S_n.

173. Prove that for $n \geq 3$ group S_n is not commutative.

A permutation can interchange some elements and fix the others. It may also happen that the permuted elements change their position cyclicly. For example, the permutation

$$\begin{pmatrix} 1 & 2 & 3 & 4 & 5 & 6 & 7 \\ 4 & 2 & 6 & 3 & 5 & 1 & 7 \end{pmatrix}$$

fixes the elements 2, 5 and 7, and permutes the other elements cyclicly: $1 \to 4$, $4 \to 3$, $3 \to 6$, $6 \to 1$. Permutations of this kind are called *cyclic permutations*, or simply *cycles*. For cyclic permutations we will even use another notation. For example, the expression (1436) will denote the permutation sending $1 \to 4$, $4 \to 3$, $3 \to 6$, $6 \to 1$ and fixing the other elements of the set we deal with. So if our permutation has degree 7 then it coincides with the permutation we had above considered.

Permutations are not all cyclic. For example, the permutation

$$\begin{pmatrix} 1 & 2 & 3 & 4 & 5 & 6 \\ 3 & 5 & 4 & 1 & 2 & 6 \end{pmatrix}$$

is not cyclic, but can be represented as product of two cycles:

$$\begin{pmatrix} 1 & 2 & 3 & 4 & 5 & 6 \\ 3 & 5 & 4 & 1 & 2 & 6 \end{pmatrix} = (134) \cdot (25).$$

The cycles obtained permute different elements. Cycles of such a kind are said to be *independent*. It is easy to see that the product of two

[9] By our definition of product of transformations (§1.2) the multiplications of permutations are carried out from right to left. Sometimes one considers the multiplications from left to right. The groups obtained with the two multiplication rules are isomorphic.

independent cycles does not depend on the order of the factors. If we identify those products of independent cycles that are distinguished only by the order of their factors, then the following proposition holds.

174. Every permutation can be uniquely represented (up to different orderings of factors) by a product of independent cycles. Prove this proposition.

A cycle of type (i, j), permuting only two elements, is called a *transposition.*

175. Prove that every cycle can be represented as a product of transpositions (not necessarily independent).

The transpositions $(1,2),(2,3),(\ldots),(n-1,n)$ are called *elementary transpositions.*

176. Prove that every transposition can be represented as product of elementary transpositions.

From the results of Problems **174**–**176** it follows that every permutation of degree n can be represented as a product of elementary transpositions. In other words, the following theorem holds.

THEOREM 4. *If a subgroup of group S_n contains all elementary transpositions, then it coincides with the whole group S_n.*

Suppose that the numbers $1, 2, \ldots, n$ are written on a row in an arbitrary order. We say that the pair i, j is an inversion in this row if $i < j$ but j appears before i in this row. The number of inversions in a row characterizes the disorder with respect to the usual order $1, 2, \ldots, n$.

177. Find the number of inversions in the row $3, 2, 5, 4, 1$.

In the sequel we shall no longer be interested in the number of inversions, but in its parity.

178. Prove that the parity of the number of inversions in a row changes if one exchanges any two numbers.

DEFINITION. The permutation $\begin{pmatrix} 1 & 2 & \ldots & n \\ i_1 & i_2 & \ldots & i_n \end{pmatrix}$ is called *even* or *odd* according to the parity of the number of inversions in the lower row. For example, the identical permutation $\begin{pmatrix} 1 & 2 & \ldots & n \\ 1 & 2 & \ldots & n \end{pmatrix}$ is even because the number of inversions in the lower row is zero.

179. Determine the parity of the permutation $\begin{pmatrix} 1 & 2 & 3 & 4 & 5 \\ 2 & 5 & 4 & 1 & 3 \end{pmatrix}$.

180. Prove that by multiplying an even permutation by an arbitrary transposition one obtains an odd permutation, and, conversely, by multiplying an odd permutation with an arbitrary transposition one obtains an even permutation.

181. Prove that an even permutation can be decomposed only into a product of an even number of transpositions, and an odd permutation only into an odd number of transpositions.

182. Determine the parity of an arbitrary cycle of length: a) 3; b) 4; c) m.

183. Prove that the result of the multiplication of two permutations of the same parity is an even permutation, whereas the result of the multiplication of two permutations of opposite parities is an odd permutation.

184. Let a be an arbitrary permutation. Prove that a and a^{-1} have the same parity.

From the results of Problems **183** and **184** it follows that the set of all the even permutations form a subgroup of group S_n.

DEFINITION. The group of all even permutations of degree n is called the *alternating group of degree n* and it is denoted by A_n.

185. Prove that for $n \geq 4$ A_n is not commutative.

186. Prove that the alternating group A_n is a normal subgroup of the symmetric group S_n and find the partition of S_n by A_n.

187. Calculate the number of elements of the group A_n.

188. Prove that the groups S_2, S_3, and S_4 are soluble.

We now prove that the alternating group A_5 is not soluble. One of the possible proofs uses the following construction. We inscribe in the dodecahedron five regular tetrahedra, numbered by the numbers 1, 2, 3, 4 and 5 in such a way that to every rotation of the dodecahedron there corresponds an even permutation of the tetrahedra, and that to different rotations there correspond different permutations. So we have defined an isomorphism between the group of rotations of the dodecahedron and the group A_5 of the even permutations of degree 5. The non-solubility of the group A_5 will thus follow from the non-solubility of the group of rotations of the dodecahedron.

189. Inscribe in the dodecahedron five tetrahedra as explained above[10].

Another proof of the non-solubility of the group A_5 consists in repeating the argument of the proof of the non-solubility of the group of rotations of the dodecahedron. To do this one must solve the next problem.

190. Prove that every even permutation of degree 5, different from the identity, can be decomposed into independent cycles in just one of the following ways: a) $(i_1i_2i_3i_4i_5)$; b) $(i_1i_2i_3)$; c) $(i_1i_2)(i_3i_4)$.

191. Let N be a normal subgroup of group A_5. Prove that if N contains at least one permutation which splits into independent cycles in one of the ways indicated in Problem **190**, then N contains all the permutations splitting into independent cycles in this way.

192. Prove that the group A_5 does not contain normal subgroups except the identity and the whole group.

From the results of Problems **192**, **161** and from the group A_5 being not commutative, it follows that the group A_5 is not soluble.

193. Prove that the symmetric group S_n for $n \geq 5$, contains a subgroup isomorphic to A_5.

From the results of Problems **193** and **162** we obtain the following theorem.

THEOREM 5. *For $n \geq 5$ the symmetric group S_n is not soluble.*

The proof of this theorem, as well as the other results of this chapter, will be needed in the next chapter to demonstrate the non-solvability by radicals of algebraic equations of degree higher than four[11].

---

[10] To inscribe the 5 tetrahedra inside the dodecahedron one can start from the 5 Kepler cubes. For their description and their relation with the tetrahedra see the footnote of the solution of Problem **189**. (*Translator's note*)

[11] The following books are indicated to students who desire to study the theory of groups more deeply: Kargapolov M.I., Merzlyakov Y.I., (1972), *Fundamentals of the Theory of Groups*, Graduate Texts in Mathematics, (Springer-Verlag: New York); Vinberg. E.B., (2003), *A Course in Algebra*, Graduate Studies in Mathematics, v. 56, (AMS).

Chapter 2

The complex numbers

When studying mathematics in school, the set of numbers considered was progressively extended. The reason for this was based on these extensions allowing us to operate on numbers with more freedom. So on passing from the natural numbers to the integers it became possible to subtract any two numbers; on passing to the rational numbers it became possible to divide any two numbers, etc.. But the most important result of such extensions consists in the properties of the extended system often allowing us to discover some new properties of the initial system. For example, many difficult problems of number theory, concerning only integers, were solved using the real numbers as well as the complex numbers.

Historically, the complex numbers appeared just as a way of solving certain problems in the real numbers. So, for example, the Italian mathematician Cardano (1501–1576) devised a correct procedure for determining the roots of the equation of third degree using, in intermediate steps of calculations, the 'non-existing' roots of negative numbers.

Afterwards the complex numbers played an increasingly important role in mathematics and applications. They were introduced for the first time in the theory of algebraic equations, because the domain of complex numbers turned out a more convenient setting for the study of such equations.

For example, every algebraic equation of degree n ($n \geq 1$) with real or complex coefficients has at least one complex root (see below the 'fundamental theorem of algebra' §2.8) whereas not all algebraic equations with real coefficients have at least one real root.

Since an interpretation of complex numbers was found in terms of vectors in the plane, geometrical notions such as that of continuity and geometrical transform became applicable to the study of complex numbers. The relation between complex numbers and vectors also allows us to rewrite several problems of mechanics in terms of complex numbers and their equations — in particular, in hydrodynamics and aerodynamics, the theory of electricity, thermodynamics, etc..

2.1 Fields and polynomials

Real numbers can be added, multiplied, and the inverse operations are also allowed: the subtraction and the division (the latter, however, not by zero). In any addition of several numbers the terms can be permuted in any way, and they can be collected arbitrarily within brackets without changing the result. The same holds for the factors of any product. All these properties, as well as the relation between the addition and the multiplication, can be summarized as it follows:

The real numbers possess the three following properties:

1) They form a commutative group (see §1.3) under addition (the unit element of this group is denoted by 0 and is called the zero).

2) If one excludes 0 then the real numbers form a commutative group under multiplication.

3) The addition and the multiplication are related by distributivity: for any numbers a, b, c

$$a(b + c) = ab + ac.$$

The existence of these three properties is very important because they allow us to simplify the arithmetic of algebraic expressions, to solve equations, etc.. The set of real numbers is not the only set to possess these three properties. In order to single out all these sets the following notion is introduced.

DEFINITION. A set in which two binary operations (addition and multiplication) possessing the above properties are defined is called a *field.*

194. Verify whether the following subsets of the real numbers set with the usual operations of addition and multiplication are a field: a) all the natural numbers; b) all the integer numbers; c) all the rational numbers; d) all the numbers of the type $r_1 + r_2\sqrt{2}$, where r_1 and r_2 are two arbitrary rational numbers.

195. Prove that in every field the identity $a \cdot 0 = 0 \cdot a = 0$ holds for any element a.

196. Prove that in every field: 1) $(-a) \cdot b = a \cdot (-b) = -(a \cdot b)$; 2) $(-a) \cdot (-b) = a \cdot b$ for any elements a and b.

197. Let a, b be two elements of an arbitrary field and $a \cdot b = 0$. Prove that either $a = 0$ or $b = 0$.

EXAMPLE 14. Suppose that in the set $\{0, 1, 2, ..., n-1\}$, besides the operation of addition modulo n (see example 9, §1.4), there is also given the *multiplication modulo n* which associates to two numbers the remainder of the division by n of their usual product.

198. Construct the tables of multiplication modulo 2, 3 and 4.

199. Prove that the remainders modulo n with the operations of addition and multiplication modulo n form a field if and only if n is a prime number.

DEFINITION. By the *difference* $(b-a)$ of the elements a and b in an arbitrary field one denotes the element which solves the equation $x+a = b$ (or $a + x = b$). One calls the *quotient* of the division of the element b by a, for $a \neq 0$, (denoted by b/a) the element which solves the equation $ya = b$ (or $ay = b$).

From the result of Problem **24** and from addition and multiplication in a field being commutative, it follows that the elements $b - a$ and b/a (for $a \neq 0$) are uniquely defined in all fields.

Since a field is a group under addition as well as if one excludes the zero, under multiplication, the equation $x + a = b$ is equivalent to the equation $x = b + (-a)$ and the equation $ya = b$, for $a \neq 0$, is equivalent to the equation $y = ba^{-1}$. Hence $b - a = b + (-a)$ and $b/a = ba^{-1}$.

The reader may easily prove that the operations of addition, subtraction, multiplication, and division in an arbitrary field possess all the basic properties which these operations possess in the field of real numbers. In particular, in any field the two members of an equation can be multiplied or divided by the same non-zero number; every term can be transported from one member to the other reversing its sign, etc.. Consider, for instance, the property which relates subtraction and multiplication.

200. Prove that in any field $(a - b)c = ac - bc$ for any three elements a, b, c.

If K is a field then it is possible, as for the field of the real numbers, to consider the polynomials with coefficients in the field K, or, in other words, the polynomials over K.

DEFINITION. An expression like (n being a natural number)

$$a_0x^n + a_1x^{n-1} + \ldots + a_{n-1}x + a_n = 0, \tag{2.1}$$

where $a_0, a_1, \ldots, a_n$ are elements of the field K, and $a_0 \neq 0$, is called a *polynomial of degree n in one variable x over K.*

If a is an element of the field K the expression a is itself considered as a polynomial over K, and if $a \neq 0$ it represents a polynomial of degree zero, whereas if $a = 0$ the degree of this polynomial is considered to be undefined.

The elements $a_0, a_1, \ldots, a_n$ are called the *coefficients* of the polynomial (2.1) and a_0 the *leading coefficient.*

Two polynomials in one variable x are considered to be equal if and only if the coefficients of the terms of the same degree in both polynomials coincide. Let

$$P(x) = a_0x^n + a_1x^{n-1} + \ldots + a_{n-1}x + a_n.$$

If in the second member of this equation one replaces x with an element a of the field K and one carries out the calculations indicated, i.e., the operations of addition and multiplication in the field K, one obtains as a result some element b of the field K. One thus writes $P(a) = b$. If $P(a) = 0$, where 0 is the zero element of field K, one says that a is a *root of the equation* $P(x) = 0$; one also says that a is a *root of the polynomial* $P(x)$.

The polynomials on any field can be added, subtracted, and multiplied.

The *sum* of two polynomials $P(x)$ and $Q(x)$ is a polynomial $R(x)$ in which the coefficient of x^k ($k = 0, 1, 2, \ldots$) is equal to the sum (in the field K) of the coefficients of x^k in the polynomials $P(x)$ and $Q(x)$. In the same way one defines the *difference* of two polynomials. It is evident that the degree of the sum or of the difference of two polynomials is not higher than the maximum of the degree of the given polynomials.

To calculate the *product* of the polynomials $P(x)$ and $Q(x)$ one must multiply every monomial ax^k of the polynomial $P(x)$ by every monomial bx^l of the polynomial $Q(x)$ according to the rule $ax^k \cdot bx^l = abx^{k+l}$, where ab is the product in K, and $k+l$ is the usual sum of integer numbers. Afterwards one must sum all the obtained expressions, collecting the monomial where the variable x has the same degree, and replacing the sum $d_1x^r + d_2x^r + \ldots + d_sx^r$ by the expression $(d_1 + d_2 + \ldots + d_s)x^r$. If

$$\begin{aligned} P(x) &= a_0x^n + a_1x^{n-1} + \ldots + a_{n-1}x + a_n, \\ Q(x) &= b_0x^m + b_1x^{m-1} + \ldots + b_{m-1}x + b_m, \end{aligned}$$

thus[1]

$$\begin{aligned} P(x) \cdot Q(x) &= a_0b_0x^{n+m} + (a_0b_1 + a_1b_0)x^{n+m-1} \\ &\quad + (a_0b_2 + a_1b_1 + a_2b_0)x^{n+m-2} + \ldots + a_nb_m. \end{aligned}$$

Since $a_0 \neq 0$ and $b_0 \neq 0$ (cf., **197**) the degree of the product $P(x) \cdot Q(x)$ is equal to $n+m$, i.e., the degree of the product of two polynomials (non-zero) is equal to the sum of the degrees of the given polynomials.

Taking into account that the operations of addition and multiplication of the elements of the field K possess the commutative, associative, and distributive properties, it is not difficult to verify that the introduced operations of addition and multiplication of polynomials also possess all these properties.

If

$$P(x) + Q(x) = R_1(x), \quad P(x) - Q(x) = R_2(x),$$
$$P(x) \cdot Q(x) = R_3(x)$$

and a is any element of the field K, one obtains

$$P(a) + Q(a) = R_1(a), \quad P(a) - Q(a) = R_2(a),$$
$$P(a) \cdot Q(a) = R_3(a).$$

The polynomials on an arbitrary field K can be divided by one another with a remainder. Dividing the polynomial $P(x)$ by the polynomial $Q(x)$ with a remainder means finding the polynomials $S(x)$ (quotient) and $R(x)$ (remainder) such that

$$P(x) = S(x) \cdot Q(x) + R(x);$$

[1] The coefficient of x^{n+m-k} in the product $P(x) \cdot Q(x)$ is equal to $a_0b_k + a_1b_{k-1} + \ldots + a_{k-1}b_1 + a_kb_0$; hence here we must impose $a_i = 0$ for $i > n$ and $b_j = 0$ for $j > n$.

moreover either the degree of the polynomial $R(x)$ must be lower than the degree of the polynomial $Q(x)$, or one must have $R(x) = 0$.

Let $P(x)$ and $Q(x)$ be any two polynomials over the field K and $Q(x) \neq 0$. We show that it is possible to divide the polynomial $P(x)$ by the polynomial $Q(x)$ with a remainder.

Let

$$\begin{aligned} P(x) &= a_0x^n + a_1x^{n-1} + \ldots + a_n, \\ Q(x) &= b_0x^m + b_1x^{m-1} + \ldots + b_m. \end{aligned}$$

If $n < m$ we put $S(x) = 0$ and $R(x) = P(x)$, and we obtain the quotient and the remainder required. If $n \geq m$ then consider the polynomial

$$P(x) - \frac{a_0}{b_0}x^{n-m}Q(x) = R_1(x).$$

The polynomial $R_1(x)$ contains no monomial in x^n because either its degree is not higher than $n-1$, or $R_1(x) = 0$. If

$$R_1(x) = c_0x^k + c_1x^{k-1} + \ldots + c_k$$

and $k \geq m$, then consider the polynomial

$$R_1(x) - \frac{c_0}{b_0}x^{k-m}Q(x) = R_2(x),$$

etc.. Since the degree of the polynomial obtained is strictly lower than the degree of the preceding polynomial, this procedure must end, i.e., at some step we obtain

$$R_{s-1}(x) - \frac{d_0}{b_0}x^{l-m}Q(x) = R_s(x),$$

where either the degree of $R_s(x)$ is lower of the degree of $Q(x)$ or $R_s(x) = 0$. We thus have

$$\begin{aligned} P(x) &= \frac{a_0}{b_0}x^{n-m}Q(x) + R_1(x) \\ &= \frac{a_0}{b_0}x^{n-m}Q(x) + \frac{c_0}{b_0}x^{k-m}Q(x) + R_2(x) \\ &\cdot \\ &\cdot \\ &\cdot \\ &= \frac{a_0}{b_0}x^{n-m}Q(x) + \frac{c_0}{b_0}x^{k-m}Q(x) + \ldots + \frac{d_0}{b_0}x^{l-m}Q(x) + R_s(x) \\ &= (\frac{a_0}{b_0}x^{n-m} + \frac{c_0}{b_0}x^{k-m} + \ldots + \frac{d_0}{b_0}x^{l-m}) \cdot Q(x) + R_s(x). \end{aligned}$$

Consequently the expression within brackets is the quotient of the division of the polynomial $P(x)$ by the polynomial $Q(x)$ and $R_s(x)$ is the remainder. The procedure of the division of two polynomials described here is called the *Euclidean algorithm.*

201. Let

$$\begin{aligned} P(x) &= S_1(x) \cdot Q(x) + R_1(x), \\ P(x) &= S_2(x) \cdot Q(x) + R_2(x), \end{aligned}$$

for which the degrees of $R_1(x)$ and $R_2(x)$ are lower than the degree of $Q(x)$ (may be $R_1(x) = 0$ or $R_2(x) = 0$). Prove that

$$S_1(x) = S_2(x), \qquad R_1(x) = R_2(x).$$

2.2 The field of complex numbers

From the solution of Problem **194** it follows that there exist fields smaller than the field of the real numbers; for example, the field of the rational numbers. We now construct a field which is bigger than the field of the real numbers: the field of the complex numbers.

Consider all the possible pairs of real numbers, i.e., the pairs of type (a, b), where a and b are two arbitrary real numbers. We will say that $(a, b) = (c, d)$ if and only if $a = c$ and $b = d$. In the set of all these pairs we define two binary operations, the addition and the multiplication, in the following way:

$$(a, b) + (c, d) = (a + c, b + d), \tag{2.2}$$

$$(a, b) \cdot (c, d) = (ac - bd, ad + bc) \tag{2.3}$$

(here within brackets in the second members of the equations the operations are the usual operations on real numbers). For example, we obtain

$$\begin{aligned} (\sqrt{2}, 3) + (\sqrt{2}, -1) &= (2\sqrt{2}, 2), \\ (0, 1) \cdot (0, 1) &= (-1, 0). \end{aligned}$$

DEFINITION. The set of all pairs of real numbers with the operations of addition and of multiplication defined by (2.2) and (2.3) is called the *set of complex numbers.*

From this definition it is clear that in the complex numbers there is nothing of the 'supernatural': the complex numbers are nothing but pairs of real numbers. However, a question may arise: is it correct to call such objects numbers? We will answer this question at the end of this section. Another question, which perhaps the reader may put, is the reason way the operations of addition and multiplication of complex numbers are defined exactly in this manner and, in particular, way the operation of multiplication is so strange. We will answer this question in §2.3.

First, we clarify the remarkable properties of the set of complex numbers which we had defined.

202. Prove that the complex numbers form a commutative group under addition. Which complex number is the unit element (zero) of this group?

In the sequel it will be convenient to denote the complex numbers by a single letter, for example by z (or w).

203. Prove that the operation of multiplication of complex numbers is commutative and associative, i.e., that $z_1 \cdot z_2 = z_2 \cdot z_1$ and $(z_1 \cdot z_2) \cdot z_3 = z_1 \cdot (z_2 \cdot z_3)$.

It is easy to verify that

$$(a, b) \cdot (1, 0) = (1, 0) \cdot (a, b) = (a, b)$$

for every complex number (a, b). The complex number $(1, 0)$ is therefore the unit element in the set of complex numbers under multiplication.

204. Let z be an arbitrary complex number and suppose that $z \neq (0, 0)$. Prove that there exists a complex number z^{-1} such that

$$z \cdot z^{-1} = z^{-1} \cdot z = (1, 0).$$

The results of Problems **203** and **204** show that complex numbers form a commutative group under multiplication.

205. Prove that the operations of addition and multiplication of complex numbers possess the distributive property, i.e., that $(z_1 + z_2) \cdot z_3 = z_1 \cdot z_2 + z_2 \cdot z_3$ for any complex numbers z_1, z_2, z_3.

From the results of Problems **202**–**205** it follows that the complex numbers with the operations of addition and multiplication defined by (2.2) and (2.3) form a field. This field is the *field of complex numbers.*

For complex numbers of type $(a, 0)$, where a is any real number, formulae (2.2) and (2.3) give

$$\begin{aligned}(a,0)+(b,0) &= (a+b,0),\\ (a,0)\cdot(b,0) &= (a\cdot b,0).\end{aligned}$$

Consequently if one associates to every complex number of type $(a, 0)$ the real number a then to the operations on numbers of type $(a, 0)$ there correspond the usual operations on real numbers. Therefore we simply identify the complex number $(a, 0)$ with the real number a and we say that the field of complex numbers contains the field of real numbers[2].

The complex number $(0, 1)$ is not real (under our definition) and we will denote it by i: $i = (0, 1)$. Since the field of complex numbers contains all real numbers and the number i, it also contains all numbers of the form $b \cdot i$ and $a + b \cdot i$, where a and b are any two real numbers and the operations of addition and multiplication are extended to operations on complex numbers.

206. Let (a, b) be a complex number. Prove that

$$(a, b) = a + b \cdot i.$$

From the result of Problem **206** we obviously obtain that $a + bi = c + di$ if and only if $a = c$ and $b = d$.

As a consequence every complex number can be represented in a unique way in the form $a + bi$, where a and b are two real numbers. If $z = a + bi$ then, following tradition, a is called the *real part* of the complex number, bi the *imaginary part*, and b the *coefficient of the imaginary part*.

The representation of a complex number z in the form $z = a + bi$ is called the *algebraic representation* of z.

For the complex numbers in algebraic representation formulae (2.2) and (2.3) read:

$$(a+bi)+(c+di) = (a+c)+(b+d)i, \tag{2.4}$$

$$(a+bi)\cdot(c+di) = (ac-bd)+(ad+bc)i. \tag{2.5}$$

207. Solve the equation (i.e., find the formula for the difference)

$$(a+bi)+(x+yi)=(c+di).$$

---

[2]In an analogous way, for example, one identifies the rational $n/1$ with the integer n.

208. Solve the equation (i.e., find the formula for the quotient)

$$(a+bi)\cdot(x+yi)=(c+di), \quad \text{where} \quad a+bi\neq 0.$$

It is easy to verify that

$$i\cdot i=(0,1)\cdot(0,1)=(-1,0)=-1,$$

i.e., $i^2=-1$. Hence in the field of complex numbers, square roots of negative numbers are well defined.

209. Calculate a) i^3; b) i^4; c) i^n.

210. Find all complex numbers $z=x+yi$ such that: a) $z^2=1$; b) $z^2=-1$; c) $z^2=a^2$; d) $z^2=-a^2$ (a is a real number).

DEFINITION. The complex number $a-bi$ is called the *conjugate* of $z=a+bi$ and it is denoted by $\overline{z}$. It is easy to verify that

$$z+\overline{z}=2a, \quad z\cdot\overline{z}=a^2+b^2.$$

211. Let z_1 and z_2 be two arbitrary complex numbers. Prove that: a) $\overline{z_1+z_2}=\overline{z_1}+\overline{z_2}$; b) $\overline{z_1-z_2}=\overline{z_1}-\overline{z_2}$; c) $\overline{z_1\cdot z_2}=\overline{z_1}\cdot\overline{z_2}$; d) $\overline{z_1/z_2}=\overline{z_1}/\overline{z_2}$.

212. Let

$$P(z)=a_0z^n+a_1z^{n-1}+\ldots+a_{n-1}z+a_n,$$

where z is a complex number and all the a_is are real. Prove that $\overline{P(z)}=P(\overline{z})$.

The passage to the complex numbers is a successive step in the series: natural numbers – integer numbers – rational numbers – real numbers – complex numbers. The reader may feel that up to the real numbers one deals with numbers, whereas the complex numbers are objects of another nature. Of course, one may use whatever terminology one wishes, but the complex numbers must, in fact, be considered as numbers.

The first objection against this is that complex numbers are not numbers, but pairs of numbers. Recall, however, that in a similar way one introduces rational numbers. A rational number is a class of equivalent fractions, and a fraction is a pair of integer numbers of the form m/n (where $n\neq 0$); in this way the operations on rational numbers are simply operations on pairs of integer numbers. Another objection should be that a number is an object which allows us to measure something. If we think that numbers are entities by which one can measure everything, then one must exclude from the set of these entities, for example, the negative numbers because there are no segments of length -3 cm, and a train cannot go for -4 hours. If, on the contrary, one thinks that numbers are objects by which it is possible (or convenient) to measure at least one quantity, then complex numbers are similar to the other numbers: with them one describes very well, for example, the potential and the resistance of alternating currents in electric circuits, which are extensively utilized in electrotechnology. Complex numbers are are successfully employed in hydro- and aerodynamics as well.

So the passage from real to complex numbers is as natural as, for example, the passage from integer to rational numbers.

2.3 Uniqueness of the field of complex numbers

Consider now the question of whether complex numbers could be defined otherwise.

In other words, the question that we answer in this section is the following: we want to obtain a field, which is an extension of the field of the real numbers: does there exist more than one field which is an extension of the field of the real numbers?

DEFINITION. We call an *isomorphic mapping* (or simply an *isomorphism*) of one field onto another one a bijective mapping ϕ which is an isomorphism with respect both to the addition, and to the multiplication, i.e., such that $\phi(a+b) = \phi(a) + \phi(b)$ and $\phi(ab) = \phi(a)\phi(b)$. Two fields between which one can define an isomorphism, are said to be *isomorphic*.

If in a field one considers exclusively the operations of addition and multiplication, then isomorphic fields all have identical properties. As a consequence, as in the case of groups, isomorphic fields cannot be distinguished.

As we have seen in the preceding section, in the field of the complex numbers there is only one element i such that $i^2 = -1$. The following problem shows that on adding this element to the field of real numbers one necessarily obtains the field of complex numbers.

213. Let M be a field containing the field of real numbers and a certain element i_0 such that $i_0^2 = -1$. Prove that M contains a field M'

isomorphic to the field of complex numbers.

We will say that a field is the *minimal field with the required properties* if it possesses these properties and it does not contain other fields with the same properties.

The result of Problem **213** can then be formulated in this way: the minimal field which contains the field of the real numbers, and an element i_0 such that $i_0^2 = -1$, is the field of complex numbers. This result proves in a certain sense the uniqueness of the field of complex numbers. However, another, much stronger, result holds. Indeed, suppose one renounces to the requirement that the field M contains an element i_0 such that $i_0^2 = -1$, and poses the problem of finding all fields that are minimal extensions of the field of real numbers. It turns out that there are only two such extensions (up to isomorphism): one of them is the field of the complex numbers. Prove this statement.

Suppose the field M contains all the real numbers, i.e., that M contains all the real numbers and that the operations on them coincide with the usual operations on the real numbers. Suppose, moreover, that M contains an element j, different from all the real numbers. Thus for all sets of n real numbers $a_1, a_2, \dots, a_n$ there exists in M an element equal to

$$j^n + a_1 j^{n-1} + a_2 j^{n-2} + \dots + a_n. \tag{2.6}$$

We call n the degree of expression (2.6).

There are two possible cases:

a) a certain expression of the form (2.6) is equal to 0 for $n \geq 1$;

b) there are no expressions of the form (2.6) equal to 0 for $n \geq 1$.

Suppose first that we are in the case (a).

DEFINITION. The polynomial with coefficients in a certain field K is said to be *reducible over* K if it can be represented as a product of two polynomials of lower degree with coefficients in K. In the opposite case it is said to be *irreducible*[3] *over* K.

For example, the polynomials $x^3 - 1$ and $x^2 - x - 1$ are reducible over the field of real numbers, because $x^3 - 1 = (x-1)(x^2+x+1)$ and $x^2 - x - 1 = (x - (1+\sqrt{5})/2)(x - (1-\sqrt{5})/2)$, whereas the polynomials $x^2 + 1$ and $x^2 + x + 1$ are irreducible over the field of real numbers. It is evident that polynomials of the first degree over any field are irreducible.

[3] Irreducible polynomials over a field K are the analogue of prime numbers in the set of integer numbers.

214. Let us choose, amongst all expressions of type (2.6), the expression of minimal degree n ($n \geq 1$) that is vanishing in M: the corresponding equation is

$$j^n + a_1 j^{n-1} + a_2 j^{n-2} + \ldots + a_n = 0.$$

Prove that the polynomial

$$x^n + a_1 x^{n-1} + a_2 x^{n-2} + \ldots + a_n$$

is not reducible over the field of real numbers.

In the sequel we will prove (cf., **272**) that every polynomial with real coefficients of degree higher than 2 is reducible over the field of real numbers. Hence in Problem **214** n cannot be higher than 2. But since $n \neq 1$ (otherwise we should have $j + a = 0$, and j should be equal to the real number $-a$) we obtain that $n = 2$.

Consequently in the case (a) there exist two real numbers p and q in M which satisfy

$$j^2 + pj + q = 0$$

and for which the polynomial $x^2 + px + q$ is irreducible over the field of real numbers.

215. Prove that in the case (a) the field M contains an element i_0 such that $i_0^2 = -1$.

From the results of Problems **215** and **213** it follows that in the case (a) the field M contains a field M' isomorphic to the field of complex numbers. Therefore if the field M is a minimal extension of the field of real numbers then the field M must coincide with M'. As a consequence, in the case (a) any minimal field which represents a minimal extension of the field of real numbers coincides (i.e., it is isomorphic) with the field of complex numbers. So in the case (a) there is only one field (up to isomorphism) which is a minimal extension of the field of real numbers, namely the field of complex numbers.

216. Find all fields that are minimal extensions of the field of real numbers in the case (b).

2.4 Geometrical descriptions of the complex numbers

Consider on the plane a system of orthogonal coordinates (x, y) and let us associate to every complex number $a + ib$ the point of the plane with coordinates (a, b). We obtain a bijective correspondence between all complex numbers and all points of the plane. This is the *first geometrical representation* of the complex numbers.

217. Which complex numbers correspond to the points shown in Figure 13?

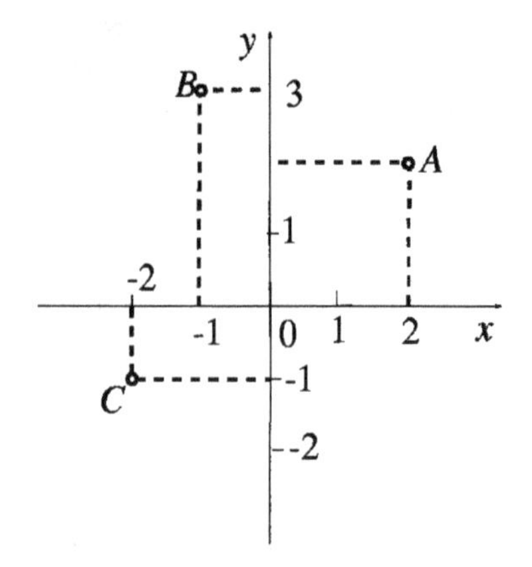

FIGURE 13

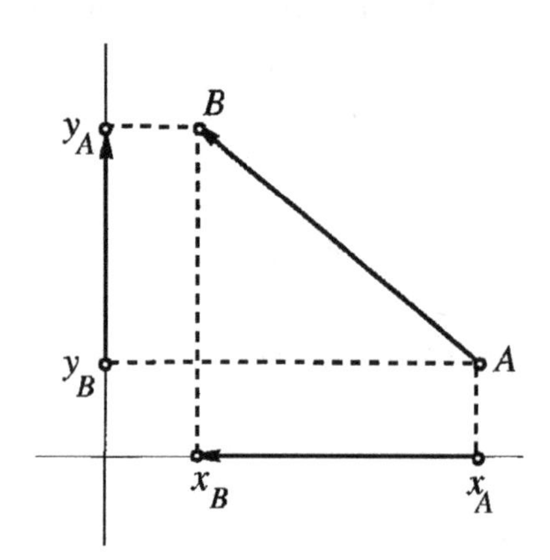

FIGURE 14

218. Let the complex numbers be represented by the points of the plane. What is the geometrical meaning of the mapping ϕ if for every complex number z: a) $\phi(z) = -z$; b) $\phi(z) = 2z$; c) $\phi(z) = \overline{z}$ ($\overline{z}$ is the conjugate of z)?

Let $A(x_A, y_A)$ and $B(x_B, y_B)$ be two points of the plane (Figure 14). The segment AB directed from A to B is called the *vector* $\overrightarrow{AB}$. The coordinates of the vector $\overrightarrow{AB}$ are by definition calculated in the following way:

$$x_{\overrightarrow{AB}} = x_B - x_A, \quad y_{\overrightarrow{AB}} = y_B - y_A$$

Two vectors are considered equal if they are parallel and have the same direction and the same length.

219. Prove that two vectors are equal if and only if their coordinates are equal.

The set of equal vectors is considered to be a unique vector, characterized by its coordinates, which is called a *free vector*. Putting into correspondence every complex number $a + bi$ with the free vector having coordinates (a, b) we obtain the *second geometrical representation* of the complex numbers.

220. Let u, v, and w be the free vectors corresponding to the complex numbers z_1, z_2, and z_3. Prove that $z_1 + z_2 = z_3$ if and only if $u + v = w$, where the vectors are added according to the parallelogram rule.

221. Prove the following relation between the two geometrical representations of the complex numbers: if z_A, z_B, $z_{\overrightarrow{AB}}$ are the complex numbers corresponding to the points A, B and to vector $\overrightarrow{AB}$, then $z_{\overrightarrow{AB}} = z_B - z_A$.

By definition, two equal vectors have equal lengths. This length is additionally assumed to be the length of the free vector corresponding to a given set of equal vectors.

DEFINITION. One calls the *modulus* of the complex number z (denoted by $|z|$) the length of the corresponding free vector[4].

222. Let $z = a + bi$. Prove that

$$|z|^2 = a^2 + b^2 = z \cdot \bar{z},$$

where $\bar{z}$ is the conjugate of z.

223. Prove the inequalities:

$$\begin{aligned} a)\quad & |z_1 + z_2| \leq |z_1| + |z_2|, \\ b)\quad & |z_1 - z_2| \geq \big||z_1| - |z_2|\big|, \end{aligned}$$

where z_1, z_2 are arbitrary complex numbers. In which case does the equality hold?

224. Prove by means of the complex numbers that in any parallelogram the sum of the squares of the lengths of the diagonals is equal to the sum of the squares of the lengths of all the sides.

[4]For real numbers (as a particular case of complex numbers) the notion of modulus introduced here coincides with the notion of ***absolute value***. Indeed, to the real number $a + 0i$ there corresponds the vector with coordinates $(a, 0)$, parallel to the x-axis, and its length is equal to $|a|$, the absolute value of the number a.

2.5 The trigonometric form of the complex numbers

Recall that the angle between the rays OA and OB is defined as the angle by which one has to turn counterclockwise the ray OA around O in order to take it over to the ray OB (if the rotation is clockwise, the angle has the sign 'minus'). So the angle is not defined uniquely, but up to rotations by $2\pi k$, where k is any integer.

DEFINITION. Let point O be the origin of the coordinates, and suppose that the vector OA with coordinates (a, b) corresponds to the complex number $z = a + bi$. One calls the *argument* of the complex number z (denoted by $\arg z$) the angle between the positive direction of the axis Ox and the ray OA (Figure 15) (if $z = 0$ then $\arg z$ is not defined).

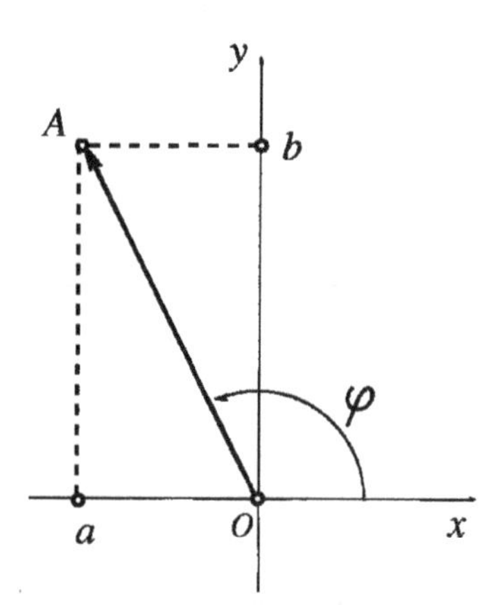

FIGURE 15

Since for a given number $z \neq 0$ the angle is not uniquely defined, by the expression $\arg z$ we mean a multi-valued function taking infinite values, between which the differences are equal to multiples of 2π.

The expression $\arg z = \varphi$ will mean that one of the values of the argument is equal to φ.

Let $z = a + bi \neq 0$ and $|z| = r$. The vector $\overrightarrow{OA}$ with coordinates (a, b) corresponds to the complex number $a + bi$, and its length is therefore equal to r. Let $\arg z = \varphi$. Thus by the definition of the trigonometric functions (see Figure 15),

$$\cos\varphi = a/r, \quad \sin\varphi = b/r,$$

and therefore

$$z = a + bi = r \cdot \cos\varphi + i \cdot r \cdot \sin\varphi = r(\cos\varphi + i\sin\varphi),$$

where $r = |z|$, $\varphi = \arg z$, and thus we have obtained the *trigonometric representation* of the complex number z.

For example, if $z = -1 + \sqrt{3}i$ then $|z| = \sqrt{1+3} = 2$ (cf., **222**) and $\cos\varphi = -1/2$, $\sin\varphi = \sqrt{3}/2$. We can assume that $\varphi = 2\pi/3$; thus $z = -1 + \sqrt{3}i = 2(\cos(2\pi/3) + i\sin(2\pi/3))$.

225. Represent in trigonometric form the following complex numbers: a) $1 + i$, b) $-\sqrt{3} - i$, c) $3i$, d) -5, e) $1 + 2i$.

226. Let $z_1 = r_1(\cos\varphi_1 + i\sin\varphi_1)$ and $z_2 = r_2(\cos\varphi_2 + i\sin\varphi_2)$. Prove that

$$\begin{aligned} z_1 \cdot z_2 &= r_1 r_2(\cos(\varphi_1 + \varphi_2) + i\sin(\varphi_1 + \varphi_2)), \\ \frac{z_1}{z_2} &= \frac{r_1}{r_2}(\cos(\varphi_1 - \varphi_2) + i\sin(\varphi_1 - \varphi_2)) \quad (z_2 \neq 0). \end{aligned}$$

Thus as a result of the multiplication the moduli of the complex numbers are multiplied and their arguments are added; as result of the division the moduli are divided and the arguments are subtracted.

227. Prove the *De Moivre formula*[5]:

$$[r(\cos\varphi + i\sin\varphi)]^n = r^n(\cos n\varphi + i\sin n\varphi)$$

for every integer $n > 0$.

228. Calculate $(1 - \sqrt{3}i)^{100}/2^{100}$.

229. Let $z = r(\cos\varphi + i\sin\varphi)$ be a given complex number and n a natural number. Find all complex numbers w satisfying the equation

$$w^n = z. \tag{2.7}$$

DEFINITION. The expression $\sqrt[n]{z}$ (nth root of z) denotes a multivalued function, which puts into correspondence with every complex number $z \neq 0$ all the n roots of equation (2.7). For $z = 0$ one has $\sqrt[n]{0} = 0$.

[5] A. de Moivre (1667–1754), French mathematician who lived in England.

230. Find all the values of the roots:
a) $\sqrt{-1}$; b) $\sqrt[3]{8}$; c) $\sqrt[4]{\cos 100^\circ + i \sin 100^\circ}$; d) $\sqrt[3]{1+i}$.

It will be convenient in the sequel to adopt the following notation

$$\epsilon_n = \cos\frac{2\pi}{n} + i\sin\frac{2\pi}{n}.$$

231. Prove that all the values of $\sqrt[n]{-1}$ are $1, \epsilon_n, \epsilon_n^2, \dots, \epsilon_n^{n-1}$.

REMARK. Since $\epsilon_n^n = 1$ the set of elements $1, \epsilon_n, \epsilon_n^2, \dots, \epsilon_n^{n-1}$ is a cyclic group under multiplication.

232. Let z_1 be one of the values of $\sqrt[n]{z_0}$. Find all the values of $\sqrt[n]{z_0}$.

We shall use the representation of complex numbers by the points of the plane, i.e., the complex number $a+bi$ will correspond to the point of the plane having the coordinates (a, b). So instead of the point corresponding to the complex number z we shall say simply point z.

233. Let the complex numbers be represented by the points of the plane. Which is the geometrical meaning of the following expressions: a) $|z|$, b) $\arg z$, c) $|z_1 - z_2|$, d) $\arg(z_1/z_2)$?

234. Find the position on the plane of the points z satisfying the following conditions (where z_0, z_1, z_2 are some given complex numbers and R is a given real number): a) $|z| = 1$; b) $|z| = R$; c) $|z - z_0| = R$; d) $|z - z_0| \leq R$; e) $|z - z_1| = |z - z_2|$; f) $\arg z = \pi$; g) $\arg z = 9\pi/4$, h) $\arg z = \varphi$.

235. How are all the values of $\sqrt[n]{z}$ distributed on the plane, z being a given complex number?

2.6 Continuity

In the sequel an important role will be played by the notion of continuity and, in particular, by that of continuous curve. If the reader does not know the precise definition of such entities he understands intuitively, however, what is a continuous curve, as well as a continuous function of one real variable (at the intuitive level one may say that it is a function whose graph is a continuous curve). But if the function of one real variable is sufficiently complicated (for example, $f(x) = (x^3 - 2x)/(x^2 - \sin x + 1)$), then saying whether it is continuous or not, using only the intuitive idea of continuity, is rather difficult. Hence we give the rigourous definition of continuity and by means of it we will prove some basic properties of continuous functions. We give the definition of continuity for functions of one real variable as well as for functions of one complex variable.

The graph of a function of one real argument can be discontinuous at some points, and at some points it can have some breaks. It is therefore natural to consider first the notion of continuity of a function at a given point rather than the general definition of continuity.

If we try to define more precisely our intuitive idea of continuity of a function $f(x)$ at a given a point x_0, we obtain that continuity means the following: under small changes of the argument near the point x_0 the value of the function changes a little with respect to value $f(x_0)$. Moreover, it is possible to obtain a variation of the function's value about $f(x_0)$ as small as we want by choosing a sufficiently small interval of the variation of the argument around x_0. One can formulate this more rigourously in the following way.

DEFINITION. Let f be a function of one real or complex variable z. One says that the function $f(z)$ is *continuous at* z_0 if for every arbitrary real number $\varepsilon > 0$ one can choose a real number $\delta > 0$ (which depends on z_0 and on ε) such that for all numbers z satisfying the condition $|z-z_0| < \delta$ the inequality[6] $|f(z) - f(z_0)| < \varepsilon$ holds.

EXAMPLE 15. Prove that the function with complex argument $f(z) = 2z$ is continuous at any point z_0. Suppose a point z_0 and a real number $\varepsilon > 0$ be given. We have to choose a real number $\delta > 0$ such that for all numbers z which satisfy the condition $|z - z_0| < \delta$ the inequality $|f(z) - f(z_0)| = |2z - 2z_0| < \varepsilon$ is satisfied. It is not difficult to see that one can choose $\delta = \varepsilon/2$ (independently of the point z_0). Indeed, from the condition $|z - z_0| < \delta$ it follows that:

$$|2z - 2z_0| = |2(z - z_0)| = (\text{cf., } \mathbf{226}) \;\; |2| \cdot |z - z_0| < 2\delta = \varepsilon,$$

i.e., $|2z-2z_0| < \epsilon$. As a consequence the function $f(z) = 2z$ is continuous at any point z_0. In particular, it is continuous for all real values of the argument z. Therefore if one restricts the function to the real values of the argument, one obtains that the function of the real variable $f(x) = 2x$ is continuous for all real values x.

[6]The geometrical meaning of the inequalities $|z - z_0| < \delta$ and $|f(z) - f(z_0)| < \varepsilon$ is given in Problems **233** and **234**.

236. Let a be a complex number (or, as a particular case, real). Prove that the complex (or real) function $f(z) \equiv a$ is continuous for all values of the argument.

237. Prove that the function of a complex argument $f(z) = z$ and the function of a real argument $f(x) = x$ are continuous for all values of their argument.

238. Prove that the function of a complex argument $f(z) = z^2$ is continuous for the all values of z.

DEFINITION. Let $f(z)$ and $g(z)$ be two functions of a complex (or real) argument. One calls the *sum* of the functions $f(z)$ and $g(z)$ the function $h(z)$ of a complex (or real) argument which satisfies at every point z_0 the equation $h(z_0) = f(z_0) + g(z_0)$. If the value $f(z_0)$ or the value $g(z_0)$ is not defined then the value $h(z_0)$ is also not defined. In the same way one defines the *difference*, the *product*, and the *quotient* of two functions.

239. Let $f(z)$ be a function of a complex or real argument and let $g(z)$ be continuous at z_0. Prove that at z_0 the functions: a) $h(z) = f(z)+g(z)$; b) $h(z) = f(z) - g(z)$; c) $h(z) = f(z) \cdot g(z)$ are continuous.

From the result of Problem **239**(b) we obtain, in particular, that if a function $f(z)$ is continuous at a point z_0 and n is an integer, then the function $[f(z)]^n$ is also continuous at the point z_0.

240. Let $f(z)$ and $g(z)$ be two functions of complex or real argument, continuous at z_0 and suppose that $g(z_0) \neq 0$. Prove that at z_0 the functions: a) $h(z) = 1/g(z)$; b) $h(z) = f(z)/g(z)$ are continuous.

DEFINITION. Let $f(z)$ and $g(z)$ be two functions of a complex or real argument. One calls the *composition* of the functions $f(z)$ and $g(z)$ the function $h(z)$ which satisfies at every point z_0 the equation $h(z_0) = f(g(z_0))$. If the value $g(z_0)$ is not defined, or the function $f(z)$ is not defined at the point $g(z_0)$, then the value $h(z_0)$ is also not defined.

241. Let $f(z)$ and $g(z)$ be two functions of complex or real argument. Let $g(z_0) = z_1$ and let functions $f(z)$ and $g(z)$ be continuous at the points z_1 and z_0 respectively. Prove that the function $h(z) = f(g(z))$ is continuous at the point z_0.

From the results of Problems **239**–**241** it follows, in particular, that any expression obtained from any functions of one complex (or real) argument, continuous for all values of the argument, by means of the operations of addition, subtraction, multiplication, division, elevation to an integer power, and composition, represents a continuous function at all points at which the denominator does not vanish.

For example, from the results of Problems **236** and **237**, we obtain that the functions $f(z) = z^n$, $f(z) = az^n$, and more generally $f(z) = a_0z^n + a_1z^{n-1} + \ldots + a_n$, are continuous functions of z for all complex numbers $a, a_0, a_1, \ldots, a_n$.

242. Prove that the functions of a real argument $f(x) = \sin(x)$ and $f(x) = \cos(x)$ are continuous for all values of x.

243. Consider for all real values $x \geq 0$ the function $f(x) = \sqrt[n]{x}$, where n is a non-zero integer and a non-negative value is chosen for $\sqrt[n]{x}$. Prove that this function is continuous for every $x > 0$.

During the study of continuity one encounters some statements which intuitively seem evident; however, their exact proofs involve serious technical difficulties and request a definition of the real numbers more strict than that learnt at school, as well as the study of the foundations of set theory and of topology.

An example of such statements can be represented by the following proposition. If a function $f(x)$ of a real argument is continuous in some interval and in this interval takes only integer values, then it takes only one value in the whole interval. Indeed, it seems intuitively evident that, as far as the point x moves in the interval, the value of the function $f(x)$ must change continuously and cannot 'jump' from one integer value to another one. Proving this proposition in an exact way is, however, rather difficult.

In the present exposition we rely on the 'intuitive evidence' of some propositions related to continuity without giving demonstrations of them. In particular, we adopt, without proof, some propositions which we have formulated in the form of examples.

2.7 Continuous curves

Suppose that the parameter t takes real values in the interval $0 \leq t \leq 1$, and that each of these values is in correspondence with some complex number

$$z(t) = x(t) + iy(t).$$

The plane on which the values of z are represented will be called simply the 'z plane'. If the functions $x(t)$ and $y(t)$ are continuous for $0 \leq t \leq 1$, then as t varies from 0 to 1, the point $z(t)$ describes a *continuous curve* in the z plane. We provide this curve with an orientation, assuming that $z_0 = z(0)$ and $z_1 = z(1)$ are the initial and the final points respectively. The function $z(t)$ is called the *parametric equation* of this curve.

EXAMPLE 16. Let $z(t) = t + it^2$. Thus $x(t) = t$, $y(t) = t^2$, and $y(t) = x^2(t)$ for every t, i.e., the point $z(t)$ lies on the parabola $y = x^2$ for every t. As t varies from 0 to 1, $x(t)$ also varies from 0 to 1 and the point $z(t)$ runs along the parabola $y = x^2$ from the point $z_0 = 0$ to the point $z_1 = 1 + i$ (Figure 16).

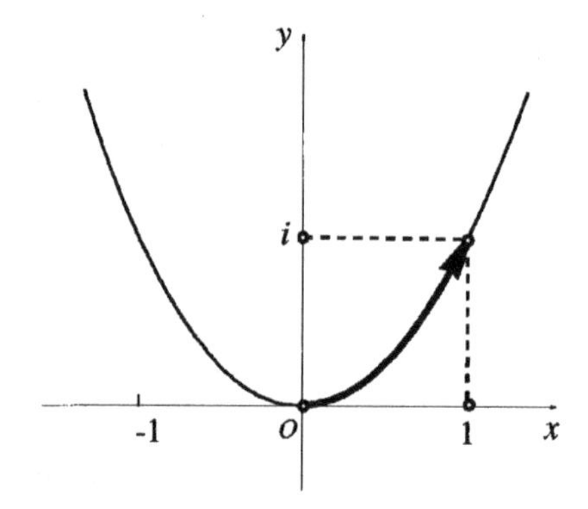

FIGURE 16

244. Trace on the z plane the curves given by the following parametric equations: a) $z(t) = 2t$; b) $z(t) = it$; c) $z(t) = it^2$; d) $z(t) = t - it$; e) $z(t) = t^2 + it$; f) $z(t) = R(\cos 2\pi t + i \sin 2\pi t)$; g) $z(t) = R(\cos 4\pi t + i \sin 4\pi t)$; h) $z(t) = R(\cos \pi t + i \sin \pi t)$;

$$\text{i)}\ z(t) = \begin{cases} \cos 2\pi t + i \sin 2\pi t & \text{for } 0 \leq t \leq 1/2, \\ 4t - 3 & \text{for } 1/2 < t \leq 1. \end{cases}$$

245. Write a parametric equation for the segment joining the points $z_0 = a_0 + b_0 i$ and $z_1 = a_1 + b_1 i$.

REMARK. In the following problems the parametric equations have some indices. These numbers have to be viewed only as labels, but all curves lie in the same z plane.

246. By means of which geometrical transformations of the curve C_1 with equation $z_1(t)$ can we obtain the curve C_2 with equation $z_2(t)$ if:

a) $z_2(t) = z_1(t) + z_0$ (z_0 being a given complex number);
b) $z_2(t) = a \cdot z_1(t)$, where a is a positive real number;
c) $z_2(t) = z_0 \cdot z_1(t)$, where $|z_0| = 1$;
d) $z_2(t) = z_0 \cdot z_1(t)$, where z_0 is an arbitrary complex number?

247. Let $z_1(t)$ be the parametric equation of curve C. What is the curve described by the equation $z_2(t)$ if $z_2(t) = z_1(1-t)$?

248. Let $z_1(t)$ and $z_2(t)$ be the parametric equations of curves C_1 and C_2 and suppose that $z_1(1) = z_2(0)$. What is the curve described by equation $z_3(t)$ if

$$z_3(t) = \begin{cases} z_1(2t) & \text{for } 0 \le t \le 1/2, \\ z_2(2t-1) & \text{for } 1/2 < t \le 1? \end{cases}$$

249. Let $z(t) = \cos \pi t + i \sin \pi t$ (Figure 17). Find all values of $\arg z(t)$ as a function of t.

250. Let $z(t) = \cos \pi t + i \sin \pi t$. How should we choose one of the values of $\arg z(t)$ for every t in such a way that the values of z vary continuously as t varies from 0 to 1, with the condition that $\arg z(0)$ is equal to: a) 0; b) 2π; c) -4π ; d) $2\pi k$ (k being a given integer)?

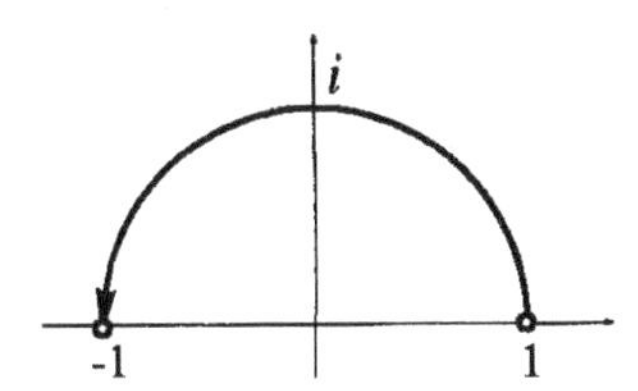

FIGURE 17

THEOREM 6. *Assume that a continuous curve C with a parametric equation $z(t)$ does not cross the origin of coordinates and that at the initial point of the curve C the argument is φ_0. Thus it is possible to choose one of the values of the argument for all points of the curve C so that along the entire curve the argument of $z(t)$ changes continuously, starting from φ_0.*

In other words, one can choose for every t one of the values $\varphi(t)$ of $\arg z(t)$ in order that the function $\varphi(t)$ be continuous for $0 \le t \le 1$ and $\varphi(0) = \varphi_0$. (See note[7]).

[7]In the book: Chinn W.G., Steenrod N.E, (1966), *First Concepts of Topology,*

251. Let $\varphi(t)$ and $\varphi'(t)$ be two functions describing the continuous variation of $\arg z(t)$ along the curve C. Prove that $\varphi(t) - \varphi'(t) = 2\pi k$, where k is a given integer which does not depend on t.

252. Prove that if one chooses a value $\varphi(0) = \varphi_0$ then the function $\varphi(t)$, which describes the continuous variation of $\arg z(t)$ along the curve C, is uniquely defined.

253. Let $\varphi(t)$ be a function describing the continuous variation of $\arg z(t)$. Prove that the function $\psi(t) = \varphi(t) - \varphi(0)$ is uniquely defined by the function $z(t)$ and does not depend on the choice of $\varphi(0)$.

From the statement of Problem **253** it follows, in particular for $t = 1$, that for a continuous curve C not passing through the point $z = 0$ the quantity $\varphi(1) - \varphi(0)$ is uniquely defined by the continuity condition of $\varphi(t)$.

DEFINITION. The difference $\varphi(1) - \varphi(0)$ is called the *variation of the argument along the curve C*.

254. Find the variation of the argument along the curves with the following parametric equations: a) $z(t) = \cos \pi t + i \sin \pi t$; b) $z(t) = \cos 2\pi t + i \sin 2\pi t$; c) $z(t) = \cos 4\pi t + i \sin 4\pi t$; d) $z(t) = (1 - t) + it$.

255. What is the variation of the argument along the curves shown in Figure 18?

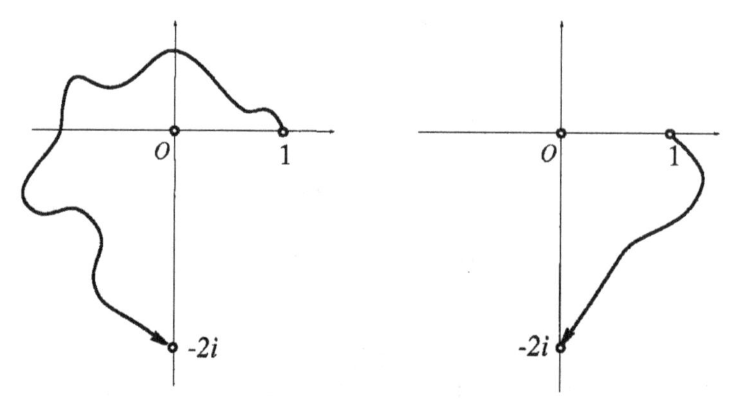

FIGURE 18

(Mathematical Association of America: Washington), (§§20–23), the angle characterized by the given curve is precisely defined. Using this angle one easily obtains the statement of Theorem 6: it suffices to put $\varphi(t) = \varphi_0 + \varphi_1(t)$, where $\varphi_1(t)$ is the angle characterized by the curve segment from $z(0)$ to $z(t)$.

REMARK. If a curve C is closed, i.e., $z(1) = z(0)$, then the quantity $\varphi(1) - \varphi(0)$ is equal to $2\pi k$, where k is an integer number.

DEFINITION. If for a continuous closed curve C not passing through a point $z = 0$ the variation of the argument is equal to $2\pi k$, then we say that the curve C turns k times around the point $z = 0$.

256. How many times do the following curves turn around the point $z = 0$: a) $z(t) = 2\cos 2\pi t + 2i \sin 2\pi t$ (Figure 19); b) $z(t) = \frac{1}{2}\cos 4\pi t - \frac{1}{2} i \sin 4\pi t$ (Figure 20); c) the curve in Figure 21; d) the curve in Figure 22?

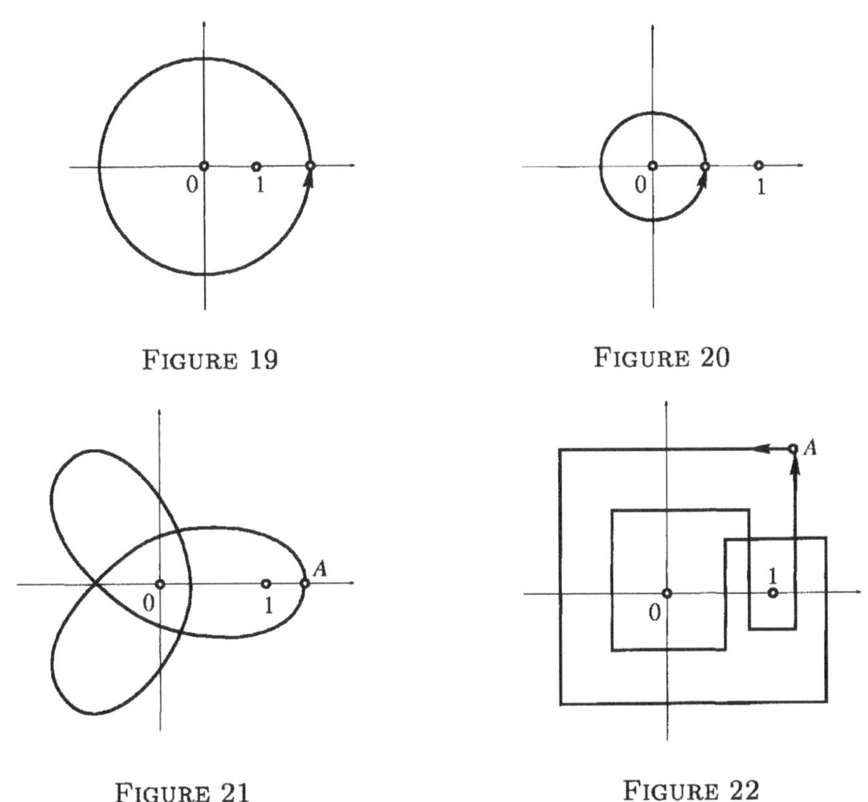

FIGURE 19

FIGURE 20

FIGURE 21

FIGURE 22

257. Prove that the number of turns of a closed continuous curve around the point $z = 0$ does not depend on the choice of the initial point, and depends only on the orientation of the curve.

258. Suppose that a curve C with equation $z_1(t)$ turns k times around the point $z = 0$. How many times does the curve with the equation $z_2(t)$

turn around the point $z = 0$ if: a) $z_2(t) = 2 \cdot z_1(t)$; b) $z_2(t) = -z_1(t)$; c) $z_2(t) = z_0 \cdot z_1(t)$, where $z_0 \neq 0$; d) $z_2(t) = \overline{z_1}(t)$, where $\overline{z}$ is the conjugate of z?

DEFINITION. Suppose that a continuous curve C with the equation $z_1(t)$ does not pass through the point $z = z_0$. We thus say that the curve C turns k times around the point $z = z_0$ if the curve with equation $z_2(t) = z_1(t) - z_0$ turns k times around the point $z = 0$ (Figure 23).

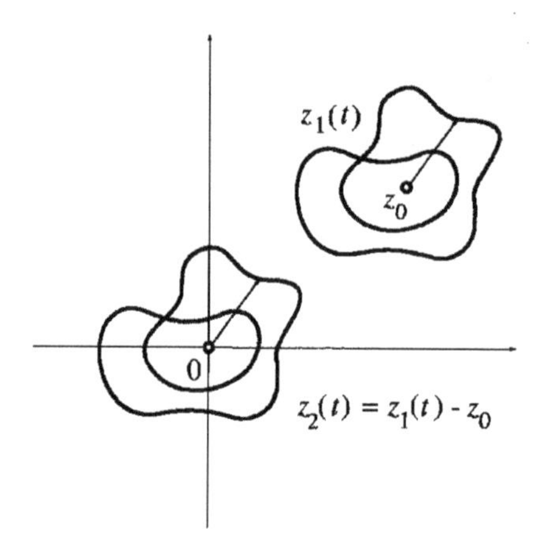

FIGURE 23

Consequently to define the number of turns of a curve around the point $z = z_0$ we have to look at the rotation of the vector $z_1(t) - z_0$, i.e., the vector joining points z_0 and $z_1(t)$ (cf., **221**).

259. How many times do the curves described in Problem **256** turn around the point $z = 1$?

260. Let $z_1(t)$ and $z_2(t)$ be the equations of two curves C_1 and C_2 not passing through the point $z = 0$. Let the variations of the argument along these curves be equal, respectively, to φ_1 and φ_2. What is the variation of the argument along the curve C with the equation $z(t)$ if: a) $z(t) = z_1(t) \cdot z_2(t)$; b) $z(t) = z_1(t)/z_2(t)$?

2.8 Images of curves: the basic theorem of the algebra of complex numbers

Consider two planes of complex numbers: the z plane and the w plane, and a function $w = f(z)$, which puts into correspondence with every value z a value w uniquely defined. If on the z plane there is a continuous curve C having equation $z(t)$ then by the function $w = f(z)$ every point of this curve is sent to a point of the w plane. Hence if the function $f(z)$ is continuous we also obtain on the w plane a continuous curve, having equation $w_0(t) = f(z(t))$. We shall denote this curve by $f(C)$.

261. What is the curve $f(C)$ if $w = f(z) = z^2$ and the curve C is
a) a quarter of a circle: $z(t) = R(\cos(\pi t/2) + i\sin(\pi t/2))$;
b) a semi-circle: $z(t) = R(\cos\pi t + i\sin\pi t)$;
c) a circle: $z(t) = R(\cos 2\pi t + i\sin 2\pi t)$?

262. Let the variation of the argument along a curve C be equal to φ. What is the variation of the argument along the curve $f(C)$ if: a) $f(z) = z^2$; b) $f(z) = z^3$; c) $f(z) = z^n$; where n is a non-zero arbitrary integer?

263. Suppose that the curve C turns k times around the point $z = z_0$. How many times does the curve $f(C)$ turn around the point $w = 0$ if $f(C) = (z - z_0)^n$?

264. Suppose that a curve C turns around the points $z = 0$, $z = 1$, $z = i$, $z = -i$ respectively k_1, k_2, k_3, k_4 times. How many times does the curve $f(C)$ turn around the point $w = 0$ if: a) $f(z) = z^2 - z$; b) $f(z) = z^2 + 1$; c) $f(z) = (z + iz)^4$; d) $f(z) = z^3 - z^2 + z - 1$?

Consider the equation

$$a_0 z^n + a_1 z^{n-1} + \ldots + a_{n-1} z + a_n = 0,$$

where all the a_is are arbitrary complex numbers, $n \geq 1$, and $a_0 \neq 0$. Our first aim is to show that this equation has at least one complex root. If $a_n = 0$ then the equation possesses the root $z = 0$. In the sequel, therefore, we will assume that $a_n \neq 0$.

Let us denote by A the maximum amongst the numbers $|a_0|$, $|a_1|, \ldots,$ $|a_n|$. Since $a_0 \neq 0$, $A > 0$. Choose two real numbers R_1 and R_2 with such conditions: let R_1 be sufficiently small to satisfy the inequalities: $R_1 \leq 1$ and $R_1 < |a_n|/(10An)$; let R_2 be sufficiently large to satisfy the inequalities: $R_2 \geq 1$ and $R_2 > 10An/|a_0|$.

265. Let $|z| = R_1$. Prove that

$$|a_0 z^n + \ldots + a_{n-1} z| < \frac{|a_n|}{10}.$$

266. Let $|z| = R_2$. Prove that

$$\left|\frac{a_1}{z} + \ldots + \frac{a_n}{z^n}\right| < \frac{|a_0|}{10}.$$

Let us denote by C_R the curve with equation $z(t) = R(\cos 2\pi t + i \sin 2\pi t)$ (i.e., the circle with radius equal to R, oriented counterclockwise). Since the curve C_R is closed ($z(1) = z(0)$), the curve $f(z) = a_0 z^n + \ldots + a_n$ is closed as well ($f(z(1)) = f(z(0))$). Let $\nu(R)$ be the number of turns of the curve $f(C_R)$ around the point $w = 0$ (if $f(C_R)$ does not pass through the point $w = 0$).

267. Calculate $\nu(R_1)$ and $\nu(R_2)$.

Let us now increase the radius R from R_1 to R_2. The curve $f(C_R)$ will consequently be deformed from $f(C_{R_1})$ to $f(C_{R_2})$. If for a value R^* the curve $f(C_{R^*})$ does not pass through the point $w = 0$, by a sufficiently small variation of R near R^* the curve $f(C_R)$ will turn out to be deformed by too small an amount for the number of its turns around the point $w = 0$ to change: the function $\nu(R)$ is indeed continuous at the value R^*. If the curves $f(C_R)$ avoid the point $w = 0$ for all values of R between R_1 and R_2, then $\nu(R)$ is a continuous function for all $R_1 \leq R \leq R_2$. Since the function $\nu(R)$ takes only integer values it can be continuous only if for all values of it $R_1 \leq R \leq R_2$ takes a unique value. But, solving Problem **267**, we have obtained that $\nu(R_1) = 0$ and $\nu(R_2) = n$. Therefore the claim that none of the curves $f(C_R)$ passes through the point $z = 0$ for all $R_1 \leq R \leq R_2$ is untrue. We thus have $f(z) = 0$ for a certain z. In this way we have proved the following theorem[8].

THEOREM 7. (The fundamental theorem of the algebra of complex numbers[9]). *The equation*

$$a_0 z^n + a_1 z^{n-1} + \ldots + a_{n-1} z + a_n = 0,$$

[8]Our reasoning contains some non-rigourous passages: it must be considered, in general, as an idea of the proof. This reasoning can, however, be made exact (though in a non-simple way). See, for example, Chinn W.G., Steenrod N.E, (1966), *First Concepts of Topology*, (Mathematical Association of America: Washington).

[9]This theorem was proved in 1799 by the German mathematician C.F. Gauss (1777–1855).

where all the a_is are arbitrary complex numbers, $n \geq 1$, and $a_0 \neq 0$, has at least one complex root.

268. Prove Bézout's theorem[10]: *If z_0 is a root of the equation $a_0z^n + \ldots + a_{n-1}z + a_n = 0$, then the polynomial $a_0z^n + \ldots + a_{n-1}z + a_n$ is divisible by the binomial $z - z_0$ without remainder.*

269. Prove that the polynomial $a_0z^n + \ldots + a_{n-1}z + a_n$, where $a_0 \neq 0$, can be represented in the form:

$$a_0z^n + \ldots + a_{n-1}z + a_n = a_0(z - z_1)(z - z_2)\cdots(z - z_n).$$

REMARK. Suppose that the polynomial $P(z)$ decomposes into factors:

$$P(z) = a_0(z_z - 1)(z_z - 2) \cdot \ldots \cdot (z - z_n).$$

The left member of the equation is equal to 0 if and only if at least one of the factors inside brackets is equal to 0 (cf., **195, 197**). Hence the roots of equation $P(z) = 0$ are the numbers $z_1, z_2, \ldots, z_n$ and them alone.

270. Let z_0 be a root of the equation

$$a_0z^n + a_1z^{n-1} + \ldots + a_{n-1}z + a_n = 0,$$

where the a_is are real numbers. Prove that the number $\overline{z_0}$, the conjugate of z_0, is also a root of that equation.

271. Suppose that the equation with real coefficients

$$a_0z^n + a_1z^{n-1} + \ldots + a_{n-1}z + a_n = 0$$

has a complex root z_0, being not a real number. Prove that the polynomial

$$a_0z^n + a_1z^{n-1} + \ldots + a_{n-1}z + a_n$$

is divisible by a polynomial of second degree with real coefficients.

272. Prove that every polynomial with real coefficients can be written in the form of a product of polynomials of first and second degree with real coefficients.

REMARK. From the result of Problem **272** it follows that the sole irreducible polynomials (cf., §2.3) over the field of real numbers are the

[10]É. Bézout (1730–1783), French mathematician.

polynomials of first degree and those of second degree with no real roots. We had used this property in §2.3 of this chapter. Over the field of complex numbers, according to the result of Problem **269**, only polynomials of first degree are irreducible.

DEFINITION. Let z_0 be a root of the equation

$$a_0z^n + a_1z^{n-1} + \ldots + a_{n-1}z + a_n = 0.$$

One says that z_0 is a *root of order* k if the polynomial $a_0z^n+\ldots+a_{n-1}z+a_n$ is divisible by $(z-z_0)^k$ and not by $(z-z_0)^{k+1}$.

273. Which is the order of the roots $z = 1$ and $z = -1$ in the equation

$$z^5 - z^4 - 2z^3 + 2z^2 + z - 1 = 0?$$

DEFINITION. One calls the *derivative of the polynomial*

$$P(z) = a_0z^n + a_1z^{n-1} + \ldots + a_kz^{n-k} + \ldots + a_{n-1}z + a_n$$

the polynomial

$$P'(z) = a_0nz^{n-1} + a_1(n-1)z^{n-2} + \ldots + a_k(n-k)z^{n-k-1} + \ldots + a_{n-1}.$$

The derivative is usually denoted by the symbol $'$ (prime).

274. Let $P(z)$ and $Q(z)$ be two polynomials. Prove that: a) $(P(z)+Q(z))' = P'(z)+Q'(z)$; b) $(c\cdot P(z))' = c\cdot P'(z)$, where c is an arbitrary complex constant; c) $(P(z)Q(z))' = P'(z)Q(z) + P(z)Q'(z)$.

275. Let $P(z) = (z-z_0)^n$ ($n \geq 1$ integer). Prove that $P'(z) = n(z-z_0)^{n-1}$.

276. Prove that if the equation $P(z) = 0$ has a root z_0 of order $k > 1$ then the equation $P'(z) = 0$ has a root z_0 of order $k-1$, and that if the equation $P(z) = 0$ has a root z_0 of first order then $P'(z_0) \neq 0$.

2.9 The Riemann surface of the function $w = \sqrt{z}$

We had considered single-valued functions for which there corresponds a unique value of the function to every value of the independent variable. In what follows we will deal mainly with multi-valued functions, for which there correspond distinct values of the function to a value of the independent variable[11]. We will explain the reason for our interest

[11]Whenever the context is sufficiently clear the term multi-valued will be omitted.

in such functions. In fact, the final aim of our study is the proof of the Abel theorem, according to which a function, expressing the roots of the general equation of fifth degree, cannot be represented by radicals. But this function is multi-valued, because an equation of fifth degree has, in general, for given coefficients, five roots. Also the functions which are represented by radicals are multi-valued.

The principal idea of the demonstration of the Abel theorem is the following. We put into correspondence with a multi-valued function of a complex variable a certain group, the so called *monodromy group*.

The monodromy group of the function expressing the roots of the general equation of fifth degree in terms of a parameter z cannot coincide with any monodromy group of functions representable by radicals, and therefore this function cannot be represented by radicals.

In order to introduce the notion of the monodromy group we consider first another notion very important in the theory of functions of one complex variable — the notion of the *Riemann*[12] *surface of a function.* We begin by the construction of the Riemann surface for the simplest example of a multi-valued function, the function $w = \sqrt{z}$.

We already know that the function $w = \sqrt{z}$ takes the single value $w = 0$ for $z = 0$ and two values for all values $z \neq 0$ (cf., **229**). Moreover, if w_0 is one of the values of $\sqrt{z_0}$ then the other value of $\sqrt{z_0}$ is $-w_0$.

277. Find all values of: a) $\sqrt{1}$; b) $\sqrt{-1}$; c) $\sqrt{i}$; d) $\sqrt{1+i\sqrt{3}}$ (here $\sqrt{3}$ is the positive value of the root).

Let us cut the z plane along the negative side of the real axis from 0 to $-\infty$, and for every z not belonging to the cut let us choose the value $w = \sqrt{z}$ which lies on the right half w plane. In this way we obtain a continuous single-valued function over the whole z plane, except the cut. This function, which we denote by ${}_1\sqrt{z}$, defines a continuous and single-valued mapping of the z plane, except the cut, on the right half w plane (Figure 24).

REMARK. If we choose $\arg z$ in such a way that $-\pi < \arg z < \pi$, then for the function ${}_1\sqrt{z}$ we obtain $\arg_1\sqrt{z} = \frac{1}{2}\arg z$ (cf., **229**). Therefore under the mapping $w = {}_1\sqrt{z}$ the z plane shrinks like a fan whose radii are shortened as its opening angle is halved.

If we now choose, for every z not lying on the cut, the value of $w = \sqrt{z}$ which lies on the left half w plane we obtain another function, still single-

[12] B. Riemann (1826–1866), German mathematician.

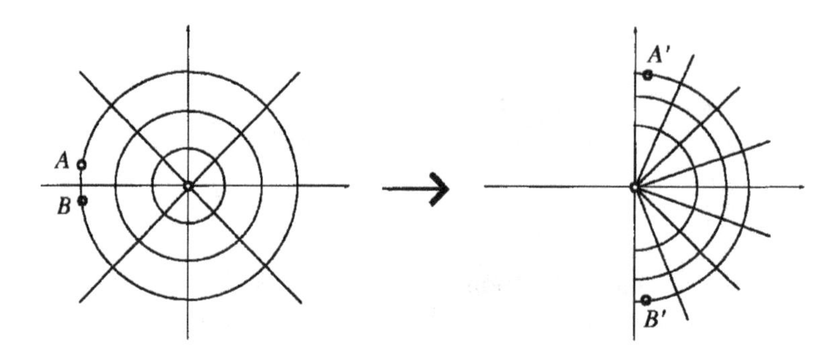

FIGURE 24

valued and continuous over the whole z plane except the cut. This function, which we denote by ${}_2\sqrt{z}$, defines a continuous single-valued mapping of the z plane except the cut, on the left half w plane (Figure 25). Here ${}_2\sqrt{z} = -{}_1\sqrt{z}$.

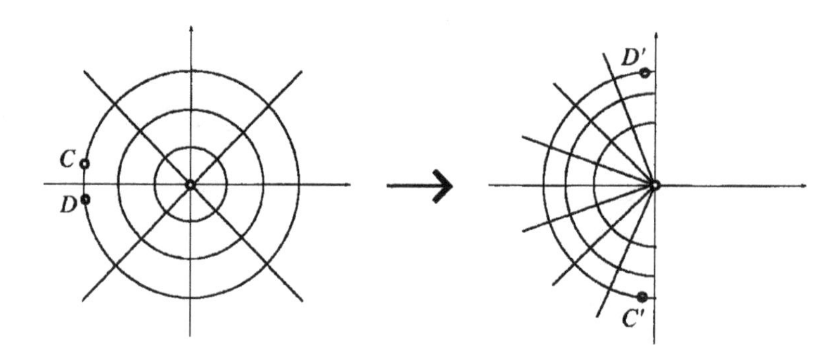

FIGURE 25

Functions ${}_1\sqrt{z}$ and ${}_2\sqrt{z}$ so defined are called the *continuous single-valued branches* of the function $w = \sqrt{z}$ (for the given cut).

Consider now two copies of the z plane, which we shall call *sheets*, and cut every sheet along the negative side of the real axis from 0 to $-\infty$ (Figure 26).

Let us take the function ${}_1\sqrt{z}$ on the first sheet and the function ${}_2\sqrt{z}$ on the second sheet. Thus we can see the functions ${}_1\sqrt{z}$ and ${}_2\sqrt{z}$ as a unique single-valued function, defined no longer on the z plane but on a more complex surface consisting of two distinct sheets.

So if a point z moves continuously on the first (or on the second) sheet,

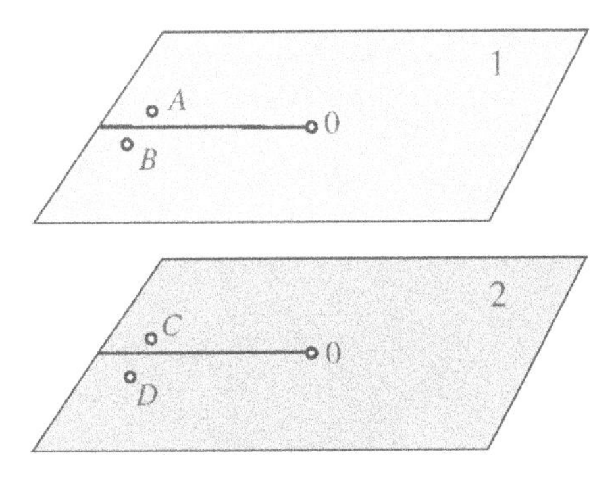

FIGURE 26

not crossing the cut, the single-valued function that we have defined varies continuously. But if the point z, moving, for example, on the first sheet, traverses the cut, then the continuity is lost. This follows from the close points A and B on the z plane being sent by the mapping ${}_1\sqrt{z}$ respectively to points A' and B', far from each other (cf., Figure 24).

On the other hand, it is easy to see in Figures 24 and 25 that the image of the point A under the mapping $w = {}_1\sqrt{z}$ (the point A') is close to the image of the point D under the mapping $w = {}_2\sqrt{z}$ (the point D').

Consequently if, traversing the cut, the point z moves from the upper side of one sheet to the lower side of the other sheet, the single-valued function we have defined varies continuously. To guarantee that the point z moves as requested, we consider the upper side of the cut on the first sheet joined to the lower side of the cut on the second sheet, and the lower side of the cut on the first sheet joined to the upper side of the cut on the second sheet (Figure 27). Furthermore, when joining the sheets we shall add between them the real negative axis from the point 0 to $-\infty$. During the first joining, for the points z which lie on this half axis we choose the values of $w = \sqrt{z}$ lying on the positive side of the imaginary axis, and during the second joining we choose the values of $w = \sqrt{z}$ which lie on the negative side of the imaginary axis.

By means of the joining explained above we had transformed the 2-valued function $w = \sqrt{z}$ into another function which is single-valued and continuous, no longer on the z plane but on a new surface. This surface is called the *Riemann surface of the function* $w = \sqrt{z}$.

Attempts to do the joining without intersections (and without reversal of the plane) are made in vain. We assume that Figure 27 represents an

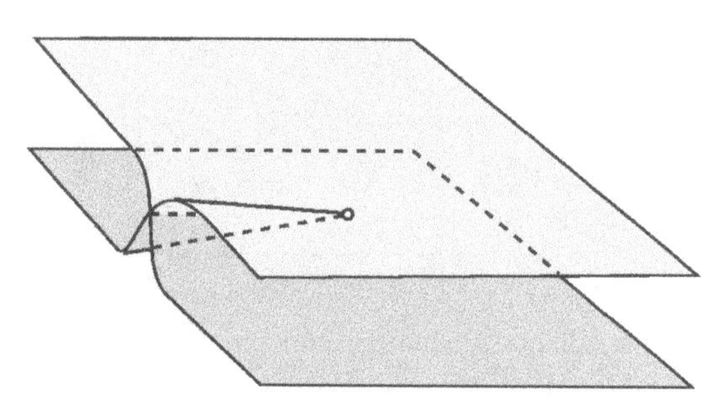

FIGURE 27

image of the Riemann surface of the function $w = \sqrt{z}$, using the further convention that the self-intersection along the negative side of the real axis is only apparent. In order to understand this, consider the following example. In Figure 7 (§1.12) we see the image of a cube. Whereas in this image some edges intersect, we know that such intersections are only apparent, and this knowledge allows us to avoid mistaken interpretations.

The Riemann surface of a multi-valued function $w(z)$ can be constructed in a way analogous to that used for the Riemann surface of the function $w = \sqrt{z}$. To do this we have first to separate the continuous single-valued branches of the function $w(z)$, excluding the points z which belong to the cuts. Afterwards we have to join the branches obtained, choosing the values on the cuts in such a way as to obtain a continuous single-valued function on the whole surface. The surface obtained is called the Riemann surface of the multi-valued function[13] $w(z)$.

It remains, therefore, to explain how to separate the continuous single-valued branches of a function $w(z)$ and how to join them. To understand this, look again more attentively at the Riemann surface of the multi-valued function $w = \sqrt{z}$.

Let $w(z)$ be a multi-valued function, and fix one of the values w_0 of the function $w(z)$ at a certain point z_0. Let $w'(z)$ be a continuous single-valued branch of the function $w(z)$ defined on some region of the z plane (for example, on the whole plane, except some cuts), and such that $w'(z_0) = w_0$. Suppose, moreover, that there exists a continuous curve C, connecting z_0 to a point z_1, lying entirely in the region of the plane considered. Thus while the point z moves continuously along the curve

[13]This type of construction is not always possible for every multi-valued function, but for all functions we shall consider in the sequel, the construction of their Riemann surfaces will, in fact, be possible.

C from z_0 to z_1, the function $w'(z)$ varies continuously from $w'(z_0)$ to $w'(z_1)$.

One can also use this property conversely, i.e., for the definition of the function $w'(z)$.

Indeed, suppose at a certain point z_0 one of the values w_0 of the function $w(z)$ be chosen. Let C be a continuous curve beginning at z_0 and ending at a certain point z_1. Moving along the curve C we choose for every point z lying on C one of the values of the function $w(z)$ in such a way that these values vary continuously while we move along the curve C starting from the value w_0. So when we arrive at the point z_1 the value $w_1 = w(z_1)$ is completely defined. We say that w_1 is the value of $w(z_1)$ *defined by continuity along the curve* C under the condition $w(z_0) = w_0$. If the values of the function $w(z)$, chosen for all points of the curve C, are represented on the w plane then we obtain a continuous curve beginning at the point w_0 and ending at the point w_1. This curve is one of the continuous images of the curve C under the mapping $w = w(z)$.

278. For the function $w(z) = \sqrt{z}$ let us choose $w(1) = \sqrt{1} = 1$. Define by continuity $w(-1) = \sqrt{-1}$ along the following curves: a) the upper semi-circle of radius 1 with centre at the origin of the coordinates; b) the lower semi-circle (Figure 28).

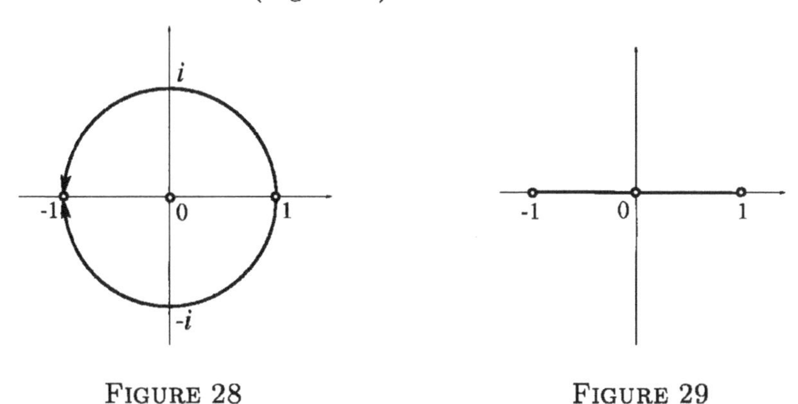

FIGURE 28 FIGURE 29

In fact, using for a function the definition by continuity along a certain curve we may encounter some difficulties. Consider the following example.

279. Find all continuous images $w_0(t)$ of a curve C with parametric equation $z(t) = 2t - 1$ (Figure 29) under the mapping $w = \sqrt{z}$, beginning: a) at the point i; b) at the point $-i$.

From the solution of Problem **279** we obtain that also by fixing the image of the initial point of the curve C the continuous image of the curve C under the mapping $w = \sqrt{z}$ may be defined non-uniquely. The uniqueness is lost when the curve C passes through the point $z = 0$. In fact, for the function $w = \sqrt{z}$ the uniqueness is lost only in this case, because only in this case do the two images of the point $z(t)$ melt into one point.

To avoid the non-uniqueness in the definition of the images of curves under the mapping $w = \sqrt{z}$ we may exclude the point $z = 0$ and forbid any curve to pass through this point. This restriction, however, does not always allows the continuous single-valued branches of the function $\sqrt{z}$ to be separated.

Indeed, if we fix at a point z_0 one of the values $w_0 = w(z_0)$ and we try to define $w(z_1)$ at a certain point z_1 by continuity along two distinct curves joining z_0 and z_1, we may obtain different values of $w(z_1)$ (for example, see **278**). Observe now how we can avoid the non-uniqueness of the obtained value.

280. Suppose the variation of the argument of $z(t)$ along a curve C be equal to φ. Find the variation of the argument of $w_0(t)$ along an arbitrary continuous image of the curve C under the mapping $w(z) = \sqrt{z}$.

281. Let $w(z) = \sqrt{z}$ and choose $w(1) = \sqrt{1} = -1$. Define the value of $w(i) = \sqrt{i}$ by continuity along: a) the segment joining points $z = 1$ and $z = i$; b) the curve with the parametric equation $z(t) = \cos \frac{3}{2}\pi t - i \sin \frac{3}{2}\pi t$; c) the curve with the parametric equation $z(t) = \cos \frac{5}{2}\pi t + i \sin \frac{5}{2}\pi t$.

282. Let $w(z) = \sqrt{z}$ and choose at the initial point of a curve C, $w(1) = \sqrt{1} = 1$. Define by continuity along the curve C the value of $w(1) = \sqrt{1}$ at the end point if the curve C has the equation: a) $z(t) = \cos 2\pi t + i \sin 2\pi t$; b) $z(t) = \cos 4\pi t - i \sin 4\pi t$; c) $z(t) = 2 - \cos 2\pi t - i \sin 2\pi t$.

283. Let C be a closed curve on the z plane (i.e., $z(0) = z(1)$). Prove that the value of the function $\sqrt{z}$ at the end point of the curve C, defined by continuity, coincides with the value at the initial point if and only if the curve C wraps around the point $z = 0$ an even number of times.

In the sequel it will be convenient to use the following notation:

DEFINITION. Let C be a continuous curve with a parametric equation $z(t)$. We shall denote by C^{-1} the curve geometrically identical to C but oriented in the opposite direction; its equation is (cf., **247**) $z_1(t) = z(1-t)$.

DEFINITION. Suppose that the initial point of a curve C_2 coincides with the final point of a curve C_1. We will denote by C_1C_2 the curve obtained by joining the end point of C_1 to the initial point of C_2 (cf., **248**).

284. Let C_1 and C_2 be two curves, joining the points z_1 and z_2 and let one of the values of $\sqrt{z_0} = w_0$ be chosen. Prove that the values of $\sqrt{z_1}$, defined by continuity along the curves C_1 and C_2, are equal if and only if curve $C_1^{-1}C_2$ (Figure 30) turns around the point $z = 0$ an even number of times.

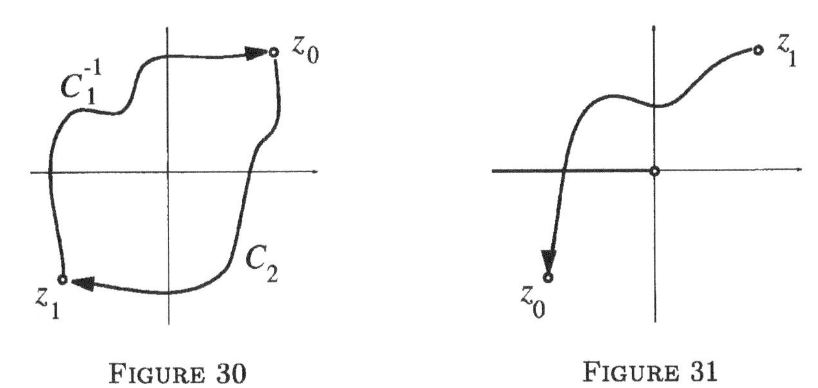

FIGURE 30 FIGURE 31

From the statement of the last problem it follows that, in particular, if the curve $C_1^{-1}C_2$ turns zero times around the point $z = 0$, then the values of the function $\sqrt{z}$ at the final points of the curves C_1 and C_2 will coincide if the values at the initial point coincide.

So to separate the single-valued continuous branches of the function $\sqrt{z}$ it suffices to take the curves C_1 and C_2 in such a way that the curve $C_1^{-1}C_2$ does not turn around the point $z = 0$. For this it suffices to make a cut from the point $z = 0$ to infinity avoiding intersecting the curve. We operated exactly in this way drawing the cut from the point $z = 0$ to $-\infty$ along the negative side of the real axis.

If, after having made the cut, one fixes at a certain point z_0 one of the values, $w_0' = \sqrt{z_0}$, and if the value at every point z_1 is defined by continuity along an arbitrary curve C joining z_0 and z_1 and not passing through the cut, then at every point of the plane (except those on the cut) a certain single-valued continuous branch ${}_1\sqrt{z}$ of the function $\sqrt{z}$ is defined. If at the point $z_0 = 0$ one fixes the other value $w_0'' = \sqrt{z_0}$ then one defines the other branch ${}_2\sqrt{z}$ of the function $\sqrt{z}$.

285. Prove that ${}_1\sqrt{z} \neq {}_2\sqrt{z}$ for every point z outside the cut.

286. Fix the value $w' = {}_1\sqrt{z'}$ at a certain point z' and define the values of function $\sqrt{z}$ at the other points of the z plane (except the cut) by continuity along the curves starting from z' and not intersecting the cut. Prove that the continuous single-valued branches so obtained coincide with the function ${}_1\sqrt{z}$ (defined by the value at point z').

It follows from the result of Problem **286** that, choosing as initial point different points of the z plane, one obtains the same splitting of the Riemann surface into single-valued continuous branches. Therefore this splitting depends only on the way in which we have made the cut.

287. Suppose that points z_0 and z_1 do not lie on the cut and that the curve C, joining them, traverses the cut once (Figure 31). Choose a value $w_0 = \sqrt{z_0}$ and by continuity along C define the value $w_1 = \sqrt{z_1}$. Prove that values w_0 and w_1 correspond to different branches of $\sqrt{z}$.

In this way, traversing the cut, one moves from one branch to the other, i.e., the branches join each other exactly as we have put them together joining the sheets (Figure 29). One obtains in this way the Riemann surface of the function $\sqrt{z}$.

We say that a certain property is satisfied by any turn around a point z_0 if it is satisfied by a simple turn counterclockwise along all circles centred on z_0 and with sufficiently small radii[14].

288. Prove that by a turn around a point z_0 one remains on the same sheet of the Riemann surface of the function $\sqrt{z}$ if $z_0 \neq 0$, and one moves onto the other sheet if $z_0 = 0$.

The following notion is very important in the sequel.

DEFINITION. Points, around which one may turn and move from one sheet to another (i.e., changing the value of the function) are called the *branch points* of the given multi-valued function.

The Riemann surface of the function $\sqrt{z}$ can be represented by a scheme (Figure 32).

This scheme shows that the Riemann surface of the function $w = \sqrt{z}$ has two sheets, that the point $z = 0$ is a branch point, and that by turning around the point $z = 0$ one moves from one of the sheets to the other. Moreover, the arrow between the two sheets in correspondence with the point $z = 0$ indicates the passage from one sheet to the other not only by a turn around the point $z = 0$, but also by crossing a point of the

[14]More precisely, this means that there exists a real number $\delta > 0$ such that the property mentioned is satisfied by any turn along any circle with centre z_0 and radius smaller than δ.

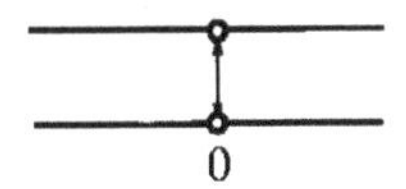

FIGURE 32

cut, joining the point $z = 0$ to infinity. We have seen that this relation between branch points and cuts coming from these points is not arbitrary.

In the sequel instead of the Riemann surfaces of multi-valued functions we will represent their schemes.

2.10 The Riemann surfaces of more complicated functions

Consider the multi-valued function $\sqrt[3]{z}$.

289. Let the variation of the argument along the curve $z(t)$ be equal to φ, and let $w_0(t)$ be the continuous image of the curve $z(t)$ under the mapping $w = \sqrt[3]{z}$. Find the variation of the argument along the curve $w_0(t)$.

290. Find the branch points of the function $\sqrt[3]{z}$.

291. Let us make a cut from the point $z = 0$ to $-\infty$ along the negative side of the real axis and assume, moreover, the single-valued continuous branches of function $w = \sqrt[3]{z}$ to be given by the conditions:

$$\begin{aligned} &f_1(1) = 1, \\ &f_2(1) = \cos(2\pi/3) + i\sin(2\pi/3) = -1/2 + i\sqrt{3}/2, \\ &f_3(1) = \cos(4\pi/3) + i\sin(4\pi/3) = -1/2 - i\sqrt{3}/2. \end{aligned}$$

Find: a) $f_1(i)$; b) $f_2(i)$; c) $f_1(8)$; d) $f_3(8)$; e) $f_3(-i)$.

292. Draw the Riemann surface (and its scheme) of the function $w = \sqrt[3]{z}$.

293. Let C be a continuous curve with parametric equation $z(t)$, and let w_0 be one of the values of $\sqrt[n]{z_0}$. Prove that there exists at least one continuous image of the curve C under the mapping $w(z) = \sqrt[n]{z}$, starting at the point w_0.

294. Suppose the variation of the argument along a curve $z(t)$ be equal to φ and $w_0(t)$ be the continuous image of the curve $z(t)$ under the mapping $w(z) = \sqrt[n]{z}$. Find the variation of the argument along the curve $w_0(t)$.

295. Find the branch points of the function $\sqrt[n]{z}$.

In §2.5 we have used the notation

$$\epsilon_n = \cos(2\pi/n) + i\sin(2\pi/n)$$

and we have considered some properties of this complex number.

296. Suppose that a curve $z(t)$ does not pass through the point $z = 0$ and that $w_0(t)$ is one of the continuous images of the curve $z(t)$ under the mapping $w = \sqrt[n]{z}$. Find all the continuous images of the curve $z(t)$ under the mapping $w = \sqrt[n]{z}$.

Suppose that two curves C_1 and C_2 join a point z_0 to a different point z_1. Exactly as in the case of function $\sqrt{z}$ (cf., **284**), one proves that if the curve $C_1^{-1}C_2$ does not turn around the point $z = 0$ then the function $\sqrt[n]{z}$ is uniquely defined by continuity along the curves C_1 and C_2. In this way if, as for the function $\sqrt{z}$, we make a cut from the point $z = 0$ to infinity the image of the function $w = \sqrt[n]{z}$ turns out to be decomposed into single-valued continuous branches.

297. Make a cut from the point $z = 0$ to infinity, not passing through the point $z = 1$, and define the single-valued continuous branches of the function $\sqrt[n]{z}$ by the conditions: $f_i(1) = \epsilon_n^i$, where i takes the integer values from 0 to $n-1$. How are the branches $f_i(z)$ expressed in terms of $f_0(z)$?

298. Draw the scheme of the Riemann surface of the function $\sqrt[n]{z}$.

299. Find the branch points and draw the scheme of the Riemann surface for the function $\sqrt{z-1}$.

300. Find the branch points and draw the scheme of the Riemann surface of the function $\sqrt[n]{z+i}$.

When a multi-valued function has several branch points, in order to separate the single-valued continuous branches we make the cuts from every branch point to infinity along lines which do not intersect each other.

In this way the scheme of the Riemann surface of a given function may depend essentially on the choice of the cuts from the branch points to infinity (some examples of such cases are given later in Problems **327** and **328**). Whenever we are in this situation, and only in this case, we will indicate how to make the cuts.

The schemes of the Riemann surfaces drawn by the reader solving the above proposed problems may differ from the schemes given in Solutions, owing to different possibilities in numbering the sheets. Yet different schemes must coincide under a suitable relabelling.

301. Let $f(z)$ be a continuous single-valued function and C be a continuous curve on the z plane, starting at the point z_0. Let w_0 be one of the values of $\sqrt[n]{f(z_0)}$. Prove that there exists at least one continuous image of the curve C under the mapping $w = \sqrt[n]{f(z)}$ starting at the point w_0.

From the result of Problem **301** it follows that it is possible to define the function $\sqrt[n]{f(z)}$ by continuity along an arbitrary curve not passing through the points at which the uniqueness of the continuous images is lost.

302. Let $f(z)$ be a continuous single-valued function and $w_0(z)$ one of the single-valued continuous branches (according to the corresponding cuts) of the function $w = \sqrt[n]{f(z)}$. Find all the continuous single-valued branches (according to the same cuts) of the function $w(z)$.

303. Find all the branch points and draw the schemes of the Riemann surface for the functions: a) $\sqrt{z(z-i)}$; b) $\sqrt{z^2+1}$.

304. Draw the schemes of the Riemann surfaces of the following functions: a) $\sqrt[3]{z^2-1}$; b) $\sqrt[3]{(z-1)^2 z}$; c) $\sqrt[3]{(z^2+1)^2}$.

305. Separate the single-valued continuous branches and draw the scheme of the Riemann surface of the function $\sqrt{z^2}$.

REMARK. From the result of Problem **305** we obtain that the point $z = 0$ is not a branch point of the function $\sqrt{z^2}$. However, the images of the curves passing through the point $z = 0$ are not uniquely defined. For example, the continuous images of the broken line AOB (Figure 33) under the mapping $w = \sqrt{z^2}$ are the broken lines COD, COF, EOD, and EOF (Figure 33). Passing through the point $z = 0$ one may either remain on the same sheet (the lines COD and EOF) or move onto the other one (the lines COF and EOD). The Riemann surface of function $w(z) = \sqrt{z^2}$ is shown in Figure 34.

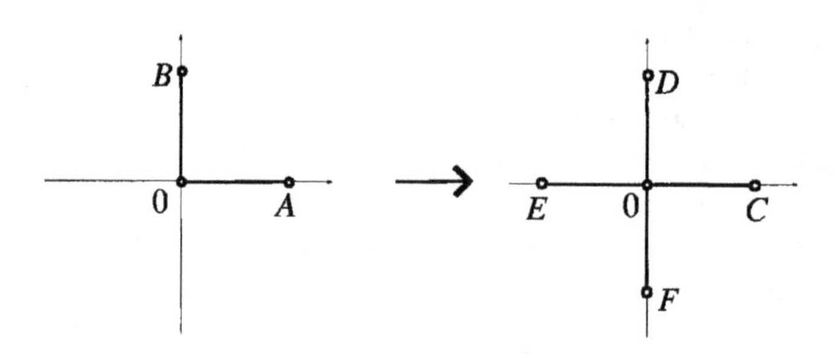

FIGURE 33

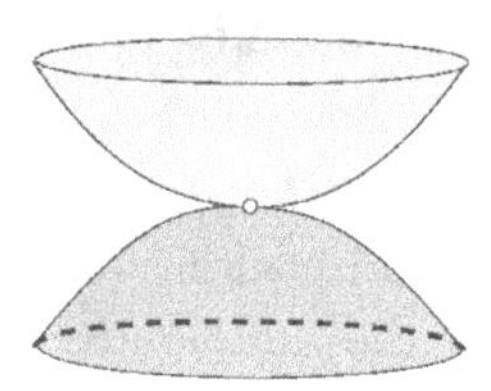

FIGURE 34

DEFINITION. Points where the uniqueness of the continuous images of the curves is lost but that are not branch points are called the *non-uniqueness points* of the given function.

When building the Riemann surfaces one should draw no cuts from the points of non-uniqueness to infinity: in drawing any curve these points must always be avoided.

306. Draw the schemes of the Riemann surfaces of the following functions: a) $\sqrt[4]{z^2+2}$; b) $\sqrt[4]{z^2}$; c) $\sqrt[4]{(z-1)^2(z+1)^3}$; d) $\sqrt[4]{(z^2-1)^3(z+1)^3}$; e) $\sqrt[4]{z(z^3-1)}$.

Later we shall also consider functions which are not defined at some points. These points may, however, be branch points.

307. Draw the scheme of the Riemann surfaces of function $\sqrt{1/z}$.

308. Draw the schemes of the Riemann surfaces of the following functions: a) $\sqrt{1/(z-i)}$; b) $\sqrt[3]{(z-1)/(z+1)}$; c) $\sqrt[4]{(z+i)^2/(z(z-1)^3)}$.

Solving the problems of this section we have always found that, after having made the cuts from all the branch points to infinity the function considered turned out to be decomposed into continuous single-valued branches which join to each other in a way defined by the cuts. This property is possessed by a quite wide class of functions. In particular, it is possessed by all functions we have considered, for instance, by all functions representable by radicals (§2.11) and by algebraic functions (§2.14) (these two classes are special cases of a wider class of functions, called *analytic*, which possess this property as well).

The proof of this statement lies outside the purposes of this book. We have therefore to accept this proposition without proof, knowing that the proof does exist[15]. The reader can thus pass directly to §2.11.

However, a sense of dissatisfaction may arise in the reader about something. Although we cannot destroy it completely, we prove that the property formulated above follows from another property, the so called *monodromy property*, which turns out to be more evident.

We know that after having decomposed the function $w(z)$ into continuous single-valued branches (in a certain region of the z plane), the function $w(z)$ is defined by continuity in the same way along two arbitrary curves C_1 and C_2 lying in this region and joining two arbitrary points z_0 and z_1. The monodromy property is related to this condition.

Suppose a multi-valued function $w(z)$ be such that if one fixes one of its values w_0 at an arbitrary point z_0 then the value of the function can be defined by continuity (possibly in a unique way) along an arbitrary curve beginning at the point z_0 (and not passing through the points at which $w(z)$ is not defined). We say that the function $w(z)$ possesses the monodromy property if it satisfies the following condition.

MONODROMY PROPERTY. Suppose that two continuous curves C_1 and C_2 on the z plane join two points z_0 and z_1 passing neither through the branch points nor through the points of non-uniqueness of the function $w(z)$. Furthermore, suppose that the curve C_1 can be transformed, varying continuously, into the curve C_2, in such a way that none of the curves during the deformation passes through the branch points, and that the ends of these curves are fixed (see Figure 35; a and b are branch points). Thus the value $w(z_1)$ is uniquely defined by continuity along the curves C_1 and C_2 (when a value $w_0 = w(z_0)$ is chosen).

309. Suppose a function $w(z)$ possess the monodromy property. On the z plane make the cuts, not intersecting each other, from the branch points of $w(z)$ to infinity. Prove that in this way the function $w(z)$ is decomposed into continuous single-valued branches.

[15]Cf., for example, Springer G., (1957), ***Introduction to Riemann Surfaces***, (Addison-Wesley: Reading, Mass).

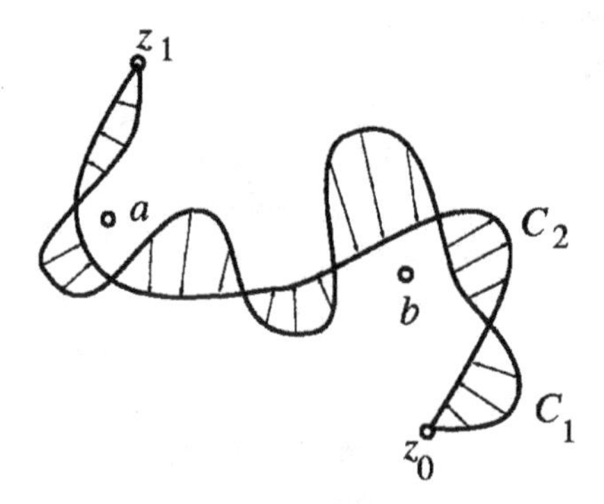

FIGURE 35

310. Suppose that in the conditions of the preceding problem the cuts do not pass through the non-uniqueness points of the function $w(z)$ and that $w(z)$ has a finite number of branch points. Prove that on traversing the cut (in a defined direction) one moves from a given branch of the function $w(z)$ to another, a well defined one, independently of the actual point at which the cut is crossed.

REMARK 1. During a turn around a branch point one traverses once the cut joining this point to infinity. Consequently by virtue of the result of Problem **310** the passages between two different branches traversing the cut at an arbitrary point coincide with the passages obtained by a turn (with the corresponding orientation) around the branch point from which the cut has been drawn, and they thus coincide with the passages indicated by the arrows in correspondence with this point in the scheme of the Riemann surface.

REMARK 2. From the results of Problems **309** and **310** it follows that if a multi-valued function $w(z)$ possesses the monodromy property then one can build its Riemann surface. In order to understand the structure of this surface it thus suffices to find the branch points of the function $w(z)$ and to define the passages between the branches corresponding to the turns around these points.

All functions which we shall consider in the sequel possess the monodromy property. We do not prove exactly this claim, because for this it would be needed to possess the precise notion of the analytic function. We give, however, the idea of the proof of the statement that a function $w(z)$ possesses the monodromy property if it is 'sufficiently good'. What this means will be clarified during the exposition of the proof's arguments.

Suppose the conditions required by the monodromy property be satisfied. Let C_1' and C_2' be the continuous images of the curves C_1 and C_2 under the mapping $w(z)$, with $w_0 = w(z_0)$ as the initial point. We have to prove that the curves C_1' and C_2' end at the same point.

Suppose first that all the curves obtained deforming C_1 into C_2 pass neither through the branch points, nor through the non-uniqueness points of the function. Let C be one of these curves. Thus there exists a unique image C' of the curve C under the mapping $w(z)$, beginning at the point $w_0 = w(z_0)$. If the function $w(z)$ is 'sufficiently good'[16] then during the continuous deformation of the curve C from C_1 to C_2 the curve C' is continuously deformed from C_1' to C_2'. The end point of the curve C' is continuously displaced as well. But the curve C ends at the point z_1, thus the final point of the curve C' must coincide with one of the images of z_1. If the function $w(z)$ takes at every z (in particular at z_1) only a finite number of values (and we consider only functions of this type), then the final point of the curve C' cannot jump from an image of the point z_1 to another, because this should destroy the continuity of the deformation. Hence the final points of the curves C' and, in particular, of C_1' and of C_2' do coincide.

Consider now what happens when the curve C passes through a non-uniqueness point (when it is not a branch point) of the function $w(z)$. Consider only the particular case in which the curve varies only in a neighbourhood of one non-uniqueness point, say a (Figure 36). If at the point z_0 one has chosen a value $w_0 = w(z_0)$ then one defines by continuity the value of $w(z)$ at the point A. Afterwards the values of $w(z)$ at the point E are uniquely defined by continuity along the curves ADE and ABE, because otherwise, making a turn along the curve $EDABE$, the value of the function $w(z)$ should change and the point a should be a branch point of the function $w(z)$. Afterwards, as we had uniquely defined the value of $w(z)$ at the point E along the two curves, we also define by continuity along the curve Ez_1 the value of $w(z)$ at the point z_1.

So, the 'obscure' point in our exposition remains the claim that all functions which we shall consider are 'sufficiently good'.

The reader either accepts this proposition or will apply himself to the

[16]The monodromy property is usually proved for arbitrary analytic functions, cf., for example, G. Springer, (1957), *Introduction to Riemann Surfaces*, (Addison-Wesley: Reading, Mass.).

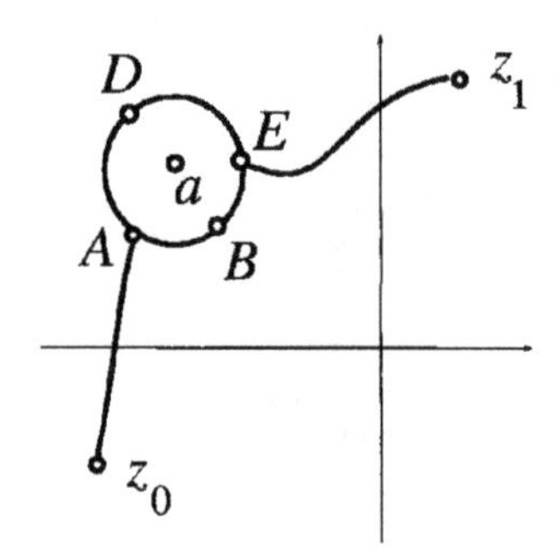

FIGURE 36

study of analytic functions[17].

2.11 Functions representable by radicals

DEFINITION. Let $f(z)$ and $g(z)$ be two multi-valued functions. By $f(z)+g(z)$ we will denote the multi-valued function whose values at a point z_0 are obtained by adding each value $f(z_0)$ to each value of $g(z_0)$. Similarly one defines the functions $f(z)-g(z)$, $f(z)\cdot g(z)$ and $f(z)/g(z)$.

By $[f(z)]^n$, where n is an arbitrary non-zero integer, we will denote a function whose values at the point z_0 are obtained raising to power n each value $f(z_0)$.

By $\sqrt[n]{f(z)}$, where n is a non-zero integer, we will denote the function whose values at a point z_0 are obtained extracting all roots of order n of each value $f(z_0)$.

311. Find all values of: a) $\sqrt[3]{-8}+\sqrt{2i}$; b) $(1-\sqrt{-2i})/\sqrt{-4}$; c) $\sqrt{i+\sqrt{-1}}$; d) $\sqrt[4]{(1+i)^2}$; e) $(\sqrt{i}+\sqrt{i})^2$.

DEFINITION. We will say that a function $h(z)$ *is representable by radicals* if it can be written in terms of the function $f(z)=z$ and of constant functions ($g(z)=a$, a being any complex number) by means of the operations of addition, subtraction, multiplication, division, raising to an integer power and extraction of a root of integer order.

[17]See, for example: Shabat B.V., (1992) *Introduction to Complex Analysis*, Pt. 2, (AMS: Providence, R.I.).

For example, the function $h(z) = (\sqrt[3]{\sqrt{z} + 3z^2} - 3/\sqrt{z})^4$ is representable by radicals. We have already seen other functions that are representable by radicals.

312. Let $h(z)$ be a function representable by radicals and let C be a continuous curve on the z plane, beginning at a point z_0 and not passing through the points at which $h(z)$ is not defined. Prove that if w_0 is one of the values $h(z_0)$ then there exists at least one continuous image of the curve C under the mapping $w = h(z)$, beginning at the point w_0. (We suppose that the parametric equation $w(t) = a$, where a is a given complex number, describes a curve degenerated to a point.)

From the result of Problem **312** one obtains that an arbitrary function representable by radicals can be defined by continuity along an arbitrary continuous curve C, not passing through the points at which $h(z)$ is not defined. Moreover, if the curve C passes neither through the branch points nor those of non-uniqueness of the function $h(z)$, then the function $h(z)$ is uniquely defined by continuity along the curve C.

We had already remarked in the preceding section that functions representable by radicals are 'sufficiently good'[18], i.e., they possess the monodromy property. Hence for every function representable by radicals one can build the Riemann surface[19] (cf., **309** and **310**). Let us analyze the structure of such Riemann surfaces.

In this section we shall consider only functions representable by radicals.

313. Let $h(z) = f(z) + g(z)$. Eliminate from the plane all points of non-uniqueness of the function $h(z)$ and make the cuts not intersect each other, starting from all branch points of $f(z)$ and of $g(z)$ and going to infinity. Let $f_1(z), \ldots, f_n(z)$ and $g_1(z), \ldots, g_m(z)$ be the continuous single-valued branches of the functions $f(z)$ and $g(z)$ defined on the plane with the cuts. Find the continuous single-valued branches of the function $h(z)$.

If, turning once around the point z_0, one moves from the branch $f_{i_1}(z)$ to the branch $f_{i_2}(z)$ and from the branch $g_{j_1}(z)$ to the branch $g_{j_2}(z)$, then evidently one moves from the branch $h_{i_1,j_1}(z) = f_{i_1}(z) + g_{j_1}(z)$ to the branch $h_{i_2,j_2}(z) = f_{i_2}(z) + g_{j_2}(z)$. This result indicates to us the formal method for drawing the scheme of the Riemann surface of the function

[18] All functions representable by radicals are analytic.

[19] Every function representable by radicals has a finite number of branch points.

$h(z) = f(z)+g(z)$ when we already have the schemes (with the same cuts) of the Riemann surfaces of the functions $f(z)$ and $g(z)$. In correspondence with every pair of branches $f_i(z)$ and $g_j(z)$ we take a sheet on which we consider defined the branch $h_{i,j}(z) = f_i(z)+g_j(z)$. If in the schemes of the Riemann surfaces of the functions $f(z)$ and $g(z)$ at the point z_0 arrows indicate, respectively, the passage from the branch $f_{i_1}(z)$ to the branch $f_{i_2}(z)$ and the passage from the branch $g_{j_1}(z)$ to the branch $g_{j_2}(z)$, then in the scheme of the Riemann surface of the function $h(z)$ we indicate by an arrow over the point z_0 the passage from the branch $h_{i_1,j_1}(z)$ to the branch $h_{i_2,j_2}(z)$.

314. Draw the schemes of the Riemann surfaces of the following functions: a) $\sqrt{z}+\sqrt{z-1}$; b) $\sqrt[3]{z^2-1}+\sqrt{1/z}$; c) $\sqrt{z}+\sqrt[3]{z}$; d)$\sqrt{z^2-1}+\sqrt[4]{z-1}$.

The formal method described above for building the scheme of the Riemann surface of the function $h(z) = f(z)+g(z)$ does not always give the correct result, because it may happen that some of the branches $h_{i,j}(z)$ coincide.

For simplicity we shall suppose that the cuts do not pass through the non-uniqueness points of the function $h(z)$. In this case, traversing a cut, by virtue of the uniqueness, we shall move from one set of sheets, corresponding to the same branch of the function $h(z)$, to a new set of sheets, all corresponding to a different branch. As a consequence, if we join the sheets corresponding to the same branches of the function $h(z)$, i.e., if we substitute every set of sheets with a single sheet, then the passages between the sheets so obtained are uniquely defined by any turn round an arbitrary branch point z_0.

315. Find all values of $f(1)$ if: a) $f(z) = \sqrt{z}+\sqrt{z}$; b) $f(z) = \sqrt{z}+\sqrt[4]{z^2}$; c) $\sqrt[3]{z}+\sqrt[3]{z}$.

316. Draw the schemes of the Riemann surfaces using the formal method and the correct schemes for the following functions:
a) $f(z) = \sqrt{z}+\sqrt{z}$; b) $f(z) = \sqrt{z}+\sqrt[4]{z^2}$; c) $\sqrt[3]{z}+\sqrt[3]{z}$.

We obtain finally that to draw the scheme of the Riemann surface of the function $h(z) = f(z)+g(z)$, starting from the schemes of the Riemann surfaces of the functions $f(z)$ and $g(z)$ (with the same cuts), it suffices to build the scheme using the formal method described above and afterwards identify the sheets corresponding to equal values.

It is easy to see that one can apply this procedure to build the schemes of the Riemann surface for the functions $h(z) = f(z) - g(z)$; $h(z) = f(z) \cdot g(z)$; $h(z) = f(z)/g(z)$.

317. Draw the schemes of the Riemann surfaces of the functions: a) $i\sqrt{z} - \sqrt[4]{z^2}$; b) $\sqrt{z-1} \cdot \sqrt[4]{z}$; c) $\sqrt{z^2-1}/\sqrt[4]{z+1}$; d) $(\sqrt{z} + \sqrt{z})/\sqrt[3]{z(z-1)}$.

318. Let $f_1(z), f_2(z), \ldots, f_m(z)$ be the single-valued continuous branches of the function $f(z)$. Find all single-valued continuous branches, having the same cuts, of the function $h(z) = [f(z)]^n$, where n is a non-zero integer.

From the result of the last problem it follows that the scheme of the Riemann surface of the function $h(z) = [f(z)]^n$ coincides with the scheme of the Riemann surface of the function $f(z)$ if all branches $h_i(z) = [f_i(z)^n]$ are distinct. But this in not always true. If one obtains some equal branches, then traversing the cuts one moves, by virtue of the uniqueness, from equal branches to equal branches.

We thus obtain that to draw the scheme of the Riemann surface of the function $h(z) = [f(z)]^n$ it suffices to consider, in the scheme of the Riemann surface of the function $f(z)$, the branches $h_i(z) = [f_i(z)]^n$ instead of the branches $f_i(z)$. If this produces equal branches it suffices to identify the corresponding sheets.

319. Build the schemes of the Riemann surfaces of the following functions: a) $(\sqrt[4]{z})^2$; b) $(\sqrt{z} + \sqrt{z})^2$; c) $(\sqrt{z} \cdot \sqrt[3]{z-1})^3$.

Let us now analyse how the Riemann surface of the function $\sqrt[n]{f(z)}$ is related to the scheme of the Riemann surface of the function $f(z)$.

320. Which points can be branch points of the function $\sqrt[n]{f(z)}$?

On the z plane make the cuts between the branch points of the function $f(z)$ and infinity in such a way that these cuts do not pass through the points at which some value of $f(z)$ vanishes, and separate the single-valued continuous branches of the function $f(z)$. Let $f_1(z), f_2(z), \ldots, f_m(z)$ be these branches. Make the cuts between the points at which one of the values of $f(z)$ is equal to 0 and infinity. Let $g(z)$ be one of the single-valued continuous branches of the function $\sqrt[n]{f(z)}$ with these cuts.

321. Prove that the function $[g(z)]^n$ coincides with one of the functions $f_i(z)$ everywhere, except on the cuts.

It follows from the result of the preceding problem that every branch of the function $\sqrt[n]{f(z)}$ corresponds to some branch of the function $f(z)$.

322. Let $g(z)$ be a single-valued continuous branch of the function $\sqrt[n]{f(z)}$, corresponding to the branch $f_i(z)$ of the function $f(z)$. Find all single-valued continuous branches of the function $\sqrt[n]{f(z)}$, corresponding to the branch $f_i(z)$.

From the result of the last problem we obtain that to every branch $f_i(z)$ of the function $f(z)$ there corresponds a 'bunch' consisting of n branches of the function $\sqrt[n]{f(z)}$. Enumerate the branches of this 'bunch' by $f_{i,0}(z), f_{i,1}(z), \ldots, f_{i,n-1}(z)$ in such a way that for every k the equation $f_{i,k}(z) = f_{i,0}(z) \cdot \epsilon_n^k$ be satisfied.

Let z_0 be a branch point of the function $f(z)$ and suppose that turning once around the point z_0 one moves from the branch $f_i(z)$ to the branch $f_j(z)$. Thus evidently for the function $\sqrt[n]{f(z)}$ we obtains that on turning around the point z_0 one moves from all branches of the bunch which corresponds to the branch $f_i(z)$ to all branches of the bunch which corresponds to the branch $f_j(z)$.

323. Let C be a curve on the z plane with a parametric equation $z(t)$ and let $w_0(t)$ be the parametric equation of the continuous image of the curve C on the w plane under the mapping $w = \sqrt[n]{f(z)}$. Prove that the curve with the equation $w_k(t) = w_0(t) \cdot \epsilon_n^k$ is also a continuous image of the curve C under the mapping $w = \sqrt[n]{f(z)}$.

324. Suppose that a curve C on the z plane avoids the branch points and the non-uniqueness points of the function $\sqrt[n]{f(z)}$. Prove that if on moving along the curve C one moves from the branch $f_{i,s}(z)$ to the branch $f_{j,r}(z)$, then one moves from the branch $f_{i,s+k}(z)$ to the branch $f_{j,r+k}(z)$, where the sums $s+k$ and $r+k$ are calculated modulo n (cf., **40**).

In this way, to define where one arrives, starting from a branch of a given bunch, on turning around a given branch point of the function $\sqrt[n]{f(z)}$, it suffices to define where one arrives from one of the branches of that bunch; for the other branches the passages turn out to be automatically defined by virtue of the result of Problem **324**.

325. Draw the scheme of the Riemann surface of the function $f(x) = \sqrt{\sqrt{z} - 1}$.

326. Draw the schemes of the Riemann surfaces of the following functions: a) $\sqrt[3]{\sqrt{z} - 2}$; b) $\sqrt{\sqrt[3]{z} - 1}$.

327. Draw the scheme of the Riemann surface of the function $f(z) = \sqrt{z^2+1} - 2$ with different cuts, respectively shown: a) in Figure 37; b) in Figure 38. In both cases say whether the points z at which $f(z) = 0$ lie on the same sheet or on different sheets.

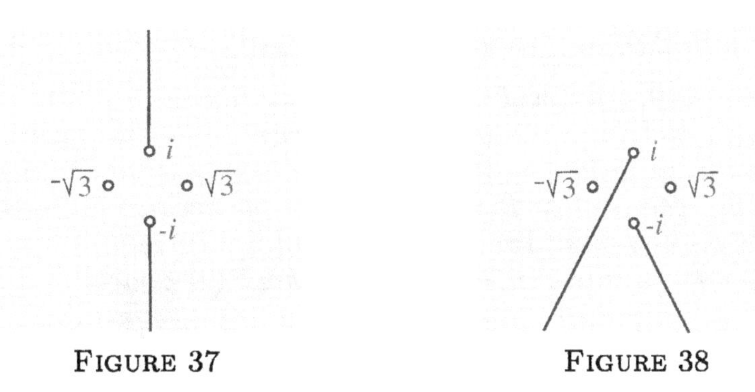

FIGURE 37 FIGURE 38

328. Draw the scheme of the Riemann surface of the function $h(z) = \sqrt{\sqrt{z^2+1}-2}$ with the cuts shown: a) in Figure 39; b) in Figure 40.

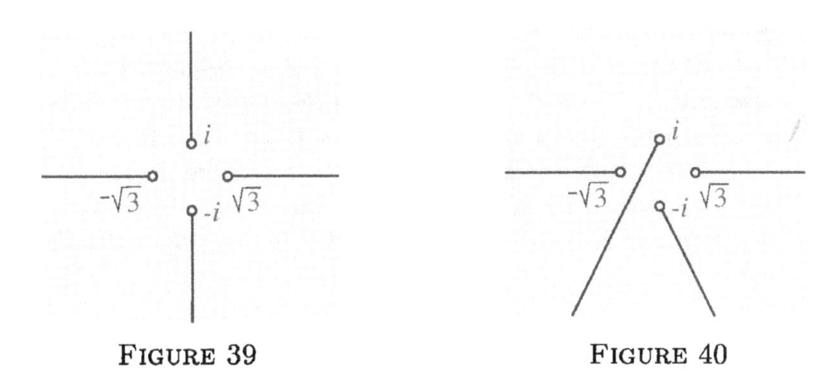

FIGURE 39 FIGURE 40

We formulate once more the results of this section which will be useful in the sequel.

THEOREM 8. *To build the schemes of the Riemann surfaces of the functions* $h(z) = f(z) + g(z)$, $h(z) = f(z) - g(z)$, $h(z) = f(z) \cdot g(z)$, $h(z) = f(z)/g(z)$ *starting from the schemes of the Riemann surfaces of the functions* $f(z)$ *and* $g(z)$, *with the same cuts, it suffices to do the following:*

a) put into correspondence with every pair of branches, $f_i(z)$ *and* $g_j(z)$, *a sheet on which the branch* $h_{i,j}(z)$, *equal respectively to* $f_i(z) + g_j(z)$ $f_i(z) - g_j(z)$, $f_i(z) \cdot g_j(z)$, $f_i(z)/g_j(z)$, *is defined;*

b) if by turning around the point z_0 one moves from the branch $f_{i_1}(z)$ to the branch $f_{i_2}(z)$ and from the branch $g_{j_1}(z)$ to the branch $g_{j_2}(z)$, then for the function $h(z)$ by the same turn one moves from the branch $h_{i_1,j_1}(z)$ to the branch $h_{i_2,j_2}(z)$;

c) identify the sheets on which the branches $h_{i,j}$ coincide.

THEOREM 9. *To build the scheme of the Riemann surface of the function $h(z) = [f(z)]^n$ starting from the scheme of the Riemann surface of function $f(z)$, defined by the same cuts, it suffices to do the following:*

a) in the scheme of the Riemann surface of the function $f(z)$ consider, instead of the branches $f_i(z)$, the branches $h_i(z) = [f_i(z)]^n$.

b) identify the sheets on which the branches $h_i(z)$ coincide.

THEOREM 10. *To build the scheme of the Riemann surface of the function $h(z) = \sqrt[n]{f(z)}$ starting from the scheme of the Riemann surface of the function $f(z)$, defined by the same cuts, it suffices to do the following:*

a) replace every sheet of the scheme of the Riemann surface of the function $f(z)$ by a 'pack' of n sheets;

b) turning around an arbitrary branch point of the function $h(z)$ one moves from all sheets of one pack to all sheets of a different pack.

c) these passages from one pack of sheets to another correspond to the passages between the sheets of the Riemann surface of the function $f(z)$;

d) if the branches in the bunches are enumerated in such a way that $f_{i,k}(z) = f_{i,0}(z)\epsilon_n^k$, then by moving from one bunch to another the sheets of the corresponding packs are not mixed, but they permute cyclically (cf., **324**.*)*

2.12 Monodromy groups of multi-valued functions

We now associate with every scheme of the Riemann surface a group of permutations.

329. Suppose that a curve C on the z plane avoids the branch points and the non-uniqueness points of the function $w(z)$. Prove that moving along the curve C, starting from distinct sheets of the scheme of the Riemann surface of function $w(z)$, one arrives at distinct sheets.

So by virtue of the result of Problem **249**, to each turn (counterclockwise) around any branch point of the function $w(z)$ there corresponds a permutation of the sheets of the scheme of the Riemann surface of the function $w(z)$.

330. Consider the schemes of the Riemann surfaces of the functions of Problem **314**, drawn as in the solution of this problem (see the chapter 'Hint, Solutions and Answers') and consider in these schemes the sheets enumerated from bottom to top by the numbers $1, 2, 3, \ldots$. For every function write the permutations of the sheets corresponding to one turn around every branch point.

331. Let $g_1, g_2, \ldots, g_s$ be some elements of a group G. Consider all elements of G which can be obtained from $g_1, \ldots, g_s$ by means of the iterated operations consisting of multiplying and of taking of the inverse of an element. Prove that the set of the elements obtained is a subgroup of the group G.

DEFINITION. The group obtained in Problem **331** is called the *subgroup generated by the elements* $g_1, \ldots, g_s$.

DEFINITION. Let $g_1, \ldots, g_s$ be the permutations of the sheets of the scheme of a Riemann surface corresponding to the turns (counterclockwise) around all the branch points. We call the subgroup generated by the elements $g_1, \ldots, g_s$ the *permutation group of the sheets of the given scheme of the Riemann surface*, for brevity the *permutation group of the given scheme.*

REMARK 1. If the number of sheets in a scheme is finite (and we consider only schemes of such a type) then to define the permutation group of this scheme it suffices to use the operation of composition, without using any permutation's inversion. Indeed, in this case every permutations of the sheets g has a finite order k: $g^k = e$, and therefore

$$g^{-1} = g^{k-1} = \underbrace{g \cdot g \cdot \ldots \cdot g}_{k-1}.$$

REMARK 2. The permutation group of the scheme we have considered is defined, as usual, up to isomorphism. The numbering of the sheets will be not important, because for different numberings one obtains different, but isomorphic, subgroups of S_n.

332. Which of the groups you already know are isomorphic to the permutation groups of the schemes of the following functions: a) $\sqrt{z}$; b) $\sqrt[3]{z}$; c) $\sqrt[n]{z}$; d) $\sqrt[3]{z^2 - 1}$ (cf., **304**); e) $\sqrt[4]{(z-1)^2(z+1)^3}$ (cf., **306**)?

333. To which of the groups you already know are the permutation groups of the schemes of the functions considered in Problems: 1) **314**; 2) **317**; 3) **319** isomorphic?

334. For the two schemes of the function $h(z) = \sqrt{\sqrt{z^2+1}-2}$, built in the solution of Problem **328**, describe the permutation groups.

Suppose that the point z_0 is neither a branch point nor a non-uniqueness point of the multi-valued function $w(z)$, and that $w_1, w_2, \ldots, w_n$ are all values of the function $w(z)$ at the point z_0. Consider a continuous curve C starting and ending at the point z_0 and not passing through any branch point and any non-uniqueness point of the function $w(z)$. Take a value $w_i = w(z_0)$ and define by continuity along the curve C a new value $w_j = w(z_0)$. Starting from distinct values w_i we obtain different values w_j (otherwise along the curve C the uniqueness would be lost). Hence to the curve C there corresponds a certain permutation of the values $w_1, w_2, \ldots, w_n$. Thus if to the curve C there corresponds the permutation g, to the curve C^{-1} there corresponds the permutation g^{-1}, and if to the curves C_1 and C_2 (both with ending points at z_0) there correspond the permutations g_1 and g_2, then to the curve C_1C_2 there corresponds the permutation g_2g_1 (recall that the permutations in a product are kept from right to left).

In this way if we consider all possible curves starting and ending at z_0, then the corresponding permutations will form a group, the *group of permutations of the values* $w(z_0)$.

335. Let G_1 be the permutation group of the values $w(z_0)$ and G_2 the permutation group of some scheme of the function $w(z)$. Prove that the groups G_1 and G_2 are isomorphic.

Notice that in the definition of the permutation group of the values $w(z_0)$ one does not use any scheme of the Riemann surface of the function $w(z)$. From the result of Problem **335** it then follows that the permutation group of the values $w(z_0)$ for an *arbitrary* point z_0 and the permutation group of an *arbitrary* scheme of the Riemann surface of the function $w(z)$ are isomorphic. Consequently the permutation group of the values $w(z_0)$ for **all** points z_0 and the permutation group of **all** the schemes of the Riemann surfaces of the function $w(z)$ are isomorphic, i.e., they represent one unique group. This group is called the *monodromy group of the multi-valued function* $w(z)$.

2.13 Monodromy groups of functions representable by radicals

In this section we prove one of the main theorems of this book.

THEOREM 11. *If the multi-valued function $h(z)$ is representable by radicals its monodromy group is soluble* (cf., §1.14).

The proof of Theorem 11 consists in the solutions of the following problems.

336. Let $h(z) = f(z) + g(z)$, or $h(z) = f(z) - g(z)$, or $h(z) = f(z) \cdot g(z)$, or $h(z) = f(z)/g(z)$, and suppose that we have built the scheme of the Riemann surface of the function $h(z)$ from the schemes of the Riemann surfaces of the functions $f(z)$ and $g(z)$ by the formal method (cf., Theorem 8 (a), §2.11). Prove that if F and G are the permutation groups of the initial schemes, then the permutation group of the scheme built is isomorphic to a subgroup of the direct product $G \times F$ (cf., Chapter 1, 1.7).

337. Let H_1 be the permutation group of the scheme built by the formal method under the hypotheses of the preceding problem, and let H_2 be the group of permutations of the real scheme of the Riemann surface of the function $h(z)$. Prove that there exists a surjective homomorphism (cf., §1.7) of the group H_1 onto the group H_2.

338. Suppose the monodromy groups of the functions $f(z)$ and $g(z)$ be soluble. Prove that the monodromy groups of the functions $h(z) = f(z) + g(z)$, $h(z) = f(z) - g(z)$, $h(z) = f(z) \cdot g(z)$, $h(z) = f(z)/g(z)$ are soluble as well.

339. Suppose the monodromy group of the function $f(z)$ be soluble. Prove that the monodromy group of the function $h(z) = [f(z)]^n$ is also soluble.

340. Let H be the permutation group of a scheme of the function $h(z) = \sqrt[n]{f(z)}$, and F the permutation group of a scheme of the function $f(z)$, made with the same cuts. Define a surjective homomorphism of the group H onto the group F.

341. Prove that the kernel of the homomorphism (cf., §1.13) defined by the solution of the preceding problem is commutative.

342. Suppose the monodromy group F of the function $f(z)$ be soluble. Prove that the monodromy group H of the function $h(z) = \sqrt[n]{f(z)}$ is also soluble.

The functions $h(z) = a$ and $h(z) = z$ are continuous single-valued functions on the entire z plane. Their Riemann surfaces thus consist of a single sheet and therefore the corresponding monodromy groups consist of a single element e and are therefore soluble. As a consequence, taking into account the definition of the functions representable by radicals (§2.11) and the results of Problems **338**, **339**, **342**, one obtains the proposition of Theorem 11.

REMARK 1. This remark is for that reader who knows the theory of analytic functions. Theorem 11 holds for a wider class of functions. For example, to define a function $h(z)$ one can be allowed to use, besides of constant functions, the identity function, the functions expressed by arithmetic operations and radicals, also all analytic single-valued functions (for example, e^z, $\sin z$, etc.), the multi-valued function $\ln z$ and some others. In this case the monodromy group of the function $h(z)$ will be soluble, though it is not necessarily finite.

2.14 The Abel theorem

Consider the equation

$$3w^5 - 25w^3 + 60w - z = 0. \tag{2.8}$$

We consider z as a parameter and for every complex value of z we look for all complex roots w of this equation. By virtue of the result of Problem **269** the given equation for every z has 5 roots (taking into account the multiplicities).

343. Which values of w can be multiple roots (of order higher than 1, cf., §2.8) of the equation

$$3w^5 - 25w^3 + 60w - z = 0?$$

For which values of z are these roots multiples?

It follows from the solution of the preceding problem that for $z = \pm 38$ and $z = \pm 16$ equation (2.8) has four distinct roots, and for the other values of z it has 5 distinct roots. Let us study the function $w(z)$.

First we prove that for small variations of the parameter z the roots of equation (2.8) vary only slightly. This property is more precisely expressed by the following problem.

344. Let z_0 be an arbitrary complex number and w_0 be one of the roots of equation (2.8) for $z = z_0$. Consider a disc of radius r arbitrarily small with its centre at w_0. Prove that there exists a real number $\rho > 0$ such that if $|z_0' - z_0| < \rho$ then in the disc considered there exists at least one root of equation (2.8) for $z = z_0'$ also.

Suppose the function $w(z)$ express the roots of equation (2.8) in terms of the parameter z and w_0 be one of the values $w(z_0)$. It follows from the result of Problem **344** that if z changes continuously along a curve, starting at the point z_0, then one can choose one of the values $w(z)$ in such a way that the point w, too, moves continuously along a curve starting from the point w_0. In other words, the function $w(z)$ can be defined by continuity along an arbitrary curve C. Therefore if the curve C avoids the branch points and the non-uniqueness points of the function $w(z)$, the function $w(z)$ is uniquely defined by continuity along the curve C.

345. Prove that points different from $z = \pm 38$ and $z = \pm 16$ can be neither branch points nor non-uniqueness points of a function expressing the roots of equation (2.8) in terms of the parameter z.

Let $w(z)$ be the function expressing the roots of equation (2.8) in terms of the parameter z. The function $w(z)$, being an algebraic function[20], is 'sufficiently good' (cf., §2.10), i.e., it possesses the monodromy property. One can therefore build for the function $w(z)$ the Riemann surface (cf., **309** and **310**). This Riemann surface evidently has 5 sheets.

By virtue of the result of Problem **345** the only possible branch points of the function $w(z)$ are the points $z = \pm 38$ and $z = \pm 16$, but it is not yet clear whether this is really the case.

346. Suppose it is known that the point $z_0 = +38$ (or $z_0 = -38$, or $z_0 = \pm 16$) is a branch point of the function $w(z)$ expressing the roots of equation (2.8) in terms of the parameter z. How do the sheets of the

[20] The multi-valued function $w(z)$ is said to be algebraic if it expresses in terms of the parameter z all the roots of some equation

$$a_0(z)w^n + a_1(z)w^{n-1} + \ldots + a_n(z) = 0$$

in which all the $a_i(z)$s are polynomials in z. All algebraic functions are analytic.

Riemann surface of the function $w(z)$ at the point z_0 (more precisely, along the cuts joining the point z_0 to infinity; cf., Remark 2, §2.10) join?

347. Let w_0 be a function expressing the roots of equation (2.8) in terms of the parameter z. Moreover, let z_0 and z_1 be two arbitrary points different from $z = \pm 38$ and $z = \pm 16$, and w_0 and w_1 be two arbitrary images of these points under the mapping $w(z)$. Prove that it is possible to draw a continuous curve joining the points z_0 and z_1, not passing through the points $z = \pm 38$ and $z = \pm 16$ and such that its continuous image, starting from the point w_0, ends at the point w_1.

348. Prove that all four points $z = \pm 38$ and $z = \pm 16$ are branch points of the function $w(z)$. How can we represent the scheme of the Riemann surface of the function $w(z)$? Draw all different possible schemes (we consider different two schemes if they cannot be obtained one from another by a permutation of the sheets and of the branch points).

349. Find the monodromy group of the function $w(z)$ expressing the roots of the equation

$$3w^5 - 25w^3 + 60w - z = 0$$

in terms of the parameter z.

350. Prove that the function $w(z)$, expressing the roots of the equation

$$3w^5 - 25w^3 + 60w - z = 0$$

in terms of the parameter z is not representable by radicals.

351. Prove that the algebraic general equation of fifth degree

$$a_0w^5 + a_1w^4 + a_2w^3 + a_3w^2 + a_4w + a_5 = 0$$

(where a_0, a_1, a_2, a_3, a_4 are complex parameters, $a_0 \neq 0$) is not solvable by radicals, i.e., that there are no formulae expressing the roots of this equation in terms of the coefficients by means of the operations of addition, subtraction, multiplication, division, elevation to an integer power and extraction of a root of integer order.

352. Considerer the equation

$$(3w^5 - 25w^3 + 60w - z)w^{n-5} = 0 \tag{2.9}$$

and prove that for $n > 5$ the general algebraic equation of degree n is not solvable by radicals.

The results of Problems **351** and **352** contain the proof of the theorem which is the subject of the present book. We have indeed proved the

ABEL'S THEOREM. *For $n \geq 5$ the general algebraic equation of degree n*

$$a_0 w^n + a_1 w^{n-1} + \ldots + a_{n-1} w + a_n = 0$$

is not solvable by radicals.

REMARK 1. In the introduction we deduced the Cardano formulae for the solution of the general algebraic equation of third degree. The roots of the equation are not given by all values expressed by these formulae, but only by those which satisfy some supplementary conditions. One may therefore pose the question of whether it is possible, also for the general equation of degree n ($n \geq 5$), to build by radicals a formula such that the roots of the equation are only a part of the values that are expressed by this formula. This is not possible even for equation (2.8).

Indeed, if the values of the function $w(z)$, expressing the roots of equation (2.8) in terms of the parameter z, are only a part of the values of a function $w_1(z)$, represented by radicals, then the Riemann surface of the function $w(z)$ is a separate part of the Riemann surface of the function $w_1(z)$. If G is the monodromy group of the function $w_1(x)$ then to every permutation of the group G there corresponds a permutation of the five sheets of the function $w(z)$. This mapping is a homomorphism of the group G into the group S_5. Since the group S_5 is not soluble then the group G is also not soluble (cf., **163**). On the other hand, the group G must be soluble, being the monodromy group of a function representable by radicals. We have thus obtained a contradiction.

REMARK 2. From Remark 1 in §2.13 it follows that the Abel theorem also holds if one is allowed to use, besides radicals, some other functions, for example all analytic functions (such as e^z, $\sin z$, etc.), the function $\ln z$, and some others.

REMARK 3. Consider equation (2.8) only in the domain of real numbers. Suppose that the function $y(x)$ expresses the real roots of the equation

$$3y^5 - 25y^3 + 60y - x = 0$$

in terms of the real parameter x. Is the function $y(x)$ representable by radicals? The answer is 'no'. To the reader who knows the theory of analytic functions we say that this follows from the theorem of the analytic continuation. Indeed, the function $w(z)$ expressing the roots of equation (2.8) in terms of parameter z is analytic. If the function $y(x)$ were representable by radicals, then the corresponding formula, considered in the domain of complex numbers, should give, by virtue of the theorem of the analytic continuation, the function $w(z)$; i.e., the function $w(z)$ would be representable by radicals.

Hence the Abel theorem remains true also if one considers only the real roots of the general equation of degree n ($n \geq 5$) for all possible real values of the coefficients. Moreover, by virtue of Remark 2 the theorem holds also if one is allowed to use, besides radicals, some other functions, for example all functions with an analytic single-valued continuation (e^x, $\sin x$, etc.), the function $\ln x$, and some others.

REMARK 4. The class of algebraic functions (cf., note 20) is sufficiently rich and interesting. In particular, one can prove that all functions representable by radicals are algebraic. We have proved that every function representable by radicals possesses a soluble monodromy group (Theorem 11). It turns out that if the analysis is restricted to algebraic functions then the converse also holds: if the monodromy group of an algebraic function is soluble then this function is representable by radicals. An algebraic function is thus representable by radicals if and only if its monodromy group is soluble.

Chapter 3

Hints, Solutions, and Answers

3.1 Problems of Chapter 1

1. *Answer.* In the cases 1 a); 1 c); 2 c); 3 a).

2. See Table 3.

Table 3

	e	a	b
e	e	a	b
a	a	b	e
b	b	e	a

3. See Table 4.

Table 4

	e	a	b	c	d	f
e	e	a	b	c	d	f
a	a	b	e	f	c	d
b	b	e	a	d	f	c
c	c	d	f	e	a	b
d	d	f	c	b	c	a
f	f	c	d	a	b	e

4. See Table 5.

Table 5

	e	a	b	c
e	e	a	b	c
a	a	e	c	b
b	b	c	a	e
c	c	b	e	a

5. See Table 6.

Table 6

	e	a	b	c	d	f	g	h
e	e	a	b	c	d	f	g	h
a	a	e	c	b	f	d	h	g
b	b	c	a	e	g	h	f	d
c	c	b	e	a	h	g	d	f
d	d	f	h	g	e	a	c	b
f	f	d	g	h	a	e	b	c
g	g	h	d	f	b	c	e	a
h	h	g	f	d	c	b	a	e

6. See table 7, where e and a are rotations of the rhombus about its centre by 0° and 180° respectively, b and c reflections of the rhombus with respect to its diagonals.

Table 7

	e	a	b	c
e	e	a	b	c
a	a	e	c	b
b	b	c	e	a
c	c	b	a	e

7. See table 7, where e and a are rotations of the rectangle about its centre by 0° and 180° respectively, b and c are reflections of the rectangle with respect to the straight lines passing through the middle points of its opposite sides.

8. No, because there are no capitals in the world whose names begin with letter X.

9. *Answer.* a) ϕ is not a mapping **onto** because, for example, an integer n such that $\phi(n) = 5$ does not exist, i.e., the number 5 has no pre-image; b) ϕ is a mapping **onto**, but not one to one because every positive integer under the mapping $\phi(n) = |n|$ has two pre-images, for example, the pre-images of 5 are 5 and -5; c) ϕ is a bijective mapping, because the numbers $0, 1, 2, \ldots$ are mapped to the numbers $0, 2, 4, \ldots$, while the numbers $-1, -2, -3, \ldots$ are mapped to the numbers $1, 3, 5, \ldots$.

10. *Answer.* $e^{-1} = e$, $a^{-1} = b$, $b^{-1} = a$, $c^{-1} = c$, $d^{-1} = d$, $f^{-1} = f$.

11. *Answer.* $g^{-1}(x) = x/2$.

12. Suppose that the given group of transformations contain the transformation g. Thus by definition it contains also the transformation g^{-1} and the transformation $gg^{-1} = e$.

13. By the definition of the transformation e and of multiplication we obtain $(eg(A) = e(g(A)) = g(A)$ and $(ge)(A) = g(e(A)) = g(A)$ for every element A. Hence $eg = g$ and $ge = g$.

14. $((g_1g_2)g_3)(A) = (g_1g_2)(g_3(A)) = g_1(g_2(g_3(A)))$; $(g_1(g_2g_3)(A) = g_1((g_2g_3)(A)) = g_1(g_2(g_3(A)))$.

15. 1) No. Here the only possible unit element is 1, and element 0^{-1}, i.e., an element x such that $0 \cdot x = x \cdot 0 = 1$, does not exist. 2) Yes.

16. Yes.

17. a) No. Amongst the natural numbers there are no elements x such that $n + x = x + n = n$ for every natural n. (If one considers the natural numbers with zero, then no elements except 0 will posses the opposite). b) No. The only possible unit element is 1, but thus no elements except 1 will possess the inverse.

18. Let e_1 and e_2 be two unit elements. Thus $e_1a = a$ and $ae_2 = a$ for every element a. Therefore $e_1e_2 = e_2$ and $e_1e_2 = e_1$. Hence $e_1 = e_2$.

19. Let element a have two inverse elements, a_1 and a_2. Thus $(a_1a)a_2 = ea_2 = a_2$ and $a_1(aa_2) = a_1e = a_1$. Therefore $a_1 = a_2$.

20. 1) $ee = e$. 2) $a^{-1}a = aa^{-1} = e$.

21. We prove this by induction. For $n = 3$ the statement of the problem is true, because in this case we can construct only two expressions, $(a_1a_2)a_3$ and $a_1(a_2a_3)$ and, by hypothesis, $(a_1a_2)a_3 = a_1(a_2a_3)$. Let the statement of the problem be true for all n such that $3 \leq n < k$. We prove that it is true for $n = k$ also. Let an arbitrary well arranged expression A be given, containing k factors $a_1, a_2, \ldots, a_k$. Within A there is a multiplication operation which is the last one to be carried out. Therefore the product A can be written in the form $A = (A_1)(A_2)$, where A_1 and A_2 are two well arranged expressions, containing l and $k-l$ factors, respectively. Hence $l < k$ and $k - l < k$. Since $l < k$ and $k - l < k$, by the induction hypothesis the expression A_1 gives the same element as the expression $(\dots((a_1 \cdot a_2) \cdot a_3) \cdot \dots \cdot a_{l-1}) \cdot a_l$, and A_2 gives the same element as the expression $(\dots((a_{l+1} \cdot a_{l+2}) \cdot a_{l+3}) \cdot \dots \cdot a_{k-1}) \cdot a_k$. Hence the expression A gives the same element as the product

$$(\dots((a_1 \cdot a_2) \cdot a_3) \cdot \dots \cdot a_l) \cdot (\dots((a_{l+1} \cdot a_{l+2}) \cdot a_{l+3}) \cdot \dots \cdot a_k).$$

Let

$$\begin{aligned}(\dots((a_1 \cdot a_2) \cdot a_3) \cdot \dots \cdot a_{l-1}) \cdot a_l &= a, \\ (\dots((a_{l+1} \cdot a_{l+2}) \cdot a_{l+3}) \cdot \dots \cdot a_{k-2}) \cdot a_{k-1} &= b.\end{aligned}$$

Thus A expresses the element $a \cdot (b \cdot a_k)$. But by virtue of associativity $a \cdot (b \cdot a_k) = (a \cdot b) \cdot a_k)$. The product $a \cdot b$ is a well arranged expression, because it contains $k - 1 < k$ factors. Hence by hypothesis

$$ab = (\dots((a_1 \cdot a_2) \cdot a_3) \cdot \dots \cdot a_{k-2}) \cdot a_{k-1}.$$

Therefore the expression A gives the element

$$a \cdot (b \cdot a_k) = (a \cdot b) \cdot a_k = (\dots((a_1 \cdot a_2) \cdot a_3) \cdot \dots \cdot a_{k-1}) \cdot a_k$$

as we had to prove.

22. *Answer.* 1) Yes, 2) yes, 3) no, 4) yes, 5) yes.

23. 1) $(ab)(a^{-1}b^{-1}) = a(bb^{-1})a^{-1} = aea^{-1} = aa^{-1} = e$ and $(b^{-1}a^{-1})(ab) = b^{-1}(a^{-1}a)b = b^{-1}eb = b^{-1}b = e$. 2) It is proved by induction on n: if for $n - 1$ it is already proved, then $(a_1 \cdot \dots \cdot a_n)^{-1}$ is equal by virtue of point 1) to $a_n^{-1} \cdot (a_1 \cdot \dots \cdot a_{n-1})^{-1} = a_n^{-1} \cdot a_{n-1}^{-1} \cdot \dots \cdot a_1^{-1}$.

24. Let $ax = b$ for an element x. Thus, multiplying on the left both members of the equality by a^{-1}, we obtain $a^{-1}ax = a^{-1}b$ and $x = a^{-1}b$. In this way the only possible solution of the equation $ax = b$ is the element $a^{-1}b$. This element is indeed a solution, because $a(a^{-1})b = (aa^{-1})b = b$. Exactly in the same way one proves that the equation $ya = b$ has the unique solution $y = ba^{-1}$.

25. Since $aa = e$ for every element a, then for all elements b and c we will have $(bc)(bc) = e$. Multiplying the two members of this equality by b

on the left and by c on the right, one obtains $bbcbcc = bec$. Since $bb = e$ and $cc = e$, one obtains $(ec)(be) = (be)c$, i.e., $cb = bc$. Since b and c are two arbitrary elements of the group G, this group is commutative.

26. We need to prove that $a^m \cdot (a^{-1})^m = (a^{-1})^m \cdot a^m = e$. We have $a^m \cdot (a^{-1})^m = a^{m-1} \cdot a \cdot a^{-1} \cdot (a^{-1})^{m-1} = a^{m-1} \cdot e \cdot (a^{-1})^{m-1} = a^{m-1} \cdot (a^{-1})^{m-1} = \ldots = a \cdot a^{-1} = e$. Exactly in the same way one proves that $(a^{-1})^m \cdot a^m = e$. (For a more rigourous proof use the induction method).

27. We will consider some cases: a) if $m > 0$, $n > 0$, then

$$a^m \cdot a^n = \underbrace{a \cdot a \cdot \ldots \cdot a}_{m} \cdot \underbrace{a \cdot a \cdot \ldots \cdot a}_{n} = \underbrace{a \cdot a \cdot \ldots \cdot a}_{m+n} = a^{m+n};$$

b) if $m < 0$, $n < 0$, $m = -k$ $(k > 0)$, $n = -l$ $(l > 0)$, then $a^m \cdot a^n = a^{-k} \cdot a^{-l} = (a^{-1})^k \cdot (a^{-1})^l =$ (see the case(a)) $= (a^{-1})^{k+l} = a^{-(k+l)} = a^{m+n}$; c) if $m > 0$, $n < 0$, $m + n \geq 0$, then $a^m \cdot a^n =$ (see the case(a)) $= (a^{m+n} \cdot a^{-n}) \cdot a^{-(-n)} = a^{m+n} \cdot a^{-n} \cdot (a^{-n})^{-1} = a^{m+n}$; d) if $m > 0$, $n < 0$, $m + n < 0$, then $a^m \cdot a^n =$ (see the case(b)) $= a^m \cdot (a^{-m} \cdot a^{m+n}) = a^m \cdot (a^m)^{-1} \cdot a^{m+n} = a^{m+n}$; the case $m < 0$, $n > 0$ is treated in the same way as the cases (c) and (d). The cases $m = 0$ or $n = 0$ are easily verified.

28. We will consider some cases: a) if $n > 0$, then

$$(a^m)^n = \underbrace{a^m \cdot a^m \cdot \ldots \cdot a^m}_{n} = \text{(see \textbf{27})} = a^{\overbrace{m + m + \ldots + m}^{n}} = a^{mn};$$

b) if $n < 0$, $m > 0$ $n = -l$ $(l > 0)$, then $(a^m)^n = ((a^m)^l)^{-1} =$ (see the case(a)) $= (a^{ml})^{-1} =$ (since $ml > 0$) $= a^{-ml} = a^{mn}$; c) if $m < 0$, $n < 0$, $m = -k$ $(k > 0)$, $n = -l$ $(l > 0)$, then $(a^m)^n = (((a^k)^{-1})^{-1})^l =$ (see **20**) $= (a^k)^l =$ (see the case(a)) $= a^{kl} = a^{mn}$. The cases $m = 0$ or $n = 0$ are easily verified.

29. *Answer.* In the group of symmetries of the triangle (see **3**) a and b are of order 3, c and d of order 2; for the square (see **5**) b and c are of order 4, a,d,f,g,h of order 2; for the rhombus (see solution **6**) all elements (except e) have order 2.

30. 1) Let $a^k = a^l$, where $0 \leq k \leq n-1$ and $k > l$. If we multiply on the right the two members of the equality by a^{-l} we obtain $a^k \cdot a^{-l} = a^l \cdot a^{-l}$ and (see **27**) $a^{k-l} = e$. Since $0 < k - l \leq n - 1$ we obtain a contradiction of the hypothesis that the element a has order n.

2) any integer m can be written in the form $m = nt + r$, where $0 \leq r \leq n-1$ and t is some integer. Thus $a^m = a^{nt+r} =$ (see **27**) $= a^{nt} \cdot a^r =$ (see **28**) $= (a^n)^t \cdot a^r =$ (since $a^n = e$) $= a^r$, where $0 \leq r \leq n-1$.

31. *Hint.* The generator is a rotation of order $2\pi/n$.

32. *Answer.* In the group of rotations of the triangle the generators are: a, rotation by 120° and b, rotation by 240°; in the group of rotations of the square they are: a, rotation by 90° and c, rotation by 270°.

33. Let $m = nd + r$, where $0 \leq r \leq n-1$. Thus (see solution **30**-2)) $a^m = a^r$. But $a^r = e$ if and only if (see **30**-1)) $r = 0$. Therefore $a^m = e$ if and only if $m = nd$.

34. $(a^m)^p = a^{mp} = (a^p)^m = e^m = e$. Hence (see **33**) the order of the element a^m must divide the number p. Since p is prime the statement follows.

35. Since the numbers n/d and m/d are non-zero integers, $(a^m)^{n/d} = a^{mn/d} = (a^n)^{m/d} = e^{m/d} = e$. If k is an integer such that $(a^m)^k = a^{mk} = e$ then (see **33**) mk is divisible by n and $m/d \cdot k$ is divisible by n/d. Since the numbers m/d and n/d are relatively prime k is divisible by n/d. Hence the smallest non-zero integer k such that $(a^m)^k = e$ is $k = n/d$.

36. Let a be a rotation counterclockwise by an angle equal to $2\pi/12$. The elements of the group considered are thus $e, a, a^2, \ldots, a^{11}$. The element a^m, in order to be a generator, must have its order equal to 12, and therefore the numbers m and 12 (see **35**) must be relatively prime. Hence a^m will be a generator whenever $m = 1, 5, 7, 11$. *Answer.* The generators are the rotations by angles $\pi/6, 5\pi/6, 7\pi/6, 11\pi/6$.

37. Let $a^k = a^l$, and $k > l$. Thus $a^k \cdot a^{-l} = a^l \cdot a^{-l}$ and $a^{k-l} = e$, contradicting the hypothesis that a is an element of infinite order.

38. *Hint.* If one considers the group under addition, then by a^m one indicates the sum $a + a + \ldots + a$ (m terms) and by a^{-1} one indicates $-a$. The generators are 1 and -1.

39. a) See Table 8; b) see Table 9; c) Tee table 10.

Table 8

+	0	1
0	0	1
1	1	0

Table 9

+	0	1	2
0	0	1	2
1	1	2	0
2	2	0	1

Table 10

+	0	1	2	3
0	0	1	2	3
1	1	2	3	0
2	2	3	0	1
3	3	0	1	2

40. We prove that all properties of a group are satisfied: 1) for three arbitrary remainders a, b, c we have $(a+b)+c = a+(b+c)$ modulo n, because in both terms of the equality we have the remainder of the division by n of the number $a+b+c$; 2) the unit element is 0 because $m+0 = 0+m = m$ for every remainder m; 3) if $m \neq 0$ then the inverse element (the opposite element) of m is $n-m$, because modulo n one has $m+(n-m) = (n-m)+m = 0$; the inverse element of element 0 is 0 itself. This group is cyclic with generator 1, because the smallest integer k such that $\underbrace{1+1+\ldots+1}_{k} = 0$ modulo n, is equal to n.

41. Since $a^m \cdot a^r = a^k$ one has $a^m \cdot a^r \cdot a^{-k} = a^k \cdot a^{-k} = e$ and $a^{(m+r)-k} = e$. Hence (see **33**) $(m+r)-k$ is divisible by n, i.e., $m+r$ and k are equal modulo n.

42. *Answer.* They are isomorphic: the groups (1) and (4) (consider the mapping $\phi : \phi(e) = 0$, $\phi(a) = 2$, $\phi(b) = 1$, $\phi(c) = 3$), and the groups (2) and (3) (consider the mapping $\phi : \phi(e) = e$, $\phi(a) = a$, $\phi(b) = b$, $\phi(c) = c$ (see solutions **6** and **7**)).

43. Since ϕ is a bijective mapping, ϕ^{-1} exists and is bijective. Let c and d be two arbitrary elements of the group G_2. In the group G_1 there are two (unique) elements a and b such that $\phi(a) = c$ and $\phi(b) = d$. Since ϕ is an isomorphism, $\phi(ab) = \phi(a)\phi(b) = cd$ (the products are taken in the corresponding groups). Therefore $\phi^{-1}(cd) = ab = \phi^{-1}(c)\phi^{-1}(d)$. Since c and d are two arbitrary elements of the group G_2, ϕ^{-1} is an isomorphism.

44. Since the mappings $\phi_1 : G_1 \to G_2$ and $\phi_2 : G_2 \to G_3$ are bijective, $\phi_3 : G_1 \to G_3$ is also bijective. Let a and b be two arbitrary elements of the group G_1. Thus $(\phi_2\phi_1)(ab) = \phi_2(\phi_1(ab)) = \phi_2(\phi_1(a) \cdot \phi_1(b)) = \phi_2(\phi_1(a)) \cdot \phi_2(\phi_1(b)) = ((\phi_2\phi_1)(a)) \cdot ((\phi_2\phi_1)(b))$ and hence $\phi_2\phi_1$ is an isomorphism (the products are taken in the corresponding groups).

45. *Hint.* Use the result of Problem **41**. The isomorphism is $\phi(a^m) = m$.

46. *Hint.* Use the result of Problem **27**. The isomorphism is $\phi(a^m) = m$.

47. Let $\phi(e_G) = x$. Thus $x \cdot x = \phi(e_G) \cdot \phi(e_G) =$ (since ϕ is an isomorphism) $= \phi(e_G \cdot e_G) = \phi(e_G) = x$, i.e., $x^2 = x$. Multiplying the two members of this equality by x^{-1} in the group F one obtains $x^2x^{-1} = xx^{-1}$ and $x = e_F$.

48. $\phi(g)\phi(g^{-1}) =$ (since ϕ is an isomorphism) $= \phi(g \cdot g^{-1}) = \phi(e_G) =$ (see **47**) $= e_F$. It follows that $\phi(g^{-1}) = [\phi(g)]^{-1}$.

49. Let n be a non-zero integer. Thus $\phi(g^n) = \phi(\underbrace{g \cdot \ldots \cdot g}_{n}) =$ (since ϕ is an isomorphism) $= \underbrace{\phi(g) \cdot \ldots \cdot \phi(g)}_{n} = h^n$. Let n_1 and n_2 be the orders of the elements g and h. Thus $\phi(g^{n_2}) = h^{n_2} = e_F$ and (see **47**) $g^{n_2} = e_G$. Therefore $n_1 \leq n_2$. On the other hand, $h^{n_1} = \phi(g^{n_1}) = \phi(e_G) = e_F$, so $n_2 \leq n_1$. Hence $n_1 = n_2$.

50. *Answer.* a) The sole group is $\mathbb{Z}_2$, b) The sole group is $\mathbb{Z}_3$. *Solution.* a) Let a and e be the elements of the group and e the unit element. Thus $e \cdot e = e$, $a \cdot e = e \cdot a = a$ and only the product $a \cdot a$ remains unknown. If $a \cdot a = a = a \cdot e$, then (see **24**) one obtains the contradiction $a = e$. It follows that $a^2 = e$ and that there exists only one group, containing 2 elements. It is the cyclic group $\mathbb{Z}_2$.

b) Let e, a, b be the elements of the group and e the unit element. We need to know to which elements the products ab, ba, aa, bb correspond. If $ab = a$ one obtains the contradiction $a = e$; if $ab = b$ one obtains the contradiction $b = e$. It follows that $ab = e$. Consequently $ba = e$. If $a^2 = a$, one has the contradiction $a = e$. If $a^2 = e = ab$ one has the contradiction $a = b$. It follows that $a^2 = b$. Thus $b^2 = a$, and there exists only one group containing 3 elements. Since $a^3 = a^2 \cdot a = ba = e$ this group with elements $e, a, b = a^2$ is the cyclic group $\mathbb{Z}_3$.

51. *Answer.* For example, the group of rotations of the square is not isomorphic to the group of symmetries of the rhombus, because in the former there is an element of order 4, whereas in the latter there is not (see **49**).

52. Consider the map $\phi(x) = 2^x$. If x takes all the real values, 2^x takes exactly once all the real positive values. Hence ϕ is a bijective mapping of the real numbers into the real positive numbers. For any two positive numbers x and y we have $\phi(x+y) = 2^{x+y} = 2^x \cdot 2^y = \phi(x)\cdot\phi(y)$ and it follows that ϕ is an isomorphism of the group of the real numbers under addition in the group of the positive real numbers under multiplication.

53. Let y be an arbitrary element of group G. Thus $\phi_a(a^{-1}y) = aa^{-1}y = y$. Hence ϕ_a is a surjective mapping of the group G onto the group G. If $\phi_a(x) = \phi_a(y)$ then $ax = ay$ and $x = y$; hence ϕ_a is a bijective mapping of the group G into itself.

54. $(\phi_a\phi_b)(x) = \phi_a(\phi_b(x)) = \phi_a(bx) = abx = \phi_{ab}(x)$ for every element x, therefore $\phi_a\phi_b = \phi_{ab}$. Since $\phi_a(a^{-1}x) = aa^{-1}x = x$, $(\phi_a)^{-1}(x) = a^{-1}x = \phi_{a^{-1}}(x)$ for every element x; therefore $(\phi_a)^{-1} = \phi_{a^{-1}}$. Hence we have a group of transformations (see §1.2).

55. Consider a mapping ψ such that $\psi(a) = \phi_a$. Since $\phi_a(e) = ae = a$ and $\phi_b(e) = be = b$, $\phi_a \neq \phi_b$ if $a \neq b$. The mapping ψ is therefore bijective. Moreover, $\psi(ab) = \phi_{ab} =$ (see solution **54**) $= \phi_a\phi_b = \psi(a)\psi(b)$. It follows that ψ is an isomorphism.

56. a) Let e_G and e_H be the unit elements respectively in the group G and in the subgroup H. In the subgroup H one has the identity $e_He_H = e_H$. By the definition of subgroup this identity also holds in the group G. Moreover, in the group G we also have the identity $e_Ge_H = e_H$, from which we obtain that in G $e_He_H = e_Ge_H$ and (see **24**) $e_H = e_G$.

b) Let a be any element of the subgroup H, and let a_G^{-1} and a_H^{-1} be its inverse elements respectively in the group G and in the subgroup H. In the subgroup H we thus have $a_H^{-1}a = e_H =$ (see (a)) $= e_G$. By the definition of subgroup this identity also holds in the group G. Moreover, in the group G we have $a_G^{-1}a = e_G$, from which we obtain that in G $a_H^{-1}a = a_G^{-1}a$ and $a_H^{-1} = a_G^{-1}$.

57. The *necessity* follows from the result of Problem **56** and from the definition of subgroup.

Sufficiency. From property 1 the binary operation of the group G is also the binary operation of H. The element e_G, which belongs to H by property 2, is the unit element in H because $e_Ga = ae_G = a$ for every element a of the group G and, in particular, for all elements of H. If a is an arbitrary element of H then the element a_G^{-1}, which belongs to H by property 3, is the inverse element of a in H because $a_G^{-1}a = aa_G^{-1} = e_G = e_H$. The associativity is obviously satisfied: H is thus a subgroup of the group G.

58. *Answer.* (For notations see Examples 1–4). 1) The subgroup of rotations $\{e, a, b\}$, three subgroups generated by the three reflections with respect to the altitudes: $\{e, a\}$, $\{e, d\}$, $\{e, f\}$, two trivial subgroups: $\{e\}$ and the whole group; 2) the subgroup of rotations $\{e, a, b, c\}$, the subgroup of generated by the central symmetry $\{e, a\}$, four subgroups generated by the four reflections with respect to the symmetry axes: $\{e, d\}$, $\{e, f\}$, $\{e, g\}$, $\{e, h\}$, the two subgroups: $\{e, a, d, f\}$ and $\{e, a, g, h\}$, and the two trivial subgroups: $\{e\}$ and the whole group.

59. *Answer.* Suppose that the elements of the given groups are $\{e, a, a^2, \ldots, a^{n-1}\}$ $(n = 5, 8, 15)$. Their subgroups are thus: a)$\{e\}$, $\mathbb{Z}_5$; b) $\{e\}$, $\{e, a^4\} \cong \mathbb{Z}_2$, $\{e, a^2, a^4, a^6\} \cong \mathbb{Z}_4$, $\mathbb{Z}_8$; c) $\{e\}$, $\{e, a^5, a^{10}\} \cong \mathbb{Z}_3$, $\{e, a^3, a^6, a^9, a^{12}\} \cong \mathbb{Z}_5$, $\mathbb{Z}_{15}$ (see **60**).

60. Let the subgroup H be different from $\{e\}$ and let a^d be the element with minimum positive d among all the elements of the subgroup H. Thus H also contains all the elements of the form $a^{kd} = (a^d)^k$ for every integer k. Let a^m be an arbitrary element of the subgroup H. Divide m by d with remainder r: $m = td+r$, where $0 \leq r \leq d-1$. Thus H contains element $a^m \cdot a^{-td} = a^{m-td} = a^r$. If $r > 0$ we obtain a contradiction of the hypothesis that d is the minimum. It follows that $r = 0$ and that m is divisible by d. Since the element $a^n = e$ belongs to H, d divides n. Hence subgroup H has the requested properties.

61. The solution is the same of that of Problem **60**.

62. If an element of the group has infinite order, then the cyclic infinite subgroup generated by it contains an infinite number of subgroups (see **61**), which are also subgroups of the initial group. If the orders of all elements are finite we consider the cyclic subgroups generated by the following elements: at first an arbitrary element a_1, then an element a_2 which does not belong to the subgroup generated by a_1, then an element a_3 which does not belong to the subgroups generated neither by a_1 nor by a_2, etc.. This procedure does not end because each of the subgroups so obtained contains a finite number of elements.

63. Let $H_1, H_2, \ldots, H_m$ be the subgroups of a group G and H their intersection. Thus (see **57**): 1) if a and b belong to H, then both a and b belong to all H_is. Hence ab also belongs to all H_is and thus it belongs to H; 2) e belongs to all subgroups H_is, therefore also to H; 3) if a is an arbitrary element of H, then a belongs to all H_is; thus a^{-1} also belongs to all H_is and therefore also to H. In virtue of the result of Problem **57** H is a subgroup of the group G.

64. *Answer.* a) Yes; b) yes; c) no; d) no.

65. The vertex A can be sent onto an arbitrary vertex, B onto any one of the remaining vertices, C onto any one of the remaining two vertices. *Answer.* $4 \cdot 3 \cdot 2 = 24$.

66. *Answer.* a) All symmetries fixing vertex D;
b) $\left\{e = \begin{pmatrix} A & B & C & D \\ A & B & C & D \end{pmatrix}, d = \begin{pmatrix} A & B & C & D \\ B & C & D & A \end{pmatrix}, \right.$
$\left. d^2 = \begin{pmatrix} A & B & C & D \\ C & D & A & B \end{pmatrix}, d^3 = \begin{pmatrix} A & B & C & D \\ D & A & B & C \end{pmatrix}\right\}.$

67. Let us formulate the definition of orientation in a more symmetric way. We have defined the orientation by means of the vertex D, but if the position of the triangle ABC with respect to the vertex D is given, then the position of any triangle with respect to the fourth vertex is also uniquely defined. Hence a transformation preserving the orientation of the tetrahedron preserves the position of every triangle with respect to the fourth vertex, whereas a transformation changing the orientation of the tetrahedron changes the position of every triangle with respect to the fourth vertex. It is clear that the product of two transformations preserving the orientation of the tetrahedron, as well as the product of two transformations which do not preserve it, preserves the orientation. Conversely, if a transformation preserves the orientation and another permutation changes it, their product changes the orientation. The identity e obviously preserves the orientation and, since $a^{-1}a = e$ for every transformation a if a preserves the orientation a^{-1} also preserves the orientation. It follows that (see **57**) all symmetries of the tetrahedron preserving the orientation form a subgroup in the group of all symmetries of the tetrahedron. Since the vertex D can be sent onto an arbitrary vertex, and then the triangle ABC can take one of three positions, this group contains $4 \cdot 3 = 12$ elements.

68. *Answer.* a) The subgroup of rotations about an axis through the middle points of two opposite edges; b) the subgroup of rotations about the axis through D perpendicular to the plane of triangle ABC.

69. Using the associativity in G and in H we shall prove the associativity in $G \times H$. We have

$$\begin{aligned}((g_1,h_1)(g_2,h_2))(g_3,h_3) &= (g_1g_2,h_1h_2)(g_3,h_3) = ((g_1g_2)g_3,(h_1h_2)h_3)\\ &= (g_1(g_2g_3),h_1(h_2h_3)) = (g_1,h_1)(g_2g_3,h_2h_3)\\ &= (g_1,h_1)((g_2,h_2)(g_3,h_3)).\end{aligned}$$

Moreover $(e_G,e_H)(g,h)=(e_Gg,e_Hh)=(g,h)$ and $(g,h)(e_G,e_H)=(ge_G,he_H)=(g,h)$. The pair (e_G,e_H) is therefore the unit element in $G\times H$. Moreover, $(g^{-1},h^{-1})(g,h)=(g^{-1}g,h^{-1}h)=(e_G,e_H)$ and $(g,h)(g^{-1},h^{-1})=(gg^{-1},hh^{-1})=(e_G,e_H)$. Hence (g^{-1},h^{-1}) is the inverse element of (g,h) in $G\times H$. We have proved that $G\times H$ possesses all the properties of a group.

70. *Answer.* nk.

71. *Hint.* Verify that the transformation ϕ such that $\phi((g,h))=(h,g)$ is an isomorphism between the groups $G\times H$ and $H\times G$.

72. *Answer.* All elements of the type (g,e_H) form a subgroup in $G\times H$ isomorphic to the group G. All elements of the type (e_G,h) form a subgroup in $G\times H$ isomorphic to the group H.

73. We have

$$(g_1,h_1)(g_2,h_2)=(g_1g_2,h_1h_2)=(g_2g_1,h_2h_1)=(g_2,h_2)(g_1,h_1).$$

74. 1) If (g_1,h_1) and (g_2,h_2) belong to $G\times H$ then g_1, g_2, and g_1g_2 belong to G, and h_1, h_2, and h_1h_2 belong to H. Hence $G\times H$ contains the element $(g_1g_2,h_1h_2)=(g_1,h_1)(g_2,h_2)$.

2) Since G_1 and H_1 are subgroups respectively of G and of H, then e_G belongs to G_1 and e_H belongs to H_1. Hence $G_1\times H_1$ contains the element (e_G,e_H), the unit element of the group $G\times H$.

3) If the element (g,h) belongs to $G_1\times H_1$, then g belongs to G_1 and h belongs to H_1. Since G_1 and H_1 are subgroups g_G^{-1} belongs to G_1 and h_H^{-1} belongs to H_1. Hence $G_1\times H_1$ contains (g_G^{-1},h_H^{-1}), the inverse element of (g,h) in the group $G\times H$.

By virtue of the result of Problem **57** $G_1\times H_1$ is a subgroup of the group $G\times H$.

75. No. Consider the following example. Let $G=\{e_1,c\}$ and $H=\{e_2,d\}$ be two cyclic groups of order two. Thus $\{(e_1,e_2),(c,d)\}$

is a subgroup of the group $G \times H$, which cannot be represented in the form requested by the problem.

76. Let A, B, C, D be the vertices of the given rhombus. Let $G = \{e_1, g\}$ be the group of permutations of the elements A and C, and $H = \{e_2, h\}$ the group of permutations of the elements B and D. Thus the mapping ϕ such that (for the notations see solution **6**)

$$\phi(e) = (e_1, e_2), \quad \phi(a) = (g, h), \quad \phi(b) = (e_1, h), \quad \phi(c) = (g, e_2)$$

is, as one can easily verify, an isomorphism of the group of symmetries of the rhombus in the group $G \times H \cong \mathbb{Z}_2 \times \mathbb{Z}_2$.

77. 1) Let $\{e_1, g\}$ and $(\{e_2, h, h^2\}$ be the two given groups. We look for the order of the element $a = (g, h)$: $a^2 = (g^2, h^2) = (e_1, h^2) \neq (e_1, e_2)$, $a^3 = (g, e_2) \neq (e_1, e_2)$, $a^4 = (e_1, h) \neq (e_1, e_2)$, $a^5 = (g, h^2) \neq (e_1, e_2)$, $a^6 = (e_1, e_2)$. Since the group $\mathbb{Z}_2 \times \mathbb{Z}_3$ contains only 6 elements we have $\mathbb{Z}_2 \times \mathbb{Z}_3 \cong \mathbb{Z}_6$.

2) Let (g, h) be an arbitrary element of the group $\mathbb{Z}_2 \times \mathbb{Z}_4$. Thus $(g, h)^4 = (g^4, h^4) = (e_1, e_2)$. Hence $\mathbb{Z}_2 \times \mathbb{Z}_4$ does not contain any element of order 8 and thus is not isomorphic to the group $\mathbb{Z}_8$.

78. Let g be a generator in $\mathbb{Z}_m$, h the generator in $\mathbb{Z}_n$ and r the order of the element (g, h) in the group $\mathbb{Z}_m \times \mathbb{Z}_n$. Since $(g, h)^{mn} = (g^{mn}, h^{mn}) = (e_1, e_2)$ it follows that $r \leq mn$. But since $(g^r, h^r) = (g, h)^r = (e_1, e_2)$ it follows that (see **33**) r is divisible by m and by n. If m and n are relatively prime we obtain that $r = mn$ and (g, h) is the generator of the group $\mathbb{Z}_m \times \mathbb{Z}_n$. Hence $\mathbb{Z}_m \times \mathbb{Z}_n \cong \mathbb{Z}_{mn}$.

If, on the contrary, m and n are not relatively prime, then their lowest common multiple, k, is less than mn. Let $k = mk_1$ and $k = nk_2$. If g and h are any two elements of the groups $\mathbb{Z}_m$ and $\mathbb{Z}_n$, then $g^m = e_1$ and $h^n = e_2$. Hence $(g, h)^k = (g^{mk_1}, h^{nk_2}) = (e_1, e_2)$. Since $k < mn$ we obtain that the group $\mathbb{Z}_m \times \mathbb{Z}_n$ contains no elements of order mn, and therefore is not isomorphic to the group $\mathbb{Z}_{mn}$.

79. *Answer.* (see §1.1, Examples 1 and 2): a) $\{e, a, b\}$; b) $\{e, c\}$, $\{a, f\}$, $\{b, d\}$.

80. Since $x = xe$ and e belongs to H the element x belongs to the coset xH.

81. By hypothesis $y = xh_1$, where h_1 is an element of the subgroup H. Hence $x = yh_1^{-1}$. Let h be an arbitrary element of the subgroup

H. The elements h_1h and $h_1^{-1}h$ thus belong to H. Hence the element $yh = (xh_1)h = x(h_1h)$ belongs to xH, and the element $xh = (yh_1^{-1})h = y(h_1^{-1}h)$ belongs to yH. Since every element of yH belongs to xH, and vice versa, it follows that $xH = yH$.

82. Suppose that the element z belongs to xH and to yH. Thus (see **81**) $xH = zH$ and $yH = zH$. Therefore $xH = yH$.

83. *Hint.* The order of an element is equal to the order of the cyclic subgroup that it generates. Afterwards apply the Lagrange theorem.

84. *Hint.* If the order p of a group is prime then the order of every element different from e is equal to p (see **83**).

85. Apply the Lagrange theorem. *Answer.* Two: $\{e\}$ and the whole group.

86. *Hint.* Use the results of Problems **84** and **45**.

87. Let G be the given group of order m and $n = md$. *Answer.* $G \times \mathbb{Z}_d$ (see **72**).

88. *Answer.* It is possible. For example, in the group of rotations of the tetrahedron, containing 12 elements (see **67**), there are no subgroups containing 6 elements. *Proof.* The group of rotations of the tetrahedron contains 12 elements (see **67**): the identity e, 8 rotations (by 120° and 240°) about the altitudes perpendicular to the 4 triangular faces, and 3 rotations (by 180°) about the axes through the middle points of opposite edges. Suppose that the group of rotations of the tetrahedron contains a subgroup of 6 elements. This subgroup must obviously contain at least one rotation a about one altitude, for example, that from the vertex A. If a is a rotation by 120° (or by 240°) then a^2 is a rotation by 240° (by 120°). Therefore our subgroup must contain both rotations about the altitude drawn from vertex A. Since one has only 3 rotations (including the identity), fixing vertex A our subgroup must contain another rotation b, sending the vertex A to a different vertex, for example to B. Thus this subgroup also contains the element bab^{-1}. This rotation sends the vertex B to B and one has, moreover, $bab^{-1} \neq e$ (otherwise $a = b^{-1}b = e$). Consequently our subgroup must contain at least one, and therefore both, rotations about the altitude drawn from the vertex B. These rotations send the vertex A to C and to D. We obtain again that our subgroup must contain all rotations about the altitudes from C and from D. In this way we have, with e, 9 elements. This is in contradiction with the hypothesis. The group of rotations of the tetrahedron therefore does not contain any subgroup of order 6.

89. *Answer.* a) The left and right partitions coincide: $\{e, a, b\}$, $\{c, d, f\}$;

b) left partition: $\{e, c\}$, $\{a, f\}$, $\{b, d\}$; right partition: $\{e, c\}$, $\{a, d\}$, $\{b, f\}$.

90. *Answer.* a) The two partitions coincide: $\{e, a\}$, $\{b, c\}$, $\{d, f\}$, $\{g, h\}$;

b) left partition: $\{e, d\}$, $\{b, g\}$, $\{a, f\}$, $\{c, h\}$; right partition: $\{e, d\}$, $\{b, h\}$, $\{a, f\}$, $\{c, g\}$.

91. *Answer.* The two partitions coincide and contain three cosets: 1) all numbers of type $3k$ ($k = 0, \pm 1, \pm 2, \ldots$), 2) all numbers of type $3k + 1$ ($k = 0, \pm 1, \pm 2, \ldots$), 3) all numbers of type $3k + 2$ ($k = 0, \pm 1, \pm 2, \ldots$).

92. *Answer.* a) Two groups: $\mathbb{Z}_4$ and $\mathbb{Z}_2 \times \mathbb{Z}_2$;

b) two groups: $\mathbb{Z}_6$ and the group of symmetries of the equilateral triangle;

c) five groups: $\mathbb{Z}_8$, $\mathbb{Z}_4 \times \mathbb{Z}_2$, $(\mathbb{Z}_2 \times \mathbb{Z}_2) \times \mathbb{Z}_2$, the group of symmetries of the square, the group of quaternions with elements $\pm 1, \pm i, \pm j, \pm k$, and the following table of multiplication (Table 11).

Table 11

	1	-1	i	$-i$	j	$-j$	k	$-k$
1	1	-1	i	$-i$	j	$-j$	k	$-k$
-1	-1	1	$-i$	i	$-j$	j	$-k$	k
i	i	$-i$	-1	1	k	$-k$	$-j$	j
$-i$	$-i$	i	1	-1	$-k$	k	j	$-j$
j	j	$-j$	$-k$	k	-1	1	i	$-i$
$-j$	$-j$	j	k	$-k$	1	-1	$-i$	i
k	k	$-k$	j	$-j$	$-i$	i	-1	1
$-k$	$-k$	k	$-j$	j	i	$-i$	1	-1

Solution. a) Let $\{e, a, b, c\}$ be the elements of the initial group. The elements a, b, c can thus be of order 2 or 4 (see **83**). Let us consider some cases.

1) Amongst a, b, c there is an element of order 4. Hence the given group is the cyclic group $\mathbb{Z}_4$.

2) The orders of elements a, b and c are equal to 2, i.e., $a^2 = b^2 = c^2 = e$. We are now looking for the element corresponding to product ab. It is possible neither that $ab = e$ (otherwise $ab = a^2$ and $b = a$), nor that $ab = a$ (otherwise $b = e$), nor that $ab = b$ (otherwise $a = e$). There remains therefore only $ab = c$. Similarly one obtains $ba = c$, $ac = ca = b$ and $bc = cb = a$. The multiplication table is complete and we obtain (see **6**) the group of symmetries of the rhombus, isomorphic to $\mathbb{Z}_2 \times \mathbb{Z}_2$ (see **76**).

b) The elements of the group can be of order 1, 2, 3 or 6 (see **83**). We will consider some cases.

1) Suppose that there exists an element of order 6. The given group is thus the cyclic group $\mathbb{Z}_6$.

2) Suppose that all elements are of order 2. Thus the group is commutative (see **25**), and if a and b are elements of the initial group the elements $\{e, a, b, ab\}$ form a subgroup of it. This is not possible (see Lagrange's theorem) and this case must be excluded.

3) All elements are of order 2 or 3, and there exists an element of order 3. Let a be the element of order 3 and c an element which is not a power of a. Thus $\{c, ca, ca^2\}$ is a left cosets of the subgroup $\{e, a, a^2\}$ and the six elements e, a, a^2, c, ca, ca^2 are thus all distinct (see **82**). We will prove that in this set of 6 elements there exists only one way to define a multiplication table. At first we prove that $c^2 = e$. In fact, it is not possible that $c^2 = ca^k$ (otherwise $c = a^k$). If $c^2 = a$ (or $c^2 = a^2$), then we should have $c^3 = cc^2 = ca \neq e$ (or $c^3 = ca^2 \neq e$), but we have supposed that all elements have order 2 or 3. Thus $c^2 = e$. Since c is an arbitrary element which does not belong to the subgroup $\{e, a, a^2\}$, we also have that $(ca)^2 = e$ and $(ca^2)^2 = e$. In this way the product of all pairs of the 6 elements of the group has been defined. Indeed, $a^k a^l = a^{k+l}$, $(ca^k)a^l = ca^{k+l}$, $(ca^k)(ca^l) = (ca^k)(ca^k)a^{l-k} = ea^{l-k} = a^{l-k}$, $a^k(ca^l) = c(ca^k)(ca^l) = ca^{l-k}$. Consequently there is only one possibility of constructing a multiplication table in order to obtain a group. Hence there exists only one group of 6 elements, with orders equal to 1, 2, and 3. We know this group: it is the group of symmetries of the equilateral triangle.

c) the elements of the group can have order 1, 2, 4, or 8 (see **83**). Consider some cases.

1) Suppose that there exists an element of order 8. The given group is thus the cyclic group $\mathbb{Z}_8$.

2) Suppose that all the elements different from the unit have order 2.

The given group is thus commutative (see **25**). In this case let a and b be two distinct elements of order 2 of the initial group. Thus $\{e, a, b, ab\}$ form a subgroup. If the element c does not belong to this subgroup then the elements $\{c, ac, bc, abc\}$ form a right coset of the subgroup $\{e, a, b, ab\}$ and the 8 elements $e, a, b, ab, c, ac, bc, abc$ are thus all distinct. The products of these elements are well defined, because the group is commutative and $a^2 = b^2 = c^2 = e$ (for example, $(ac)(abc) = a^2bc^2 = b$). Therefore if all elements have order 2 then there is only one possible group. This group is, in fact, $(\mathbb{Z}_2 \times \mathbb{Z}_2) \times \mathbb{Z}_2$.

3) Suppose the element a have order 4 and that amongst the elements, different from the powers of a, there is an element b of order 2, i.e., $b^2 = e$. In this case $\{e, a, a^2, a^3\}$ and $\{b, ba, ba^2, ba^3\}$ are the two left cosets of the subgroup $\{e, a, a^2, a^3\}$, and the 8 elements that they contain are thus all distinct. We look for which of these elements may be equal to product ab. It is possible neither that $ab = a^k$ (otherwise $b = a^{k-1}$) nor $ab = b$ (otherwise $a = e$). If $ab = ba^2$ then $ab^2 = ba^2b$ and (since $b^2 = e$) $a = ba^2b$. Thus one obtains the contradiction $a^2 = (ba^2b)(ba^2b) = ba^2a^2b = bb = e$. This means that either $ab = ba$, or $ab = ba^3$.

Let us consider the two cases:

α) $ab = ba$. The table of multiplication is in this case uniquely defined. Indeed, $a^ka^l = a^{k+l}$, $a^k(ba^l) = ba^{k+l}$, $(ba^k)a^l = ba^{k+l}$, $(ba^k)(ba^l) = b^2a^{k+l} = a^{k+l}$. We can therefore have only one group: this group, in fact, does exist: it is the group $\mathbb{Z}_4 \times \mathbb{Z}_2$. If e_1 and g are the unit and the generator of $\mathbb{Z}_4$, e_2 and h the unit and the generator of $\mathbb{Z}_2$, then it suffices to write $a = (g, e_2)$, $b = (e_1, h)$ for all the properties listed above being satisfied.

β) $ab = ba^3$. Also in this case the table of multiplication is uniquely defined. Indeed, $a^ka^l = a^{k+l}$, $(ba^k)a^l = ba^{k+l}$, $a^kb = ba^{3k}$, $a^k(ba^l) = ba^{3k+l}$, $(ba^k)(ba^l) = b(a^kba^l) = b^2(a^{3k+l}) = a^{3k+l}$. In this case we therefore have only one group. In fact, this group does exist: it is the group of symmetries of the square. It suffices to put a = rotation by 90°, b = reflection with respect to a diagonal, for all the properties listed above being satisfied.

4) Suppose that there exists an element a of order 4 and that all elements different from e, a, a^2, a^3 are of order 4 as well. Let b be an arbitrary element different from e, a, a^2, a^3. Thus the elements e, a, a^2, a^3 b, ba, ba^2, ba^3 are all distinct. We look for which of these elements is equal to product bb. It is possible neither that $b^2 = ba^k$ (otherwise $b = a^k$) nor $b^2 = e$ (because the order of b is 4). If $b^2 = a$ (or $b^2 = a^3$) one has the

contradiction $b^4 = a^2 \neq e$. Hence $b^2 = a^2$. Since b is an arbitrary element different from e, a, a^2, a^3, also $(ba)^2 = (ba^2)^2 = (ba^3)^2 = a^2$. From the equality $baba = a^2 = b^2$ it follows that $aba = b$, $aba^4 = ba^3$ and $ab = ba^3$. The table of multiplication is now uniquely defined. Indeed, $a^k a^l = a^{k+l}$, $(ba^k)a^l = ba^{k+l}$, $a^k b = ba^{3k}$, $a^k(ba^l) = ba^{3k+l}$, $(ba^k)(ba^l) = b^2 a^{3k+l} = a^2 a^{3k+l} = a^{3k+l+2}$. Hence in this case we can have only one group. We can verify that our multiplication table in fact defines a group. This group is called the group of quaternions. It is better to denote its elements by the following notations: instead of e, a, a^2, a^3 b, ba, ba^2, ba^3, we have, in that order, 1, i, -1, $-i$ j, $-k$, $-j$, k. Thus the multiplication by 1 and -1 and the operations with signs are the same as in ordinary algebra. Moreover, one has $i^2 = j^2 = k^2 = -1$, $ij = k$, $ji = -k$, $jk = i$, $kj = -i$, $ki = j$, $ik = -j$. The multiplication table for the group of quaternions is shown in Table 11.

93. Consider the vertex named A by the new notation. Its old notation was thus $g^{-1}(A)$. By the action of the given transformation this vertex is sent onto a vertex named, in the old notation, $hg^{-1}(A)$ and, in the new notation, $ghg^{-1}(A)$. Similarly in the new notations the vertex B is sent onto $ghg^{-1}(B)$ and the vertex C onto $ghg^{-1}(C)$. Hence to this transformation there corresponds in the new notation the permutation ghg^{-1}.

94. $ghg^{-1} = h_1$ if and only if $h = g^{-1}h_1 g$. Hence every element h_1 has, under the mapping ϕ_g, one and only one image. It follows that $\phi_g(h) = ghg^{-1}$ is a bijective mapping of the group into itself. Moreover, $\phi_g(h_1 h_2) = g(h_1 h_2)g^{-1} = gh_1(g^{-1}g)h_2 g^{-1} = (gh_1 g^{-1})(gh_2 g^{-1}) = \phi_g(h_1)\phi_g(h_2)$. Thus ϕ_G is an isomorphism.

95. *Answer.* The reflections with respect to all altitudes.

96. *Answer.* The rotations by 120° and 240°.

97. *Answer.* Let us subdivide all the elements of the group of symmetries of the tetrahedron in the following classes: 1) e; 2) all rotations different from e about the altitudes; 3) all rotations by 180° about the axes through the middle points of opposite edges; 4) all reflections with respect to the planes through any vertex and the middle point of the opposite edge; 5) all transformations generated by a cyclic permutation of the vertices (for example, $\begin{pmatrix} A & B & C & D \\ B & C & D & A \end{pmatrix}$). Thus two elements can be transformed one into another by an internal automorphism of the group of symmetries of the tetrahedron if and only if they belong to the same class.

In the group of rotations of the tetrahedron classes 4 and 5 are absent, whereas class 2 splits into two subclasses: 2a) all the clockwise rotations by 120° about the altitudes (looking on the base of the tetrahedron from the vertex from which the altitude is drawn); 2b) all rotations by 240°.

Solution. Let the elements of the group of symmetries of the tetrahedron be divided into classes as explained above. Such classes are characterized by the following properties: all elements of class 2 have order 3 and preserve the orientation of the tetrahedron; all elements of class 3 have order 2 and preserve the orientation; all elements of class 4 have order 2 and change the orientation; all elements of the class 5 have order 4 and change the orientation. Since an internal automorphism is an isomorphism (see **94**) two elements of different order cannot be transformed one into the other (see **49**). Moreover, h and ghg^{-1} either both change the orientation or both preserve it (it suffices to consider two cases: when g changes and when g preserves the orientation). Consequently two elements of distinct classes cannot be transformed one into the other by an internal automorphism.

Let h_1 and h_2 be two rotations by 180° about two axes through the middle points of two opposite edges and let g be a rotation sending the first axis onto the other. Thus the rotation gh_1g^{-1} sends the second axis into itself without reversing it. Moreover, $gh_1g^{-1} \neq e$ (otherwise $h_1 = g^{-1}eg = e$). Hence gh_1g^{-1} coincides with h_2. Therefore any two elements of class 3 can be transformed one into the other by an internal automorphism in the group of rotations (and therefore in the group of symmetries) of the tetrahedron.

Let h_1 and h_2 be two reflections of the tetrahedron with respect to two planes of symmetry and let g be a rotation sending the first plane onto the second one. Thus as before we have $gh_1g^{-1} = h_2$.

If $g_1hg_1^{-1} = h_1$ and $g_2hg_2^{-1} = h_2$ then $h = g_1^{-1}h_1g_1$ and $g_2(g_1^{-1}h_1g_1)g_2^{-1} = h_2$. Hence $(g_2g_1^{-1})h_1(g_2g_1^{-1})^{-1} = h_2$. It follows that if h can be transformed either into h_1 or into h_2, then h_1 and h_2 can be transformed one into the other. Therefore it suffices to show that any element of a given class can be sent into all the other elements of the same class.

Let $a = \begin{pmatrix} A & B & C & D \\ B & C & D & A \end{pmatrix}$ be an element of the class 5 and let g_i $(i = 1, \dots, 5)$ be the rotations such that $g_1 = \begin{pmatrix} A & B & C & D \\ A & C & D & B \end{pmatrix}$,

$$g_2 = \begin{pmatrix} A & B & C & D \\ A & D & B & C \end{pmatrix}, \quad g_3 = \begin{pmatrix} A & B & C & D \\ C & B & D & A \end{pmatrix},$$
$$g_4 = \begin{pmatrix} A & B & C & D \\ B & D & C & A \end{pmatrix}, \quad g_5 = \begin{pmatrix} A & B & C & D \\ B & A & D & C \end{pmatrix}.$$

Hence (verify) the elements $g_i a g_i^{-1}$ $(i = 1, \ldots, 5)$ form, together with the element a, the entire class 5.

Let $b = \begin{pmatrix} A & B & C & D \\ A & C & D & B \end{pmatrix}$ be the rotation of the tetrahedron by 120° about the altitude drawn from vertex A. We will prove that in the group of symmetries of the tetrahedron this rotation can be sent by internal automorphisms to all the other rotations about the altitudes. Because of symmetry it suffices to prove that b can be transformed into the second rotation about the same altitude: $b^2 = \begin{pmatrix} A & B & C & D \\ A & D & B & C \end{pmatrix}$ as well as into an arbitrary rotation about a different altitude, for example $c = \begin{pmatrix} A & B & C & D \\ D & B & A & C \end{pmatrix}$. Let g_1 be the symmetry $\begin{pmatrix} A & B & C & D \\ A & B & D & C \end{pmatrix}$ and let g_2 be the rotation $= \begin{pmatrix} A & B & C & D \\ B & C & A & D \end{pmatrix}$.

Thus $g_1 b g_1^{-1} = b^2$ and $g_2 b g_2^{-1} = c$.

However, if we take as g only one rotation of the tetrahedron, then it is easy to verify that a counterclockwise rotation by 120° about one altitude (looking on the base of the tetrahedron from the vertex from which this altitude is drawn) cannot be transformed into the rotation by 240° about the same altitude. Indeed, the rotations by 120° about the altitudes can be transformed only into rotations of 120° about altitudes, and the rotations by 240° can be transformed only into rotations by 240°.

98. $\phi_b(ab) = b(ab)b^{-1} = ba$. Since ϕ_b is an isomorphism (see **94**), ab and ba have the same order (see **49**).

99. *Hint.* In this case for every element a of the subgroup N and every element g of the group G the element $gag^{-1} = agg^{-1} = a$ belongs to N.

100. *Answer.* Yes. Verify that for every element g of the group of symmetries of the square $geg^{-1} = e$ and $gag^{-1} = a$.

101. Suppose that the left and the right partitions coincide. Let a be an arbitrary element of N and g an arbitrary element of the group G. Since the classes gN and Ng have the element g in common, they must coincide. Hence the element ga, which belongs to gN, also belongs to

Ng, i.e., there exists an element b in N such that $ga = bg$. Hence the element gag^{-1} belongs to N and therefore N is a normal subgroup of the group G.

Let now N be a normal subgroup of the group G. We will prove that $gN = Ng$ for every element g of the group G. Let ga be an arbitrary element of gN. Thus $gag^{-1} = b$, where b is an element of N. Thus $ga = bg$ and it follows that ga (and therefore the entire gN) belongs to Ng. Now let cg be an arbitrary element of Ng. Thus $(g^{-1})c(g^{-1})^{-1} = d$, where d is an element of N. Hence $cg = gd$ and cg (and therefore the entire Ng) belongs to gN. So gN and Ng coincide.

102. *Hint.* In this case both the left and the right partitions contain two classes: one is the given subgroup, the other contains all the remaining elements. See also Theorem 2 (§1.10).

103. Let $N_1, N_2, \ldots, N_s$ be the given normal subgroups of the group G and N their intersection. If a is an arbitrary element of N then a belongs to all the N_is. Hence if g is an arbitrary element of the group G then gag^{-1} belongs to all the N_is and therefore to N. This means that N is a normal subgroup of G.

104. Let g be an arbitrary element of the group G. Since $eg = ge$, e belongs to the centre. If a belongs to the centre then $ag = ga$. Multiplying both members of this equality by a^{-1} on the left one obtains $ga^{-1} = a^{-1}g$. Thus a^{-1} also belongs to the centre. If a and b belong to the centre then $ag = ga$ and $bg = gb$. Therefore $g(ab) = (ga)b = a(gb) = (ab)g$ and thus ab also belongs to the centre. By virtue of the result of Problem **57** the centre is a subgroup.

Let a be any element of the centre and g an arbitrary element of the group G. Thus the element $gag^{-1} = agg^{-1} = a$, too, belongs to the centre. The centre is therefore a normal subgroup.

105. Let h_1, h_2 be two arbitrary elements of N_1 and N_2, and g_1, g_2 be two arbitrary elements of G_1 and G_2, respectively. Thus the element $g_1h_1g_1^{-1}$ belongs to N_1, whereas the element $g_2h_2g_2^{-1}$ belongs to N_2. Consequently the element $(g_1, g_2)(h_1, h_2)(g_1^{-1}, g_2^{-1}) = (g_1h_1, g_2h_2)(g_1^{-1}, g_2^{-1}) = (g_1h_1g_1^{-1}, g_2h_2g_2^{-1})$ belongs to $N_1 \times N_2$. It follows that $N_1 \times N_2$ is a normal subgroup of $G_1 \times G_2$.

106. Since x_1 belongs to the class x_1N, x_2 also (by hypothesis) belongs to the class x_1N. This means that there exists in N an element h_1 such that $x_2 = x_1h_1$. Similarly we obtain that there exists in N an element

h_2 such that $y_2 = y_1h_2$. Since N is a normal subgroup $Ny_1 = y_1N$. Hence there exists in N an element h_3 such that $h_1y_1 = y_1h_3$. Therefore $x_2y_2 = x_1h_1y_1h_2 = x_1y_1h_3h_2$. Since the element h_3h_2 belongs to N x_1y_1 and x_2y_2 belong to the same coset x_1y_1N.

107. Let a, b, and c be the elements arbitrarily chosen in T_1, T_2, and T_3, respectively. By the definition of multiplication of cosets, $(T_1T_2)T_3$ and $T_1(T_2T_3)$ are the cosets containing the elements $(ab)c$ and $a(bc)$, respectively. Since $(ab)c = a(bc)$, $(T_1T_2)T_3 = T_1(T_2T_3)$.

108. *Hint.* Take e as a representant of class E.

109. *Hint.* Let a be an arbitrary element of class T. Take as the class T^{-1} the coset containing the element a^{-1}.

110. It is easy to verify (see table 2, §1.11) that $A^2 = B^2 = C^2 = E$. Hence this quotient group is isomorphic to the group of symmetries of the rhombus.

111. We will show only the normal subgroups different from $\{e\}$ and the whole group:

a) see **58** (1), **95**, **96**, **102**. *Answer.* The normal subgroup is the group of rotations of the triangle; the corresponding quotient group is isomorphic to $\mathbb{Z}_2$.

b) see **99**, **74**, **75**. Let $\{e_1, c\} \times \{e_2, d\}$ be the given group. *Answer.* The normal subgroups are: 1)$\{(e_1, e_2), (c, e_2)\}$; 2)$\{(e_1, e_2), (e_1, d)\}$; 3) $\{(e_1, e_2), (c, d)\}$. The corresponding quotients groups are isomorphic to $\mathbb{Z}_2$.

c) for the notations see examples 3, 4 (§1.1). If a normal subgroup of the group of symmetries of the square contains the element b or the element c, then it contains every subgroup of the group of rotations of the square. We obtain in this case the normal subgroup $\{e, a, b, c\}$ (see **102**), and the quotient group $\mathbb{Z}_2$.

We have $bdb^{-1} = f$ and $bfb^{-1} = d$. Thus if one of the elements d, f belongs to the normal subgroup the other one also does. Since $df = a$, in this case the element a also belongs to the normal subgroup. We obtain the normal subgroup $\{e, a, d, f\}$ (see **102**), and the quotient group $\mathbb{Z}_2$.

Since $bgb^{-1} = h$, $bhb^{-1} = g$ and $hg = a$, we obtain, as before, the normal group $\{e, a, g, h\}$, and the quotient group $\mathbb{Z}_2$.

If, on the contrary, the normal subgroup does not contain the elements b, c, d, f, g, h, then it coincides with the normal subgroup $\{e, a\}$, and the corresponding quotient group is isomorphic to $\mathbb{Z}_2 \times \mathbb{Z}_2$ (see **100, 110**).

Answer. The normal subgroups are: 1) $\{e, a, b, c\}$; 2) $\{e, a, d, f\}$; 3) $\{e, a, g, h\}$; 4) $\{e, a\}$. The quotient group in the cases 1–3 is isomorphic to $\mathbb{Z}_2$, in the case 4 to $\mathbb{Z}_2 \times \mathbb{Z}_2$.

d) Let $\{1, -1, i, -i, j, -j, k, -k\}$ be the given group. If h is an arbitrary element, different from 1 and -1, then $h^2 = -1$. Hence every normal subgroup different from $\{1\}$ contains the element -1. We obtain the first normal subgroup, which consists of elements $\{1, -1\}$. The partition by this subgroup is shown in table 12.

Table 12

1 -1	i $-i$	j $-j$	k $-k$
E	A	B	C

Since $i^2 = j^2 = k^2 = -1$, one has $A^2 = B^2 = C^2 = E$ and therefore the quotient group is isomorphic in this case to $\mathbb{Z}_2 \times \mathbb{Z}_2$. Since the element -1 belongs to every (non-trivial) subgroup both elements i and $-i$ either belong or do not belong to a normal subgroup. This also holds for j and $-j$, as well as for k and $-k$. Since a (non-trivial) normal subgroup in the group of quaternions can contain only 2 or 4 elements (see Lagrange's theorem), we obtain again only 3 normal subgroups (see **102**): $\{1, -1, i, -i\}$, $\{1, -1, j, -j\}$, $\{1, -1, k, -k\}$. The quotient groups in these cases are isomorphic to $\mathbb{Z}_2$.

Answer. The normal subgroups are: 1) $\{1, -1\}$; 2) $\{1, -1, i, -i\}$; 3) $\{1, -1, j, -j\}$; 4) $\{1, -1, k, -k\}$. The quotient group in the case 1 is isomorphic to $\mathbb{Z}_2 \times \mathbb{Z}_2$ and in the cases 2–4 is isomorphic to $\mathbb{Z}_2$.

112. a) See **99**, **60**. Let $n = dk$. The partition of the group $\mathbb{Z}_n = \{e, a, a^2, \ldots, a^{n-1}\}$ by the subgroup $\{e, a^d, a^{2d}, \ldots, a^{(k-1)d}\}$ is represented in Table 13 (l takes all values from 0 to $k-1$). Element a belongs to class A_1 and the minimal positive integer m such that a^m belongs to class E is equal to d. Hence the order of the element A_1 in the quotient group is equal to d and therefore the quotient group is isomorphic to $\mathbb{Z}_d$.

Table 13

a^{dl}	a^{dl+1}	a^{dl+2}	$\ldots$	$a^{dl+(d-1)}$
E	A_1	A_2	$\ldots$	A_{d-1}

b) See **99**, **61**. Table 13 shows the partition of the group $\mathbb{Z} = \{\ldots, a^{-2}, a^{-1}, e, a^1, a^2, \ldots\}$ by the subgroup $\{\ldots, a^{-2d}, a^{-d}, e, a^d, a^{2d}, \ldots\}$

$(l = 0, \pm 1, \pm 2, \dots)$. As in the case (a) one obtains that the quotient group is isomorphic to $\mathbb{Z}_d$.

113. See **97**. A set of rotations is a normal subgroup of the group of rotations of the tetrahedron if and only if it consists of some one of the classes given in the solution of Problem **97** (for the group of rotations) and it is a subgroup. If a normal subgroup contains one rotation (by 120° or by 240°) about one altitude of the tetrahedron, then it also contains the other rotation about the same altitude and hence it contains all rotations about all altitudes of the tetrahedron. If $a = \begin{pmatrix} A & B & C & D \\ A & C & D & B \end{pmatrix}$ and $b = \begin{pmatrix} A & B & C & D \\ C & B & D & A \end{pmatrix}$ are two rotations about the altitudes from the points A and B, respectively, then $ab = \begin{pmatrix} A & B & C & D \\ D & C & B & A \end{pmatrix}$ is a rotation by 180° about the axis through the middle points of the edges AD and BC. Hence in this case the normal subgroup contains all rotations by 180° about the axes through the middle points of opposite edges, and therefore it coincides with the entire group of rotations of the tetrahedron.

In this way in the group of rotations of the tetrahedron there is only one normal subgroup (non-trivial): it consists in the identity and in the three rotations by 180° about the axes through the middle points of the opposite edges. The quotient group by this normal subgroup contains 3 elements, and is thus isomorphic to $\mathbb{Z}_3$ (see **50**).

114. Let (g_1, g_2) be an arbitrary element of $G_1 \times G_2$ and (g_3, e_2) an arbitrary element of $G_1 \times \{e_2\}$. Thus the element $(g_1, g_2) \cdot (g_3, e_2) \cdot (g_1, g_2)^{-1} =$ (see solution **69**) $= (g_1 g_3, g_2) \cdot (g_1^{-1}, g_2^{-1}) = (g_1 g_3 g_1^{-1}, e_2)$ belongs to $G_1 \times \{e\}$. Hence $G_1 \times \{e\}$ is a normal subgroup of $G_1 \times G_2$.

Let (g_1, a) be an arbitrary element of the group $G_1 \times G_2$. We look for which of the cosets of the normal subgroup $G_1 \times \{e\}$ contains this element. If we multiply the element (g_1, a) by all elements of the normal subgroup $G_1 \times \{e\}$ (for example, on the right) we obtain all elements of the type (g, a), where g runs over all the elements of the group G_1. We denote this coset by T_a. In this way the cosets of the normal subgroup $G_1 \times \{e_2\}$ are the cosets of type T_a, where a runs over all the elements of the group G_2. Since (e_1, a), (e_1, b) and (e_1, ab) belong to classes T_a, T_b and T_{ab}, respectively, and we have that $(e_1, a) \cdot (e_1, b) = (e_1, ab)$, then $T_a \cdot T_b = T_{ab}$. The bijective mapping ϕ of the group G_2 in the quotient group such that $\phi(a) = T_a$ for every a in G_2 is an isomorphism, because

$\phi(ab) = T_{ab} = T_a \cdot T_b = \phi(a)\phi(b)$. In this way the quotient group of the group $G_1 \times G_2$ by the normal subgroup $G_1 \times \{e\}$ is isomorphic to G_2.

115. See **57**. The commutant obviously possesses the property 1 stated in Problem **57**. 2) $eee^{-1}e^{-1} = e$, therefore e belongs to the commutant. 3) If k is the commutator $aba^{-1}b^{-1}$, then $k^{-1} = (aba^{-1}b^{-1})^{-1} =$ (see **23**) $= bab^{-1}a^{-1}$, i.e., k^{-1} is a commutator. By the definition of the commutant, each one of its elements a can be written in the form $a = k_1 \cdot k_2 \cdot \ldots \cdot k_n$, where all the k_is are commutators. Thus $a^{-1} = (k_1 \cdot k_2 \cdot \ldots \cdot k_n)^{-1} = k_n^{-1} \cdot \ldots \cdot k_2^{-1} \cdot k_1^{-1}$. Since all the k_i^{-1}s are commutators a^{-1} belongs to the commutant.

116. If g is an arbitrary element of the group and k is the commutator $aba^{-1}b^{-1}$, then gkg^{-1} is a commutator. indeed:

$$\begin{aligned} gkg^{-1} &= gaba^{-1}b^{-1}g^{-1} = ga(g^{-1}g)b(g^{-1}g)a^{-1}(g^{-1}g)b^{-1}g^{-1} \\ &= (gag^{-1})(gbg^{-1})(gag^{-1})^{-1}(gbg^{-1})^{-1}. \end{aligned}$$

If a is an arbitrary element of the commutant, then $a = k_1 \cdot k_2 \cdot \ldots \cdot k_n$, where all the k_is are commutators. Thus $gag^{-1} = g(k_1 \cdot k_2 \cdot \ldots \cdot k_n)g^{-1} = gk_1(g^{-1}g)k_2(g^{-1}g) \cdot \ldots \cdot (g^{-1}g)k_n g^{-1} = (gk_1g^{-1})(gk_2g^{-1}) \cdot \ldots \cdot (gk_ng^{-1})$ is a product of commutators and therefore it belongs to the commutant. Since g is an arbitrary element of the group we obtain that the commutant is a normal subgroup of the group.

117. *Hint.* Show that $aba^{-1}b^{-1} = e$ if and only if $ab = ba$.

118. a) Since the group of symmetries of the triangle is not commutative its commutant is different from $\{e\}$. If g is any transformation of the triangle then both g and g^{-1} either reverse or do not reverse the triangle. Hence the product $g_1g_2g_1^{-1}g_2^{-1}$ contains either 0, or 2, or 4 factors reversing the triangle, and therefore the element $g_1g_2g_1^{-1}g_2^{-1}$ never reverses the triangle, i.e., it is a rotation. The commutant thus contains only rotations. Since the commutant is different from $\{e\}$ and it is a subgroup, it follows that (see **58**) the commutant in the group of symmetries of the triangle coincides with the subgroup of all rotations.

b) As in the case (a) we obtain that the commutant is different from $\{e\}$ and it contains only the rotations of the square. If g is any transformation of the square, then both g and g^{-1} either exchange the diagonals, or fix them. It follows that the element $g_1g_2g_1^{-1}g_2^{-1}$ fixes the diagonals. Moreover, each commutator, being a rotation of the square, coincides either with e or with the central symmetry a. Thus the commutant can contain only the elements e and a; being different from e, it thus coincides with the subgroup of central symmetries $\{e, a\}$.

c) The elements 1 and -1 commute with all the others elements of the group of quaternions. Hence if one of the elements g_1, g_2 coincides with 1 or -1, then $g_1 g_2 g_1^{-1} g_2^{-1} = 1$. If g is an arbitrary element different from 1 and -1, then $g \cdot (-g) = -g^2 = -(-1) = 1$, i.e., $g^{-1} = -g$. Therefore if g_1, g_2 are any two elements different from of 1 and -1, $g_1 g_2 g_1^{-1} g_2^{-1} = g_1 g_2 (-g_1)(-g_2) = (g_1 g_2)^2$. But the square of every element in the group of the quaternions is equal to 1 or to -1. Hence the commutant can contain only the elements 1 and -1; because the group is not commutative, it must be different from 1. It follows that the commutant is $\{1, -1\}$.

119. By the same arguments used to solve Problems **118** (a),(b) we obtain that the commutant in the group of symmetries of the regular n-gon contains only the rotations.

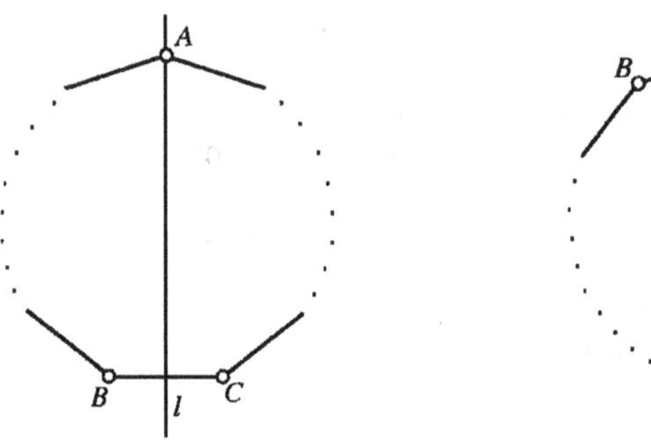

FIGURE 41 FIGURE 42

Let n be odd and let a be the reflection of the n-gon with respect to axis l (Figure 41), b the counterclockwise rotation of the n-gon by the angle $\pi - \pi/n$ (sending A onto B). Thus $aba^{-1}b^{-1}$ is the rotation of the n-gon sending (verify) B on C, i.e., the counterclockwise rotation by $2\pi/n$. Since the commutant is a subgroup we obtain that for n odd it contains the rotations by all the angles which are a multiple of $2\pi/n$. Since the commutant contains only rotations of the regular n-gon, for n odd it coincides with the subgroup of all rotations of the regular n-gon, isomorphic to $\mathbb{Z}_n$ (see **31**).

Now let $n = 2k$. Inscribe in the regular n-gon a k-gon, joining the even vertices. Joining the odd vertices, we obtain a second regular k-gon. If g is an arbitrary vertex of the n-gon, both transformations g and g^{-1} either exchange or fix the two k-gons. Therefore the element $g_1 g_2 g_1^{-1} g_2^{-1}$

sends every k-gon into itself. Consequently for $n = 2k$ the commutant can contain only rotations by angles which are a multiple of $2\pi/k$. Let c be the rotation of the regular n-gon by $2\pi/n$, d the reflection with respect to the axis m (Figure 42).

Thus $cdc^{-1}d^{-1}$ is the rotation sending (verify) the vertex C onto B, i.e., the counterclockwise rotation by $4\pi/n = 2\pi/k$. Hence the commutant contains all rotations by angles which are a multiple of $2\pi/k$, and only them. It is the subgroup of rotations of the plane sending every regular k-gon into itself. This group is isomorphic to $\mathbb{Z}_k = \mathbb{Z}_{n/2}$.

120. Let k, l, m be the axes through the middle points (K, L, M, respectively) of the opposite edges of the tetrahedron. Join the points K, L, M so obtaining the regular triangle KLM. For an arbitrary rotation of the tetrahedron, either each one or none of the three axes k, l, m is fixed (verify). If we put the permutation of the axes k, m, n in correspondence with the permutation of the vertices K, M, L of the triangle KLM, we obtain that to every rotation of the tetrahedron there corresponds a transformation of the triangle KLM, which is in fact a rotation of the triangle. Hence to every commutator in the group of rotations of the tetrahedron there corresponds a commutator in the group of rotations of the triangle KLM. Since the group of rotations of the triangle is commutative, every commutator in it is equal to e. Thus every commutator in the group of rotations of the tetrahedron must fix every one of the three axes k, l, m. Therefore the commutant in the group of rotations of the tetrahedron can contain only the identity and the rotations by 180° about the axes through the middle points of the opposite edges. Since the group of rotations of the tetrahedron is not commutative, its commutant is different from $\{e\}$; the commutant being a normal subgroup, by **113** it coincides with the subgroup containing the identity and all the rotations by 180° about the axes through the middle points of the opposite edges of the tetrahedron.

121. See solution **113**.

122. Both symmetries of the tetrahedron g and g^{-1} either change or fix the orientation of the tetrahedron (see solution **67**). Thus every commutator $g_1 g_2 g_1^{-1} g_2^{-1}$ preserves the orientation of the tetrahedron. Consequently the commutant in the group of symmetries of the tetrahedron contains only the rotations of the tetrahedron. If $a = \begin{pmatrix} A & B & C & D \\ A & C & D & B \end{pmatrix}$

and $b = \begin{pmatrix} A & B & C & D \\ A & B & D & C \end{pmatrix}$ are two symmetries of the tetrahedron, then $aba^{-1}b^{-1} = \begin{pmatrix} A & B & C & D \\ A & D & B & C \end{pmatrix}$ is a rotation about the axis through the vertex A. Since the commutant is a normal subgroup (see **116** and **121**) the commutant in the group of symmetries of the tetrahedron coincides with the subgroup of rotations.

123. *Answer.* 24. For the cube: 1) the identity; 2) the rotations (they are 9) by 90°, 180°, and 270° about the axes through the centres of opposite faces; 3) the rotations (6 in total) by 180° about the axes through the middle pints of opposite edges; 4) the rotations (8 in total) by 120° and 240° about of the axes through opposite vertices.

124. If we join the centres of the adjacent faces of the cube we obtain an octahedron. Thus to every rotation of the cube there corresponds a rotation of the octahedron and vice versa. Moreover, to every composition of two rotations of the cube there corresponds a composition of two rotations of the octahedron and we obtain an isomorphism of the group of rotations of the cube in the group of rotations of the octahedron.

125. If we fix the position of the cube and we consider as different two colourings for which at least one face takes a different colour, then there are in all $6 \cdot 5 \cdot 4 \cdot 3 \cdot 2 = 720$ colourings: indeed, by the first colour one can colour any one of the 6 faces, by the second, any one of the remaining 5, and so on. Since for every colouring one obtains 24 distinct colourings by means of rotations of the cube (see **123**), we have in all $720/24 = 30$ ways of colouring the cube.

There exist only 4 rotations transforming a box of matches into itself: the identity and the three rotations by 180° about the axes through the centres of opposite faces. Hence we have $720/4 = 180$ ways of colouring a box of matches with 6 colours.

126. *Answer.* The group of symmetries of the rhombus and the group $\mathbb{Z}_2 \times \mathbb{Z}_2$.

127. *Hint.* a) See **57**. b) Use the result that both g and g^{-1} either exchange the tetrahedra or fix them.

128. The rotations g and g^{-1} of the cube either both exchange the tetrahedra ACB_1D_1 and A_1C_1BD (see Figure 8) or fix them. Thus each commutator transforms every tetrahedron into itself. Hence to every element of the commutant of the group of rotations of the cube there corresponds a rotation of the tetrahedron ACB_1D_1.

Let a be a rotation of the cube by 90° about the axis through the centres of the faces $ABCD$ and $A_1B_1C_1D_1$ and such that the vertex B is sent onto A. Let b be the rotation of the cube by 120° about the axis through the vertices A_1 and C_1 and such that the vertex A is sent onto D_1. Thus the rotation $aba^{-1}b^{-1}$ sends the vertex A onto itself (verify) and the vertex A_1 onto D, i.e., it is a non-trivial rotation of the cube about the axis through the vertices A and C_1. This rotation is also a rotation of the tetrahedron ACB_1D_1 about the axis through the vertex A. Now it is easy to show (see **121**) that the commutant in the group of rotations of the cube contains all the rotations fixing the tetrahedron. But since it contains only these rotations, the commutant in the group of rotations of the cube is isomorphic to the group of rotations of the tetrahedron.

129. Let A and B be two arbitrary cosets and a, b their representant. Since the element $aba^{-1}b^{-1}$ belongs to the commutant, $ABA^{-1}B^{-1} = E$ and therefore $AB = BA$.

130. Let a, b be two arbitrary elements of the group and let A and B be the cosets to which they belong. Since $AB = BA$ then $ABA^{-1}B^{-1} = E$. The commutator $aba^{-1}b^{-1}$ therefore belongs to a normal subgroup N. In this way all the commutators belong to N and hence N is the commutant.

131. Let h_1, h_2 be two arbitrary elements of N and g an arbitrary element of the group G. Since N is a normal subgroup the elements gh_1g^{-1} and gh_2g^{-1} belong to N. Thus $g(h_1h_2h_1^{-1}h_2^{-1})g^{-1} = gh_1(g^{-1}g)h_2(g^{-1}g)h_1^{-1}(g^{-1}g)h_2^{-1}g^{-1} = (gh_1g^{-1})(gh_2g^{-1})(gh_1^{-1}g^{-1})^{-1} \times (gh_2^{-1}g^{-1})^{-1}$ is a commutator in the normal subgroup N, i.e., it belongs to $K(N)$. Hence $K(N)$ is a normal subgroup of the group G.

132. Let f_1 and f_2 be two arbitrary elements of the group F. Since ϕ is a surjective homomorphism of the group G onto the group F there exist two elements g_1 and g_2 of the group G such that $\phi(g_1) = f_1$ and $\phi(g_2) = f_2$. Thus one has $f_1f_2 = \phi(g_1)\phi(g_2) = \phi(g_1g_2) = \phi(g_2g_1) = \phi(g_2)\phi(g_1) = f_2f_1$. This means that the group F is commutative.

The converse proposition is not true: see Example 12 in §1.13.

133. Let $\phi(e_G) = x$. Thus $x \cdot x = \phi(e_G)\phi(e_G) = \phi(e_Ge_G) = \phi(e_G) = x$. Therefore $x \cdot x = x$ and $x = e_F$.

134. $\phi(a)\phi(a^{-1}) = \phi(aa^{-1}) = \phi(e_G) =$ (see **133**) $= e_F$. Hence $\phi(a^{-1}) = [\phi(a)]^{-1}$.

135. Let a and b be two arbitrary elements of the group G. Thus $(\phi_2\phi_1)(ab) = \phi_2(\phi_1(ab)) = \phi_2(\phi_1(a) \cdot \phi_1(b)) = \phi_2(\phi_1(a)) \cdot \phi_2(\phi_1(b)) = ((\phi_2\phi_1)(a)) \cdot ((\phi_2\phi_1)(b))$.

136. If $\phi(a) = A$ and $\phi(b) = B$, then $\phi(a) \cdot \phi(b) = A \cdot B =$ (by the definition of cosets) $= \phi(ab)$.

137. See **57**. 1) If a and b belong to $\ker\phi$, then $\phi(a) = e_F$, $\phi(b) = e_F$ and $\phi(ab) = \phi(a)\phi(b) = e_F e_F = e_F$. Therefore ab also belongs to $\ker\phi$. 2) $\phi(e_G) =$ (see **133**) $= e_F$, so e_G belongs to $\ker\phi$. 3) If $\phi(a) = e_F$ then $\phi(a^{-1}) =$ (see **134**) $= [\phi(a)]^{-1} = e_F^{-1} = e_F$. Hence if a belongs to $\ker\phi$ a^{-1} also belongs to it.

138. Let a be an arbitrary element of the kernel $\ker\phi$, and g an arbitrary element of the group G. Thus $\phi(a) = e_F$ and $\phi(gag^{-1}) = \phi(g)\phi(a)\phi(g^{-1}) =$ (see **134**) $= \phi(g) \cdot e_F \cdot [\phi(g)]^{-1} = e_F$. Hence the element gag^{-1} also belongs to $\ker\phi$, and therefore $\ker\phi$ is a normal subgroup of the group G.

139. Suppose that the two elements g_1 and g_2 belong to the same coset $g\ker\phi$. Thus there exist in $\ker\phi$ two elements r_1 and r_2 such that $g_1 = gr_1$ and $g_2 = gr_2$. Thus $\phi(g_1) = \phi(gr_1) = \phi(g)\phi(r_1) = \phi(g)e_F = \phi(g)\phi(r_2) = \phi(gr_2) = \phi(g_2)$. Vice versa let $\phi(g_1) = \phi(g_2)$. In this case we have $\phi(g_1^{-1}g_2) = \phi(g_1^{-1})\phi(g_2) =$ (see **134**) $= [\phi(g_1)]^{-1}\phi(g_1) = e_F$. Hence $g_1^{-1}g_2 = r$, where r is an element of the kernel $\ker\phi$. It follows that $g_2 = g_1 r$ and both elements g_1 and g_2 belong to the coset $g_1\ker\phi$.

140. Let f be an arbitrary element of F. Since ϕ is a surjective mapping, there exists an element a of the group G such that $\phi(a) = f$. Let A be the coset containing a. Thus by definition $\psi(A) = \phi(a) = f$.

141. Let $\psi(A) = \psi(B)$ and let a, b be two elements, representant of the classes A and B. Thus $\phi(a) = \psi(A) = \psi(B) = \phi(b)$. It follows (see **139**) that $A = B$.

142. Let A and B be two arbitrary cosets and a and b their representant elements. Thus element ab belongs to the coset AB. Applying the definition of the mapping ψ we obtain $\psi(AB) = \phi(ab) =$ (since ϕ is a homomorphism) $\phi(a) \cdot \phi(b) = \psi(A) \cdot \psi(B)$. The mapping ψ, being bijective (see **141**), is an isomorphism.

143. Let k, l, m be the three axes through the middle points of opposite edges of the tetrahedron. By every symmetry of the tetrahedron these axes are permuted in some way, i.e., we obtain a mapping ϕ_1 of the group of symmetries of the tetrahedron in the group of permutations of the three axes k, l, m. This mapping is a mapping onto the entire group of these permutations, because (verify) every permutation can be obtained by a suitable symmetry of the tetrahedron. It is easy to see that for any two symmetries g_1 and g_2 the permutation of the axes k, l, m corresponding to the symmetry $g_1 g_2$ is the composition of the permutations corresponding to symmetries g_1 and g_2, i.e., $\phi_1(g_1 g_2) = \phi_1(g_1)\phi_2(g_2)$. It follows that ϕ_1 is an isomorphism.

We put every symmetry of the equilateral triangle KLM in correspondence with every permutation of the axes k, l, m in a natural way. We obtain an isomorphism ϕ_2 of the group of the permutations of the axes k, l, m into the group of symmetries of the triangle KLM.

The mapping $\phi_2\phi_1$ is a surjective homomorphism (see **135**) of the group of symmetries of the tetrahedron onto the group of symmetries of the triangle KLM. The kernel of this homomorphism contains all symmetries of the tetrahedron that send each of the axes k, l, m into itself. Such symmetries are the identity and the rotations by 180° about the axes k, l, m. From the result of Problem **138** and Theorem 3 we obtain that these symmetries form a normal subgroup in the group of symmetries of the tetrahedron and that the corresponding quotient group is isomorphic to the group of symmetries of the triangle.

144. Let k, l, m be the axes through the centres of opposite faces of the cube. As in the solution of Problem **143** we define an isomorphism of the group of rotations of the cube in the group of symmetries of the equilateral triangle KLM. The kernel of this homomorphism contains all rotations of the cube that send each of the axes k, l, m into itself. Such rotations are only the identical transformation and the rotations by 180° about the axes k, l, m. From the result of Problem **138** and Theorem 3 we obtain that these four rotations form a normal subgroup in the group of rotations of the cube and the corresponding quotient group is isomorphic to the group of symmetries of the triangle.

145. Denote by r_α the counterclockwise rotation of the plane around the point O by an angle α. The mapping $\phi(r_\alpha) = r_{n\alpha}$ is a homomorphism of the group R into itself, because

$$\phi(r_\alpha r_\beta) = \phi(r_{\alpha+\beta}) = r_{n(\alpha+\beta)} = r_{n\alpha} r_{n\beta} = \phi(r_\alpha)\phi(r_\beta)$$

and for every rotation r_α there exists a rotation $r_{\alpha/n}$ such that $\phi(r_{\alpha/n}) = r_\alpha$. The kernel of the homomorphism ϕ contains all rotations such that $n\alpha = 2\pi k$, i.e., $\alpha = 2\pi k/n$. They are the rotations of the plane that send the regular n-gon onto itself and only them. From the result of Problem **138** and Theorem 3 we obtain the proposition that we had to prove.

146. Let ϕ_1 and ϕ_2 be the natural homomorphisms of the groups G_1 and G_2 onto the quotient groups G_1/N_1 and G_2/N_2, respectively. Let ϕ be the surjective mapping of the group $G_1 \times G_2$ onto the group $(G_1/N_1)\times(G_2\times N_2)$ such that $\phi((g_1, g_2)) = (\phi_1(g_1), \phi_2(g_2))$. This mapping is an homomorphism; indeed:

$$\begin{aligned} \phi((g_1, g_2) \cdot (g_3, g_4)) &= \phi((g_1g_3, g_2g_4)) = (\phi_1(g_1g_3), \phi_2(g_2g_4)) \\ &= (\phi_1(g_1)\phi_1(g_3), \phi_2(g_2)\phi_2(g_4)) \\ &= (\phi_1(g_1), \phi_2(g_2)) \cdot (\phi_1(g_3), \phi_2(g_4)) \\ &= \phi((g_1, g_2)) \cdot \phi((g_3, g_4)). \end{aligned}$$

The kernel of the homomorphism ϕ contains all pairs (g_1, g_2) such that $\phi((g_1, g_2)) = (E_1, E_2)$, where E_1 and E_2 are the unit elements in the quotient groups G_1/N_1 and G_2/N_2, respectively. Since $\phi((g_1, g_2)) = (\phi_1(g_1), \phi_2(g_2))$ the kernel of the homomorphism ϕ is the subgroup $N_1 \times N_2$. From the result of Problem **138** and from Theorem 3 we obtain the statement which we had to prove.

147. Yes, it is possible. For example, the group $\mathbb{Z}_4 = \{e, a, a^2, a^3\}$ and the group $\mathbb{Z}_2 \times \mathbb{Z}_2$ contain the normal subgroups $\{e, a^2\}$ and $\mathbb{Z}_2 \times \{e_2\}$, respectively, which are isomorphic to group $\mathbb{Z}_2$. The corresponding quotient groups, too, are isomorphic.

148. Yes, it is possible. For example, the group $\mathbb{Z}_4 \times \mathbb{Z}_2$. The group $\mathbb{Z}_4 = \{e, a, a^2, a^3\}$ contains two isomorphic normal subgroups: $\{e_1\} \times \mathbb{Z}_2$ and $\{e_1, a^2\} \times \{e_2\}$, and the corresponding quotient groups are respectively $(\mathbb{Z}_4 \times \mathbb{Z}_2)/(\{e_1\} \times \mathbb{Z}_2) \cong (\mathbb{Z}_4/\{e_1\}) \times (\mathbb{Z}_2/\mathbb{Z}_2) \cong \mathbb{Z}_4$ and $(\mathbb{Z}_4 \times \mathbb{Z}_2)/(\{e_1, a^2\} \times \{e_2\}) \cong \mathbb{Z}_2 \times \mathbb{Z}_2$ (see **146**).

149. Yes, it is possible. An example of an infinite group of this type is given in Problem **145**, where $R/\mathbb{Z}_n \cong R$ and, evidently, $R/\{e\} \cong R$.

An example of a finite group is given by the group $\mathbb{Z}_4 \times \mathbb{Z}_2$, which contains two normal subgroups of the type $\mathbb{Z}_4 \times \{e_2\}$ and $\mathbb{Z}_2 \times \mathbb{Z}_2$, whose corresponding quotient groups are isomorphic to $\mathbb{Z}_2$ (see **146**).

150. See **57**. Suppose that f_1 and f_2 belong to $\phi(H)$. This means that in H there exist two elements h_1 and h_2 such that $\phi(h_1) = f_1$ and $\phi(h_2) = f_2$. Thus the element h_1h_2 belongs to H and $\phi(h_1h_2) =$ (since ϕ is a homomorphism) $= \phi(h_1)\phi(h_2) = f_1f_2$. It follows that f_1f_2 also belongs to $\phi(H)$.

2) Since e_G belongs to H and $\phi(e_G) = e_F$ (see **133**) e_F belongs to the image $\phi(H)$ of the subgroup H.

3) Suppose that the element f belongs to $\phi(H)$. This means that in H there exists an element h such that $\phi(h) = f$. Thus h^{-1} belongs to H and $\phi(h^{-1}) =$ (see **134**) $= [\phi(h)]^{-1} = f^{-1}$. Hence f^{-1} also belongs to $\phi(H)$.

151. See **57**. 1) Suppose that g_1 and g_2 belong to $\phi^{-1}(H)$. This means that the elements $\phi(g_1) = h_1$ and $\phi(g_2) = h_2$ belong to H. Thus h_1h_2 also belongs to H and $\phi(g_1g_2) = \phi(g_1)\phi(g_2) = h_1h_2$. Hence g_1g_2 belongs to $\phi^{-1}(H)$.

2) Since $\phi(e_G) = e_F$ (see **133**) and e_F belongs to H, then e_G belongs to $\phi^{-1}(H)$.

3) Suppose that g belongs to $\phi^{-1}(H)$. This means that the element $\phi(g) = h_1$ belongs to H. Thus the element h^{-1} also belongs to H and $\phi^{-1}(g) =$ (see **134**) $= [\phi(g)]^{-1} = h^{-1}$. Hence g^{-1} belongs to $\phi^{-1}(H)$.

152. Let a be an arbitrary element of $\phi^{-1}(N)$. This means that the element $\phi(a) = h$ belongs to N. If g is an arbitrary element of the group G and $\phi(g) = f$, then $\phi(g^{-1}) =$ (see **134**) $= [\phi(g)]^{-1} = f^{-1}$. Thus the element $\phi(gag^{-1}) = \phi(g)\phi(a)\phi(g^{-1}) = fhf^{-1}$ belongs to N, because N is a normal subgroup of the group F. The element gag^{-1} therefore belongs to $\phi^{-1}(N)$, and $\phi^{-1}(N)$ is thus a normal subgroup of the group G.

153. If g_1 and g_2 are two arbitrary elements of the group G and $\phi(g_1) = f_1$, $\phi(g_2) = f_2$, then (see **134**) $\phi(g_1^{-1}) = f_1^{-1}$, $\phi(g_2^{-1}) = f_2^{-1}$. It follows that $\phi(g_1g_2g_1^{-1}g_2^{-1}) = \phi(g_1)\phi(g_2)\phi(g_1^{-1})\phi(g_2^{-1}) = f_1f_2f_1^{-1}f_2^{-1}$, i.e., the image of every commutator of the group G is a commutator in F. Every element of the commutant K_1 can be written in the form $k_1 \cdot k_2 \cdot \ldots \cdot k_n$, where each k_i is a commutator. The element $\phi(k_1 \cdot k_2 \cdot \ldots \cdot k_n) = \phi(k_1) \cdot \phi(k_2) \cdot \ldots \cdot \phi(k_n)$ is a product of commutators in the group F, and therefore it belongs to the commutant K_2. It follows also that K_1 is contained in $\phi^{-1}(K_2)$.

154. Let a be an arbitrary element of $\phi(N)$. This means that in N there exists an element h such that $\phi(h) = a$. Let f be an arbitrary element of the group F. Since ϕ is a surjective homomorphism there exists in the group G an element g such that $\phi(g) = f$. Thus $\phi(g^{-1}) = f^{-1}$ (see **134**) and $\phi(ghg^{-1}) = faf^{-1}$. Since N is a normal subgroup of the group G the element faf^{-1} belongs to $\phi(N)$ and $\phi(N)$ is thus a normal subgroup of the group F.

155. Let $f_1f_2f_1^{-1}f_2^{-1}$ be an arbitrary commutator in F. Since ϕ is a surjective homomorphism there exist in the group G two elements g_1 and g_2 such that $\phi(g_1) = f_1$, and $\phi(g_2) = f_2$. Thus (see **134**) $\phi(g_1^{-1}) = f_1^{-1}$, $\phi(g_2^{-1}) = f_2^{-1}$ and $\phi(g_1g_2g_1^{-1}g_2^{-1}) = f_1f_2f_1^{-1}f_2^{-1}$. Since the element $g_1g_2g_1^{-1}g_2^{-1}$ belongs to K_1 the commutator $f_1f_2f_1^{-1}f_2^{-1}$ belongs to $\phi(K_1)$. But since $\phi(K_1)$ is a subgroup of F (see **150**) and contains all commutators, $\phi(K_1)$ contains the entire commutant K_2. On the other hand, $\phi(K_1)$ is contained in K_2 (see **153**). Hence $\phi(K_1) = K_2$.

The equality $K_1 = \phi^{-1}(K_2)$ does not hold in general. For example, the mapping ϕ of the group $\mathbb{Z}_2 = \{e_1, a\}$ into the group $\{e_2\}$ consisting of a sole element, such that $\phi(e_1) = e_2$ and $\phi(a) = e_2$, is a surjective homomorphism. One has $K_1 = \{e_1\}$, $K_2 = \{e_2\}$, and $\phi^{-1}(K_2) = \{e_1, a\} \neq K_1$.

156. *Answer.* a) Yes: the group $\mathbb{Z}_n$ is commutative; b) yes (see **118**); c) yes (see **118**); d) yes (see **118**); e) yes (see **120**); f) yes (see **122** and (e)); g) yes (see **118** and (e)).

157. A given face can be sent onto any one of the 12 faces in 5 distinct ways.

Answer. 60.

158. *Answer.* 1) 1; 2) 24; 3) 20; 4) 15.

159. Let l_1 and l_2 be two axes of the same type, (i.e., two axes either through the centres of opposite faces, or through opposite vertices, or through the centres of opposite edges). There exists a rotation g of the dodecahedron which sends the axis l_1 to the axis l_2. If a is any (non-trivial) rotation about l_1 then the rotation gag^{-1} sends (verify) the axis l_2 to itself, without reversing it. The element gag^{-1} is not the identity, otherwise $a = gg^{-1} = e$ should be the identical rotation about the axis l_2. Hence if a normal subgroup in the group of rotations contains one rotation about one axis then it contains at least one rotation about every one of the axes of the same type. A subgroup of rotations of the dodecahedron about one axis have an order (according to the type of axis) equal to 5, 3, or 2. Since 5, 3, and 2 are prime numbers, any element (different from e) is a generator of this subgroup (see **34**), i.e., it generates the entire subgroup. Hence if a normal subgroup of the group of rotations of the dodecahedron contains one rotation about some axis then it contains all the rotations about all the axes of that type.

160. From the result of Problem **159** it follows that a normal subgroup in the group of the rotations of the dodecahedron must consist in some one of the classes 1–4, necessarily including class 1. The order of the normal subgroup must be a divisor of the order of the group of the rotations of the dodecahedron (i.e., 60) (see Lagrange's theorem, §1.8). It is easy to verify (see **158**) that this is possible only if the normal subgroup contains only class 1 or contains all the classes.

161. Since the group G is not commutative, the commutant $K(G) \neq \{e\}$. Since $K(G)$ is a normal subgroup of G (see **116**) it follows from the hypothesis of the problem that $K(G) = G$. Therefore in the sequence $G, K(G), K_2(G) = K(K(G)), \ldots$ all groups coincide with G and this sequence cannot end with the unit group. Hence the group G is not soluble.

162. Let the group G be soluble. Thus there exists an integer n such that the subgroup $K_n(G) = \underbrace{K(K \ldots (K}_{n}(G)) \ldots)$ is the unit group. If H is a subgroup of the group G then $K(H)$ is contained in $K(G)$, $K(K(H))$ is contained in $K(K(G))$, etc.. Since the subgroup $K_n(H)$ is contained in $K_n(G)$ and the subgroup $K_n(G)$ is the unit, $K_n(H)$ is also the unit subgroup. It follows that the subgroup H is soluble.

163. Denote by $K(G)$ the commutant in the group G and by $K_r(G)$ the subgroup $\underbrace{K(K \ldots (K}_{r}(G)) \ldots)$. Since ϕ is a surjective homomorphism of the group G onto F, $\phi(K(G)) = K(F)$ (see **155**), $\phi(K_2(G)) = K_2(F)$, and in general $\phi(K_r(G)) = K_r(F)$. Since the group G is soluble, for a certain n the subgroup $K_n(G)$ will be the unit group. But $\phi(K_n(G)) = K_n(F)$, and thus $K_n(F)$ will be also the unit group. Hence the group F is soluble.

164. *Answer.* For example, $\phi : G \to \{e\}$, where G is the group of the rotations of the tetrahedron.

165. *Hint.* Consider the natural homomorphism of the group G into the quotient group G/N (§1.13). Use later the result of Problem **163**.

166. Let $K(G)$ be the commutant in the group G and

$$K_r(G) = \underbrace{K(K \ldots (K}_{r}(G)) \ldots).$$

Consider the natural homomorphism ϕ (see §1.13) of the group G into the quotient group G/N. Thus (see **155**) $\phi(K(G)) = K(G/N)$, $\phi(K_2(G)) = K_2(G/N)$ and in general $\phi(K_r(G)) = K_r(G/N)$. Since by hypothesis the group G/N is soluble, for a certain n the subgroup $K_n(G/N)$ is the unit group, i.e., $K_n(G/N) = \{E\}$. Since $\phi(K_n(G)) = K_n(G/N) = \{E\}$, the subgroup $K_n(G)$ is contained in the normal subgroup N. The group N being soluble, for a certain s the subgroup $K_s(N)$ is the unit group. Since $K_n(G)$ is contained in N the subgroup $K_{n+s}(G)$ is contained in $K_s(N)$, and it is thus the unit group. It follows that the group G is soluble.

167. From Problem **146** we obtain $(G \times F)/(G \times \{e_2\}) \cong \{e\} \times F$. Since the groups $G \times \{e_2\}$ and $\{e\} \times F$ are isomorphic to the groups G and F, respectively, and they are therefore soluble, the group $G \times F$ (see **166**) is also soluble.

168. Since the group G is soluble, in the sequence of commutants $K(G)$, $K_2(G), \ldots$ we will have, for a certain n, $K_n(G) = \{e\}$. Consider the sequence of groups G, $K(G)$, $K_2(G), \ldots, K_n(G)$. This sequence of groups is the sequence we seek, because every group (after the first one) is the commutant of the preceding group, and is therefore a normal subgroup (see **116**) of it. Moreover, all the quotient groups $K_i(G)/K_{i+1}(G)$, as well as $G/K(G)$, are commutative (see **129**); the group $K_n(G) = \{e\}$, too, is commutative.

169. Since by hypothesis the quotient group G_{i-1}/G_i is commutative, the commutant $K(G_{i-1})$ is contained in G_i (see **130**). It follows that the subgroup $K_2(G_{i-1})$ is contained in $K(G_i)$, and in general the subgroup $K_r(G_{i-1})$ is contained in $K_{r+1}(G_i)$ for every $r \geq 1$ and $1 \leq i \leq n$. The subgroup $K_{n-1}(G_0)$ is contained in the subgroup $K_n(G_1)$, which is contained in $K_{n-1}(G_2)$, etc., up to $K(G_n)$. It follows that the subgroup $K_{n+1}(G)$ is contained in $K(G_n)$. But $K(G_n) = \{e\}$ because the group G_n is by hypothesis commutative. Hence $K_{n+1}(G) = \{e\}$, i.e., the group G is soluble.

We can obtain the result of Problem **169** also by induction, going from G_n to G_{n-1}, later to G_{n-2}, etc. and using the result of Problem **166**.

170. By hypothesis the group G is soluble. This means that for a certain n the subgroup $K_n(G)$ is the unit group and therefore the subgroup

$K_{n-1}(G)$ is commutative. Since $K_{n-1}(G)$ is a normal subgroup of the group G (see **131**), we may consider the quotient group $G_1 = G/K_{n-1}(G)$. We prove that the subgroup $K_{n-2}(G_1)$ is commutative. Consider the natural homomorphism (see §1.13) $\phi : G \to G_1$ with kernel $K_{n-1}(G)$. Since ϕ is surjective (see **155**) $\phi(K(G)) = K(G_1)$. Thus $\phi(K_2(G)) = K_2(G_1)$, etc.. Consequently we have $K_{n-1}(G_1) = \phi(K_{n-1}(G)) =$ (because $K_{n-1}(G)$ is the kernel of the homomorphism ϕ)$=\{e_G\}$. Since $K_{n-1}(G_1) = K(K_{n-2}(G_1)) = \{e_{G_1}\}$ the subgroup $K_{n-2}(G_1)$ (see **117**) is commutative. Denote by G_2 the quotient group $G_1/K_{n-2}(G_1)$. As before, we can prove that the subgroup $K_{n-3}(G_2)$ is commutative. Put $G_3 = G_2/K_{n-3}(G_2)$, etc.. The group $G_{n-1} = G_{n-2}/K(G_{n-2})$ is commutative (see **129**). The sequence of groups $G, G_1, G_2, \ldots, G_{n-1}$, with normal subgroups $K_{n-1}(G), K_{n-2}(G_1), \ldots, K(G_{n-2})$, is the requested sequence.

171. Let $G_0, G_1, \ldots, G_n$ be a sequence of groups with the properties mentioned in Problem **170**. We will prove that all the groups of the sequence, in particular G_0, are soluble. We shall use the induction from n to 0. The group G_n is soluble, because by hypothesis it is commutative and therefore $K(G_n) = \{e\}$. Suppose that we have already proved that the group G_i is also soluble. We will prove that the group G_{i-1}, too, is soluble. The group G_{i-1} contains by hypothesis a normal, commutative, and thus soluble, subgroup N_{i-1}, from which the quotient group $G_{i-1}/N_{i-1} \cong G_i$ is soluble by the induction hypothesis. Hence the group G_{i-1} is also soluble by virtue of the result of Problem **166**. In conclusion we can state that all the groups in the sequence $G_0, G_1, \ldots, G_n$ are soluble, and, in particular, G_0 is soluble.

172. Any permutation of order n can be written in the form

$$\begin{pmatrix} 1 & 2 & \ldots & n \\ i_1 & i_2 & \ldots & i_n \end{pmatrix},$$

where the i_ms are all distinct and take values from 1 to n. In particular, i_1 may take any one of the the n values. Later i_2 can take any one of the remaining $n-1$ values, and so on. It follows that the number of distinct permutations of degree n is equal to $n \cdot (n-1) \cdot \ldots \cdot 2 \cdot 1 = n!$

173. If $a = \begin{pmatrix} 1 & 2 & 3 & 4 & \ldots & n \\ 2 & 3 & 1 & 4 & \ldots & n \end{pmatrix}$ and $b = \begin{pmatrix} 1 & 2 & 3 & 4 & \ldots & n \\ 1 & 3 & 2 & 4 & \ldots & n \end{pmatrix}$, then $ab = \begin{pmatrix} 1 & 2 & 3 & 4 & \ldots & n \\ 2 & 1 & 3 & 4 & \ldots & n \end{pmatrix}$ and $ba = \begin{pmatrix} 1 & 2 & 3 & 4 & \ldots & n \\ 3 & 2 & 1 & 4 & \ldots & n \end{pmatrix}$, i.e., $ab \neq ba$. (Recall that in the product ab the permutation b is carried out first, and a later.)

174. Suppose that the element i_1 is sent to i_2, i_2 to i_3, etc.. Let i_r be the first returning element. If we suppose that $i_r = i_k$, where $2 \leq k \leq r-1$, then we obtain that two distinct elements, i_{k-1} and i_{r-1}, are sent by the permutation to the same element, which is a contradiction. It follows that $i_r = i_1$ and one obtains a cycle. Starting with an arbitrary element, not belonging to this cycle, one constructs the second cycle, etc.. It is easy to see that every permutation is the product of independent cycles.

Suppose now a given permutation be the product of independent cycles. If one of the cycles sends the element i_1 to i_2 and the elements i_1 and i_2 do not appear in other cycles, then all products of cycles send i_1 to i_2. Hence the element i_2, which follows i_1 in the cycle containing i_1, is uniquely defined by the given permutation. Therefore all cycles are uniquely defined. Note that if the cycles are not independent the decomposition into cycles may not be unique. For example,

$$\begin{pmatrix} 1 & 2 & 3 \\ 2 & 3 & 1 \end{pmatrix} = (1\ 2) \cdot (2\ 3) \quad \text{and} \quad \begin{pmatrix} 1 & 2 & 3 \\ 2 & 3 & 1 \end{pmatrix} = (1\ 3) \cdot (1\ 2).$$

175. *Hint.* Verify this equality

$$(i_1 i_2 \dots i_m) = (i_1 i_m) \cdot (i_1 i_{m-1}) \cdot \dots \cdot (i_1 i_3) \cdot (i_1 i_2).$$

176. *Hint.* Let $i < j$. Verify the equality

$$\begin{aligned}(i,j) = \ & (i,i+1) \cdot (i+1,i+2) \cdot \dots \cdot (j-2,j-1) \cdot (j-1,j) \\ & \cdot (j-2,j-1) \cdot (j-3,j-2) \cdot \dots \cdot (i+1,i+2) \cdot (i,i+1).\end{aligned}$$

177. The pairs corresponding to inversions are (3,2), (3,1), (2,1), (5,4), (5,1), (4,1). *Answer.* 6.

178. If the numbers i and j are interchanged and $k_1, k_2, \dots, k_s$ are the numbers between i and j, the property of being or not being an inversion changes into the opposite for the pairs of the following numbers: $(i,j), (i,k_r), (k_r,j)$, where $r = 1,2,\dots s$, i.e., for $2s+1$ pairs. Since the number $2s+1$ is odd the parity of the number of inversions does change.

179. *Answer.* The permutation is even (6 inversions).

180. Since

$$\begin{pmatrix} \dots & r & \dots & s & \dots \\ \dots & i_r & \dots & i_s & \dots \end{pmatrix} \cdot (r, s) = \begin{pmatrix} \dots & r & \dots & s & \dots \\ \dots & i_s & \dots & i_r & \dots \end{pmatrix},$$

the lower row of the product is obtained from the lower row of the initial permutation interchanging numbers i_r and i_s. By virtue of the result of Problem **178** the permutation obtained and the initial permutation have different parities.

181. Every permutation splits into a product of transpositions (see **174**, **175**). Suppose a permutation a be decomposed into a product of m transpositions: $a = a_1 \cdot a_2 \cdot \ldots \cdot a_m$. We can write $a = e \cdot a_1 \cdot a_2 \cdot \ldots \cdot a_m$, e being the identity permutation. Since the permutation e is even and is multiplied m times by a transposition, one obtains (see **180**) that if m is even a is an even permutation, and if m is odd a is an odd permutation.

182. See the hint to Solution **175**. *Answer.* a) even; b) odd; c) even for m odd, odd for m even.

183. *Hint.* Decompose the given permutations into products of transpositions (see **181**). Count the number of transpositions in these products.

184. If the permutations should have different parities, then (see **183**) the permutation $aa^{-1} = e$ should be odd, which is not true.

185. For example,

$$(1\ 2\ 3) \cdot (2\ 3\ 4) = \begin{pmatrix} 1 & 2 & 3 & 4 & 5 & \dots & n \\ 2 & 1 & 4 & 3 & 5 & \dots & n \end{pmatrix},$$

$$(2\ 3\ 4) \cdot (1\ 2\ 3) = \begin{pmatrix} 1 & 2 & 3 & 4 & 5 & \dots & n \\ 3 & 4 & 1 & 2 & 5 & \dots & n \end{pmatrix}.$$

186. If a is an even permutation then gag^{-1} is also even, independently of the parity of the permutation g. Hence A_n is a normal subgroup of the group S_n. Let b be an odd permutation. We prove that the coset bA_n contains all the odd permutations. Let c be an odd permutation. Thus $b^{-1}c$ is even. It follows that the permutation $c = b(b^{-1}c)$ belongs to the coset bA_n. Therefore the group S_n is decomposed, by the subgroup A_n, into two cosets: that of all the even permutations and that of all the odd permutations.

187. *Answer.* Since the group S_n is decomposed by the subgroup A_n into two cosets (having the same number of elements), the number of the elements of the group A_n is equal to (see **172**)

$$\frac{n!}{2} = \frac{1 \cdot 2 \cdot \ldots \cdot n}{2}.$$

188. The group S_2 contains 2 elements and is therefore isomorphic to the commutative group $\mathbb{Z}_2$. The groups S_3 and S_4 are isomorphic to the group of symmetries of the triangle and to the group of symmetries of the tetrahedron, respectively. Both these groups are soluble (see **156**).

189. Enumerate the vertices of the dodecahedron as shown in Figure 43. The tetrahedra are, for example, those with the vertices chosen in the following way[1]: (1, 8, 14, 16), (2, 9, 15, 17), (3, 10, 11, 18), (4, 6, 12, 19), (5, 7, 13, 20).

[1]The inscription of the five Kepler cubes inside the dodecahedron helps us to find the five tetrahedra.

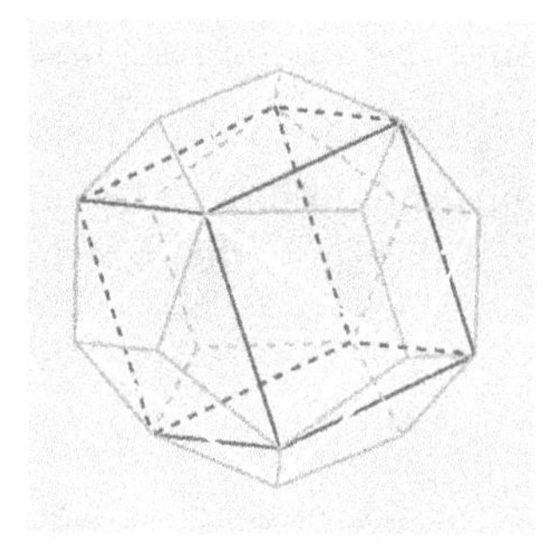

The edges of the cubes are the diagonals of the dodecahedron faces. Every pair of opposite vertices of the dodecahedron is a pair of two opposite vertices of two Kepler cubes. Each cube has thus only one pair of vertices in common with any one of the others. (Two Kepler cubes — black and white — having two opposite vertices in common are shown in the figure). In each cube one can inscribe two tetrahedra (see Problems **126** and **127**). Since each tetrahedron is defined by four vertices of the cube, and any two Kepler cubes have only 2 vertices in common, all tetrahedra inscribed in the Kepler cubes are distinct. So there are in all 10 tetrahedra, two for every vertex of the dodecahedron. Any two of such tetrahedra either have no vertices in common, or they have only one vertex in common. Indeed, if two vertices belonged to two tetrahedra, the edge of such tetrahedra joining them should be the diagonal of a face of two different cubes, but we know that any two cubes have in common only opposite vertices. There are two possible choices of 5 tetrahedra, without common vertices, inside the dodecahedron: indeed, when we choose one of them, we have to

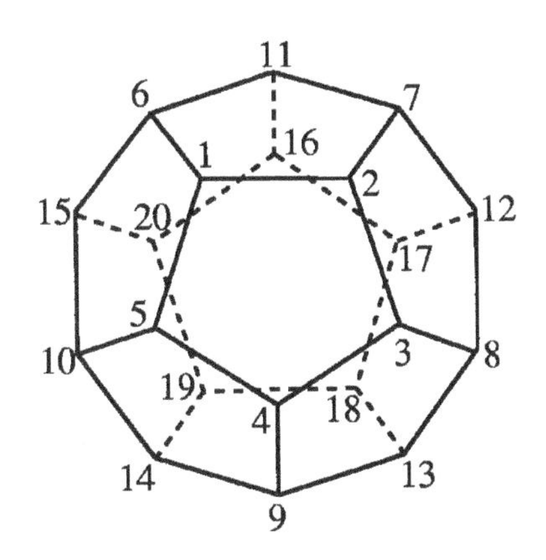

FIGURE 43

190. The permutations of degree 5 may be represented as product of independent cycles only in the following ways: a) $(i_1 i_2 i_3 i_4 i_5)$; b) $(i_1 i_2 i_3)$; c) $(i_1 i_2)(i_3 i_4)$; d) $(i_1 i_2 i_3 i_4)$; e) $(i_1 i_2 i_3)(i_4 i_5)$; f) $(i_1 i_2)$. Applying the results of Problems **182** and **183** we obtain that in the cases (a), (b), and (c) the permutations are even, whereas in the cases (d), (e) and (f) the permutations are odd.

191. Suppose that the normal subgroup N in the group A_5 contains a permutation h of the type (a) (see **190**). Without loss of generality we may suppose that $h = (1\ 2\ 3\ 4\ 5)$. We prove that any permutation $(i_1 i_2 i_3 i_4 i_5)$ of the type (a) belongs to N. If in the row i_1, i_2, i_3, i_4, i_5 there is an even number of inversions then the permutation $g = \begin{pmatrix} 1 & 2 & 3 & 4 & 5 \\ i_1 & i_2 & i_3 & i_4 & i_5 \end{pmatrix}$ is even. Hence by the definition of a normal subgroup N contains the permutation $ghg^{-1} = (i_1 i_2 i_3 i_4 i_5)$. If in the row i_1, i_2, i_3, i_4, i_5 there is an odd number of inversions then in the row i_1, i_4, i_2, i_5, i_3 there is an even number of inversions (because the order of elements is reversed in three pairs). In this case the permutation $g = \begin{pmatrix} 1 & 2 & 3 & 4 & 5 \\ i_1 & i_4 & i_2 & i_5 & i_3 \end{pmatrix}$ is even. Hence N contains the

---

reject the four tetrahedra having a vertex in common with the chosen tetrahedron. The remaining tetrahedra are five. Amongst them four tetrahedra have disjoint sets of vertices, whereas the remaining tetrahedron has one vertex in common with each of the four disjoint tetrahedra. The choice of the first tetrahedron thus forces the choice of the others, so there are in all only two choices. (*Translator's note*)

permutation $ghg^{-1} = (i_1i_4i_2i_5i_3)$ and together with it the permutation $(ghg^{-1})^2 = (i_1i_2i_3i_4i_5)$.

Suppose now that the normal subgroup N contains a permutation $h = (1\ 2\ 3)$ of the type (b) (see **190**) and that $(i_1i_2i_3)$ is an arbitrary permutation. Choose the elements i_4 and i_5, different from i_1, i_2, i_3, in the set $\{i_1, i_2, i_3, i_4, i_5\}$. Either in the row i_1, i_2, i_3, i_4, i_5 or in the row i_1, i_2, i_3, i_5, i_4 there will be an even number of inversions. Consequently one of the permutations $\begin{pmatrix} 1 & 2 & 3 & 4 & 5 \\ i_1 & i_2 & i_3 & i_4 & i_5 \end{pmatrix}$ or $\begin{pmatrix} 1 & 2 & 3 & 4 & 5 \\ i_1 & i_2 & i_3 & i_5 & i_4 \end{pmatrix}$ is even. Denoting this permutation by g we obtain that in all the cases N contains the permutation $ghg^{-1} = (i_1i_2i_3)$.

If the normal subgroup contains a permutation of the type (c) (see **190**), for example, the permutation $h = (1\ 2)(3\ 4)$, then it also contains all permutations $(i_1i_2)(i_3i_4)$ of the type (c). Indeed, in this case one of the permutations $\begin{pmatrix} 1 & 2 & 3 & 4 & 5 \\ i_1 & i_2 & i_3 & i_4 & i_5 \end{pmatrix}$ or $\begin{pmatrix} 1 & 2 & 3 & 4 & 5 \\ i_2 & i_1 & i_3 & i_4 & i_5 \end{pmatrix}$ is even. Let this permutation be denoted by g. Thus N contains the permutation $ghg^{-1} = (i_1i_2)(i_3i_4)$.

192. We will count the number of permutations of each one of types (a), (b) and (c) (see **190**).

a) There exist $5 \cdot 4 \cdot 3 \cdot 2 \cdot 1 = 120$ sequences of 5 numbers $1, 2, 3, 4, 5$. Since every permutation of the type (a) can be written in five equivalent ways (depending upon the choice of the first element) the number of permutations of the type (a) is $120/5 = 24$.

b) By the same reasoning as in the case (a) one obtains that the number of permutations of the type (b) is equal to $5 \cdot 4 \cdot 3/3 = 20$.

c) A permutation of the type (c) can be written in 8 different ways (one can choose i_1 in 4 ways and afterwards i_3 in two ways). Hence the number of permutations of type c is equal to $5 \cdot 4 \cdot 3 \cdot 2/8 = 15$.

Every normal subgroup N contains the unit element. Moreover, from the result of Problem **191** it follows that a normal subgroup of the group A_5 either contains all the permutations of a given type (see **190**) or it contains none of them. The order of a normal subgroup must divide the order of the group A_5 (60). But adding to the number 1 the numbers 24, 20, and 15 one obtains a divisor of 60 only in two cases: when one adds nothing and when one adds all the three numbers. The first case corresponds to the unit subgroup, the second one to the whole group A_5.

193. Such a subgroup is, for example, the subgroup containing all permutations of type

$$\begin{pmatrix} 1 & 2 & 3 & 4 & 5 & 6 & \dots & n \\ i_1 & i_2 & i_3 & i_4 & i_5 & 6 & \dots & n \end{pmatrix}$$

with an even number of inversions in the row i_1, i_2, i_3, i_4, i_5.

3.2 Problems of Chapter 2

194. a) *Answer.* No, because the integer numbers do not form a group under addition (cf., **17**).

b) *Answer.* No, because the integers numbers without zero do not form a group under multiplication (all numbers, except 1 and -1, have no inverse).

c) *Answer.* Yes. Use the result of Problem **57**.

Solution. Since the real numbers form a commutative group under addition, and without zero also a commutative group under multiplication, it suffices to verify that the set of rational numbers form a subgroup of the group of real numbers under addition, and without zero also under multiplication. One obtains this easily using the result of Problem **57**. Indeed: 1) if a and b are rational then $a + b$ and $a \cdot b$ are rational as well; 2) 0 and 1 are rational; 3) if a is a rational number, then $-a$ and $1/a$ (for $a \neq 0$) are also rational. The distributivity is obviously satisfied. Consequently rational numbers form a field.

d) *Answer.* Yes. Use the result of Problem **57**.

Solution. If $r_1 + r_2\sqrt{2} = 0$, where r_1 and r_2 are rational numbers and $r_2 \neq 0$, then it is not possible that $\sqrt{2} = -r_1/r_2$, because $\sqrt{2}$ is not a rational number. This means that if $r_1 + r_2\sqrt{2} = 0$, r_1 and r_2 being rational numbers, then $r_1 = r_2 = 0$. All numbers of the form $r_1 + r_2\sqrt{2}$, for different pairs (r_1, r_2), are different, because if $r_1 + r_2\sqrt{2} = r_3 + r_4\sqrt{2}$, then $(r_1 - r_3) + (r_2 - r_4)\sqrt{2} = 0$ and $r_1 = r_3$, $r_2 = r_4$. Let us now prove that all numbers of the form $r_1 + r_2\sqrt{2}$, where r_1 and r_2 are rational, form a field. To do this, we prove that the numbers of the form $r_1 + r_2\sqrt{2}$ form a subgroup under addition in the set of the real numbers, and also under multiplication without zero. Using the result of Problem **57** we obtain: 1) if $a = r_1 + r_2\sqrt{2}$ and $b = r_3 + r_4\sqrt{2}$, then $a + b = (r_1 + r_3) + (r_2 + r_4)\sqrt{2}$ and $ab = (r_1r_3 + 2r_2r_4) + (r_1r_4 + r_2r_3)\sqrt{2}$; 2) 0 and 1 belong to the set considered, because $0 = 0 + 0 \cdot \sqrt{2}$ and $1 = 1 + 0 \cdot \sqrt{2}$; 3) if $a = r_1 + r_2\sqrt{2}$, then $-a = (-r_1) + (-r_2)\sqrt{2}$ and (for $r_1 + r_2\sqrt{2} \neq 0$)

$$
\begin{aligned}
a^{-1} &= \frac{1}{r_1 + r_2\sqrt{2}} = \frac{r_1 - r_2\sqrt{2}}{(r_1 + r_2\sqrt{2})(r_1 - r_2\sqrt{2})} = \frac{r_1 - r_2\sqrt{2}}{r_1^2 - 2r_2^2} \\
&= \frac{r_1}{r_1^2 - 2r_2^2} - \frac{r_2}{r_1^2 - 2r_2^2}\sqrt{2}.
\end{aligned}
$$

Since the distributivity is satisfied by all real numbers the considered set is a field.

195. We have $a0 + a0 = a(0+0) = a0$. Since a field is a group under addition, one can add $-a0$ to both members of the equations. One obtains $0a = a$. Since the multiplication in the field is commutative one also has $a0 = 0$.

196. 1) $ab + (-a)b = (a + (-a))b = 0b =$ (cf., **195**) $= 0$. It follows that $(-a)b = -(ab)$. In the same way one proves that $a(-b) = -ab$; 2) $(-a)(-b) =$ (cf., the point (1)) $= -(a(-b)) = -(-(ab)) =$ (cf., **20**) $= ab$.

197. If $ab = 0$ and $a \neq 0$, then there exists an element a^{-1}, the inverse of the element a. Thus $a^{-1}ab = a^{-1}0 =$ (cf., **195**) $= 0$. But $a^{-1}ab = 1 \cdot b = b$. Hence $b = 0$.

198. *Answer.* See Table 14

Table 14

·	0	1
0	0	0
1	0	1

·	0	1	2
0	0	0	0
1	0	1	2
2	0	2	1

·	0	1	2	3
0	0	0	0	0
1	0	1	2	3
2	0	2	0	2
3	0	3	2	1

199. Let n be a non-prime number, i.e., $n = n_1 n_2$, where $n_1 < n$ and $n_2 < n$. Thus modulo n we have $n_1 n_2 = 0$, but $n_1 \neq 0$ and $n_2 \neq 0$. Since this is not possible in a field (cf., **197**), for non-prime n the remainders with the operations modulo n do not form a field.

Observe now the following property. Let x and y be two integers and r_1 and r_2 the remainders of their division by n, i.e., $x = k_1 n + r_1$ and $y = k_2 n + r_2$. Thus $x + y = (k_1 + k_2)n + (r_1 + r_2)$ and $xy = (k_1 k_2 n + k_1 r_2 + k_2 r_1)n + r_1 r_2$. We obtain that the numbers $x+y$ and r_1+r_2, as well as xy and $r_1 r_2$, divided by n, give the same remainder. In other words, we obtain the same result either if we first take the remainders of the division of x and y by n and afterwards their sum (or their product) modulo n, or if we first take the sum (or the product) of the integers x and y, as usual, and later on we take the remainder of the division by n of this sum (or of this product). In this way, to calculate a certain expression with the operations modulo n one may take the remainders of the division by n not after each operation, but, after having made the calculations as usual with integers, take only at the end the remainder of the division by n of the number obtained. Let us use this property. Under addition modulo n the remainders form a commutative group (see **40**). Since for all integers a, b, c the numbers $(a+b)c$ and $ac+bc$ are equal, and consequently the remainders of their division by n are also equal, the equality $(a+b)c = ac+bc$ also holds modulo n, i.e., distributivity is satisfied. In the same way one verifies the associativity and the commutativity of the multiplication modulo n: $(ab)c = a(bc)$ and $ab = ba$. The unit element for the multiplication is 1. It remains to prove that for n prime every remainder a different from 0 has an inverse, i.e., that there exists a remainder x such that $ax = 1$ modulo n.

Let $0 < a < n$. Consider the numbers $a \cdot 0, a \cdot 1, \dots, a \cdot (n-1)$ (under the usual multiplication). The difference between two of these numbers $ak - al = a(k-l)$ is not divisible by n, because n is prime and $a < n$ and $0 < |k-l| < n$. As a consequence all these n numbers divided by n give distinct remainders, and thus all the remainders possible. This means that one of these numbers divided by n gives as remainder 1, i.e., for a certain remainder x one has $ax = 1$ modulo n. Hence for n prime all the properties of fields are satisfied.

200. Since $a - b = a + (-b)$ it follows that $(a-b)c = (a+(-b))c = ac + (-b)c =$ (cf., **196**) $= ac + (-bc) = ac - bc$.

201. Subtract from

$$P(x) = S_1(x) \cdot Q(x) + R_1(x)$$

the equality

$$P(x) = S_2(x) \cdot Q(x) + R_2(x).$$

One obtains

$$0 = (S_1(x) - S_2(x))Q(x) + (R_1(x) - R_2(x)),$$

so

$$(S_1(x) - S_2(x))Q(x) = R_2(x) - R_1(x). \tag{3.1}$$

If $S_1(x) - S_2(x) \neq 0$, the degree of the polynomial in the left member of the equality (3.1) is not lower than the degree of the polynomial $Q(x)$. But the degree of the polynomial in the right member is strictly lower than the degree of the polynomial $Q(x)$. From this contradiction it follows necessarily that $S_1(x) - S_2(x) = 0$, and consequently $R_2(x) = R_1(x) = 0$, i.e., $S_1(x) = S_2(x)$ and $R_1(x) = R_2(x)$.

202. This group is the direct product of the group of the real numbers (under addition) by itself (cf., **69** and **73**). The unit element (zero) is the pair $(0, 0)$.

203. Let $z_1 = (a, b)$, $z_2 = (c, d)$, $z_3 = (e, f)$. Thus $z_1 \cdot z_2 = (ac - bd, ad + bc)$ and $z_2 \cdot z_1 = (ca - db, cb + da)$. But $ac - bd = ca - db$ and $ad + bc = cb + da$. Hence $z_1 \cdot z_2 = z_2 \cdot z_1$. Moreover we have $(z_1 \cdot z_2) \cdot z_3 = (ac - bd, ad + bc) \cdot (e, f) = (ace - bde - adf - bcf, acf - bdf + ade + bce)$ and $z_1 \cdot (z_2 \cdot z_3) = (a, b) \cdot (ce - df, cf + de) = (ace - adf - bcf - bde, acf + ade + bce - bdf)$, i.e., $(z_1 \cdot z_2) \cdot z_3 = z_1 \cdot (z_2 \cdot z_3)$.

204. Let $z = (a, b) \neq (0, 0)$ and let the required complex number be $z^{-1} = (x, y)$. Thus $z \cdot z^{-1} = z^{-1} \cdot z = (ax - by, ay + bx)$. To have $z \cdot z^{-1} = (1, 0)$, the two following equations must be satisfied:

$$\begin{cases} ax - by = 1, \\ bx + ay = 0. \end{cases}$$

This system of equations has exactly one solution:

$$x = \frac{a}{a^2 + b^2}, \qquad y = -\frac{b}{a^2 + b^2},$$

for which $a^2 + b^2 \neq 0$, because $(a, b) \neq (0, 0)$. Hence

$$(a, b)^{-1} = \left(\frac{a}{a^2 + b^2}, -\frac{b}{a^2 + b^2} \right).$$

205. Let $z_1 = (a, b)$, $z_2 = (c, d)$, $z_3 = (e, f)$. Thus $(z_1 + z_2) \cdot z_3 = (a+c, b+d) \cdot (e, f) = (ae + ce - bf - df, af + cf + be + de)$ and $z_1 \cdot z_3 + z_2 \cdot z_3 = (ae - bf, af + be) + (ce - df, cf + de) = (ae - bf + ce - df, af + be + cf + de)$, from which it results that $(z_1 + z_2) \cdot z_3 = z_1 \cdot z_3 + z_2 \cdot z_3$.

206. We have $a + bi = (a, 0) + (b, 0)(0, 1) = (a, 0) + (0, b) = (a, b)$.

207. *Answer.* $(c + di) - (a + bi) = (c - a) + (d - b)i$.

208.

$$\begin{aligned} x + yi &= \frac{c + di}{a + bi} = \frac{(c + di)(a - bi)}{(a + bi)(a - bi)} = \frac{(ac + bd) + (ad - bc)i}{a^2 + b^2} \\ &= \frac{ac + bd}{a^2 + b^2} + \frac{ad - bc}{a^2 + b^2} i \quad (\text{since } a + bi \neq 0, \ a^2 + b^2 \neq 0). \end{aligned}$$

209. *Answer.* a) $i^3 = -i$; b) $i^4 = 1$; c) $i^n = \begin{cases} 1, & \text{if } n = 4k, \\ i, & \text{if } n = 4k+1, \\ -1, & \text{if } n = 4k+2, \\ -i, & \text{if } n = 4k+3. \end{cases}$

210. *Hint.* The equation $(x+yi)^2 = (x^2 - y^2) + 2xyi = b$, where b is a real number, is equivalent to the system of equations

$$\begin{cases} x^2 - y^2 = b, \\ 2xy = 0. \end{cases}$$

Answer. a) $z = \pm 1$; b) $z = \pm i$; c) $z = \pm a$; d) $z = \pm ai$.

211. Let $z_1 = a + ib$, $z_2 = c + di$. Thus

a) $\overline{z_1 + z_2} = \overline{(a+bi)+(c+di)} = \overline{(a+c)+(b+d)i} = (a+c) - (b+d)i = (a - bi) + (c - di) = \overline{z_1} + \overline{z_2}$.

b) $(z_1 - z_2) + z_2 = z_1$. By virtue of (a) $\overline{z_1} = \overline{(z_1 - z_2) + z_2} = \overline{(z_1 - z_2)} + \overline{z_2}$. It follows that $\overline{z_1 - z_2} = \overline{z_1} - \overline{z_2}$.

c) $\overline{z_1 z_2} = \overline{(a+bi)(c+di)} = \overline{(ac - bd) + (bc + ad)i} = (ac - bd) - (bc + ad)i = (a - bi)(c - di) = \overline{z_1}\,\overline{z_2}$.

d) $\overline{z_1} = \overline{((z_1/z_2)\cdot z_2)} =$ (because of (c)) $\overline{(z_1/z_2)} \cdot \overline{z_2}$. Hence $\overline{(z_1/z_2)} = \overline{z_1}/\overline{z_2}$.

212. From the result of Problem **211** we obtains

$$\begin{aligned} \overline{P(z)} &= \overline{a_0 z^n + a_1 z^{n-1} + \ldots + a_{n-1} z + a_n} \\ &= \overline{a_0 z^n} + \overline{a_1 z^{n-1}} + \ldots + \overline{a_{n-1} z} + \overline{a_n} \\ &= \overline{a_0}\,\overline{z}^n + \overline{a_1}\,\overline{z}^{n-1} + \ldots + \overline{a_{n-1}}\,\overline{z} + \overline{a_n} \\ &\quad (\ \overline{a_i} = a_i \text{ because all the } a_i\text{s are real numbers}) \\ &= a_0 \overline{z}^n + a_1 \overline{z}^{n-1} + \ldots + a_{n-1}\overline{z} + a_n = P(\overline{z}). \end{aligned}$$

213. Since the field M contains all the real numbers and the element i_0, then it contains all possible elements of the form $a + bi_0$, where a, b are arbitrary real numbers. Let M' denote the set of all elements of the field M, represented in the form $a + bi_0$. Thus by virtue of the commutative, associative, and distributive properties of addition and multiplication in M we obtain for the elements of M':

$$(a + bi_0) + (c + di_0) = (a + c) + (bi_0 + di_0) = (a + c) + (b + d)i_0, \quad (3.2)$$

$$\begin{aligned}(a+bi_0)\cdot(c+di_0) &= ac+adi_0+bci_0+bdi_0^2 \\ (\text{since } i_0^2=-1) &= (ac-bd)+(ad+bc)i_0. \qquad (3.3)\end{aligned}$$

Let C be the field of complex numbers. Consider the mapping ϕ of the field C into the field M such that

$$\phi(a+bi) = a+bi_0.$$

Comparing formulae (3.2) and (3.3) with formulae (2.4) and (2.5) of §2.2 we obtain that ϕ is a homomorphism of C into M with respect to the addition and with respect to the multiplication. Since $\phi(C) = M'$, M' is a subgroup (cf., **150**) of M with respect to the addition and with respect to the multiplication. As the operations of addition and multiplication possess the properties of commutativity, associativity, and distributivity in M, this obviously also holds in M'. Hence M' is a field.

If $a+bi_0 = c+di_0$, then $(b-d)i_0 = c-a$. Moreover, if $b-d \neq 0$, then $i_0 = (c-a)/(b-d)$ is a real number, but this cannot be true because the square of any number is never equal to -1. Consequently $b-d=0$, and thus also $c-a=0$. Hence $b=d$ and $a=c$. Consequently the elements of the form $a+bi_0$ are different for different pairs (a,b). It follows that the mapping ϕ defined above is a bijective mapping of the field C onto the field M'. Moreover, since ϕ is a homomorphism, ϕ is an isomorphism of the field C in the field M', i.e., the field M' is isomorphic to the field of complex numbers.

214. Let the polynomial considered be reducible, i.e.:

$$\begin{aligned}&x^n + a_1x^{n-1} + a_2x^{n-2} + \ldots + a_n \\ &\qquad = (b_0x^m + b_1x^{m-1} + \ldots + b_m)\cdot(c_0x^k + c_1x^{k-1} + \ldots + c_k),\end{aligned}$$

where $m < n$, $k < n$, all the b_is and c_ls are real numbers and $b_0 \neq 0$, $c_0 \neq 0$. Put j in place of x in the first member of this equation. Since M is a field one can eliminate the brackets, carrying out the product as usual. We obtain in this way the initial polynomial, in which x is replaced by j, i.e.:

$$\begin{aligned}&j^n + a_1j^{n-1} + a_2j^{n-2} + \ldots + a_n \\ &\qquad = (b_0j^m + b_1j^{m-1} + \ldots + b_m)\cdot(c_0j^k + c_1j^{k-1} + \ldots + c_k),\end{aligned}$$

By hypothesis

$$j^n + a_1j^{n-1} + a_2j^{n-2} + \ldots + a_n = 0,$$

and thus (cf., **197**) at least one of the two polynomials above within brackets is equal to zero. By dividing this polynomial by its leading coefficient (b_0 or c_0) we obtain a vanishing expression of the type of (2.6), having a degree lower than n. We thus obtain a contradiction of n being the lowest degree for which an expression of the type of (2.6) vanishes. Therefore the claim that the polynomial considered is reducible is not true.

215. As we had proved, in the case (a) the element j of the field M satisfies the equation:

$$j^2 + pj + q = 0$$

where p and q are some real numbers, and $x^2 + px + q$ is not reducible over the field of real numbers. We have

$$x^2 + px + q = \left(x^2 + px + \frac{p^2}{4}\right) - \left(\frac{p^2}{4} - q\right) = \left(x + \frac{p^2}{2}\right)^2 - \left(\frac{p^2}{4} - q\right).$$

If $\frac{1}{4}p^2 - q \geq 0$, then $\frac{1}{4}p^2 - q = a^2$ for some real number a. Thus

$$x^2 + px + q = \left(x + \frac{p^2}{2}\right)^2 - a^2 = \left(x + \frac{p^2}{2} + a\right)\left(x + \frac{p^2}{2} - a\right),$$

i.e., the polynomial $x^2 + px + q$ should be reducible over the field of the real numbers. It follows that $\frac{1}{4}p^2 - q < 0$, i.e., $\frac{1}{4}p^2 - q = -b^2$, for a certain non-zero real real number b. Since in the field M $j^2 + pj + q = 0$, in M we have

$$j^2 + pj + \frac{p^2}{4} = \frac{p^2}{4} - q \quad \text{and} \quad \left(j + \frac{p}{2}\right)^2 = \frac{p^2}{4} - q = -b^2.$$

Consequently

$$\left(\frac{j}{b} + \frac{p}{2b}\right)^2 = \frac{-b^2}{b^2} = -1.$$

Hence the element

$$i_0 = \frac{j}{b} + \frac{p}{2b},$$

belonging to M, is the element sought.

216. *Answer.* The sole field satisfying the required properties is (up to isomorphism) the field whose elements are fractions, having polynomials in j as numerators and denominators, with the usual operations on these fractions.

217. *Answer.* A: $2+2i$, B: $-1+3i$, C: $-2-i$.

218. *Answer.* a) The symmetry with respect to the origin of the coordinates (or, equivalently, the rotation by 180° around the origin of the coordinates). b) the dilation of the plane by 2 (fixing the origin of the coordinates). c) the reflection with respect to the x axis.

219. *Hint.* Use the property of the triangles $A_1B_1C_1$ and $A_2B_2C_2$ being equal (Figure 44).

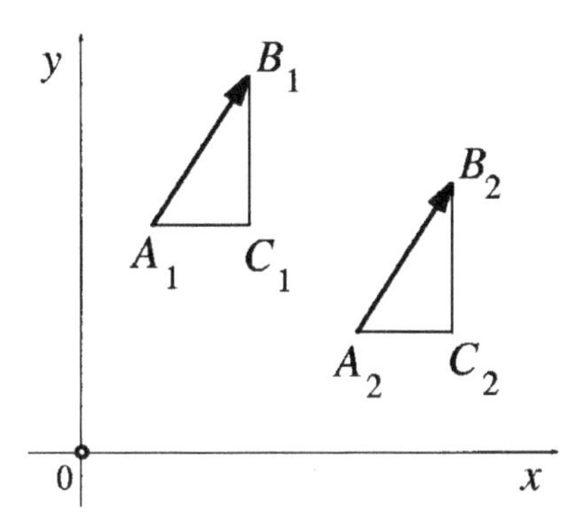

FIGURE 44

220. By hypothesis $z_1 = x_u + iy_u$, $z_2 = x_v + iy_v$, $z_3 = x_w + iy_w$. Consequently the equation $z_3 = z_1 + z_2$ is equivalent to the two equations:

$$\begin{cases} x_w = & x_u + x_v, \\ y_w = & y_u + y_v. \end{cases} \tag{3.4}$$

On the other hand, if $w = u + v$ then (cf., Figure 45) $x_w =$ (by definition) $x_C - x_A = (x_C - x_D) + (x_D - x_A) = x_u + x_v$, and similarly $y_w = y_u + y_v$. Thus the equation $w = u + v$ is equivalent to the equations (3.4).

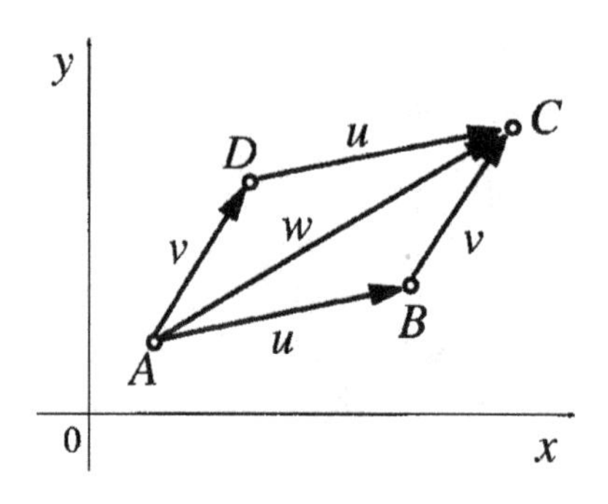

FIGURE 45

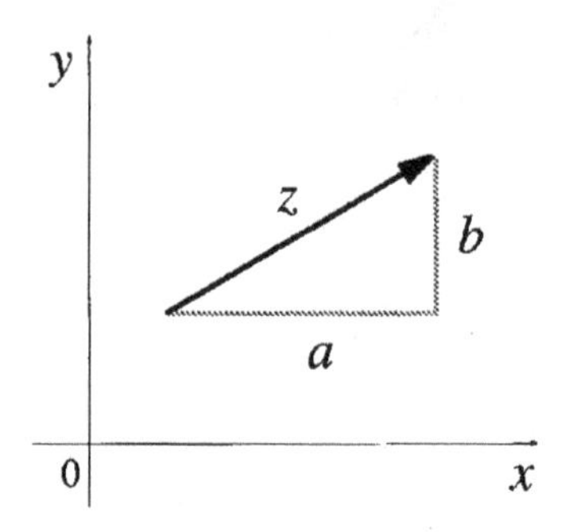

FIGURE 46

221. By definition $x_{\overrightarrow{AB}} = x_B - x_A$ and $y_{\overrightarrow{AB}} = y_B - y_A$.

222. Equation $|z|^2 = a^2 + b^2$ comes from Pythagoras' theorem (Figure 46). Equations $z \cdot \overline{z} = (a+bi)(a-bi) = a^2 + b^2$ are easily verified.

223. See Figure 47. *Hint.* The inequalities stated by the problem follow from the property that in a triangle the length of any side is smaller than the sum and longer than the difference of the lengths of the two other sides.

224. See Figure 48. $|z_1 + z_2|^2 + |z_1 - z_2|^2 =$ (cf., **222**) $= (z_1 + z_2)\overline{(z_1 + z_2)} + (z_1 - z_2)\overline{(z_1 - z_2)} = (z_1 + z_2)(\overline{z_1} + \overline{z_2}) + (z_1 - z_2)(\overline{z_1} - \overline{z_2}) = 2z_1\overline{z_1} + 2z_2\overline{z_2} = 2|z_1|^2 + 2|z_2|^2$.

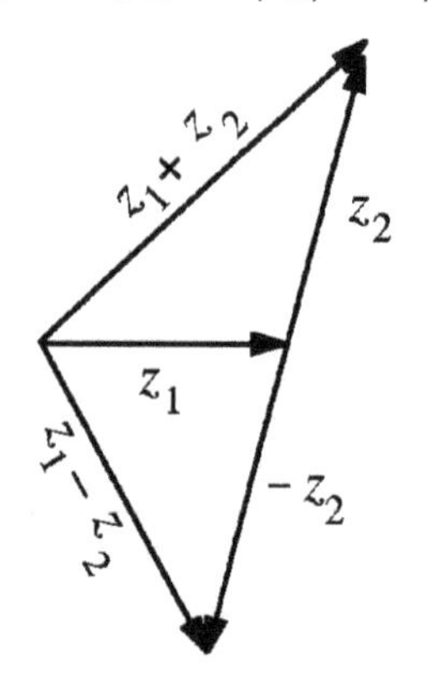

FIGURE 47

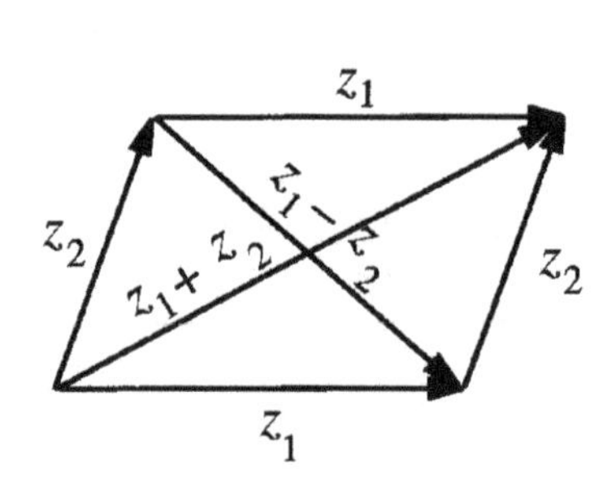

FIGURE 48

225. *Answer.* a) $\sqrt{2}(\cos(\pi/4) + i\sin(\pi/4))$; b) $2(\cos(7\pi/6) + i\sin(7\pi/6))$;

c) $3(\cos(\pi/2)+i\sin(\pi/2))$; d) $5(\cos\pi+i\sin\pi)$; e) $\sqrt{5}(\cos\varphi+i\sin\varphi)$, where $\varphi = \arctan 2$.

226. We have: $z_1z_2 = r_1r_2[(\cos\varphi_1 \cdot \cos\varphi_2 - \sin\varphi_1 \cdot \sin\varphi_2) + i(\sin\varphi_1 \cdot \cos\varphi_2 + \cos\varphi_1 \cdot \sin\varphi_2)]$ = (by the trigonometric formulae of the sum) $= r_1r_2(\cos(\varphi_1+\varphi_2)+i\sin(\varphi_1+\varphi_2))$. The second equation in the problem is equivalent to the equation

$$z_1 = z_2 \cdot \frac{r_1}{r_2}(\cos(\varphi_1 - \varphi_2) + i\sin(\varphi_1 - \varphi_2)),$$

which follows directly from the first equality.

227. *Hint.* Use the result of Problem **226**.

228. $1 - \sqrt{3}i = 2(\frac{1}{2} - \frac{\sqrt{3}}{2}i) = 2(\cos(-\pi/3) + i\sin(-\pi/3))$. So, $(1 - \sqrt{3}i)^{100} = [2(\cos(-\pi/3) + i\sin(-\pi/3))]^{100}$ = (cf., **227**) $= 2^{100} \cdot [\cos(-100\pi/3) + i\sin(-100\pi/3)] = 2^{100} \cdot (\cos(2\pi/3) + i\sin(2\pi/3))$. It follows that

$$\frac{(1-\sqrt{3}i)^{100}}{2^{100}} = \cos\frac{2\pi}{3} + i\sin\frac{2\pi}{3} = -\frac{1}{2} + \frac{\sqrt{3}}{2}i.$$

Answer. $-\frac{1}{2} + \frac{\sqrt{3}}{2}i$.

229. If $z = 0$ then also $w = 0$ (cf., **197**). If $z \neq 0$ then also $w \neq 0$. In this case let $w = \rho(\cos\psi + i\sin\psi)$. Thus by de Moivre's formula (cf., **227**)

$$w^n = \rho^n(\cos n\psi + i\sin n\psi) = r(\cos\varphi + i\sin\varphi).$$

It follows that $\rho^n = r$ and $n\psi = \varphi + 2\pi k$, where k is any integer. Therefore

$$w = \sqrt[n]{r}\left(\cos\frac{\varphi + 2\pi k}{n} + i\sin\frac{\varphi + 2\pi k}{n}\right). \tag{3.5}$$

If $k_1 - k_2 = ln$, where l is an integer number, then the quantities $(\varphi + 2\pi k_1)/n$ and $(\varphi + 2\pi k_2)/n$ differ by $2\pi l$, and the values of w expressed by formula (3.5) coincide. Formula (3.5) thus gives n distinct values of w, obtained by giving to the parameter k the values $k = 0, 1, \ldots, n-1$.

230. *Hint.* Put the expressions under the square root into trigonometric form, afterwards use formula (3.5) of the solution of Problem **229**.

Answer. a) $+i$, $-i$; b) 2, $-1+\sqrt{3}i$, $-1-\sqrt{3}i$ (here $\sqrt{3}$ is the real positive root); c) $\cos(25^\circ+90^\circ n)+i\sin(25^\circ+90^\circ n)$, where $n=0,1,2,3$; d) $\sqrt[6]{2}(\cos(15^\circ+120^\circ n)+i\sin(15^\circ+120^\circ n))$, where $n=0,1,2$ (here $\sqrt[6]{2}$ is the real positive root).

231. *Hint.* See formula (3.5) of the solution of Problem **229**; $1 = 1(\cos 0 + i\sin 0)$; $\epsilon_n^k = \cos(2\pi k/n) + i\sin(2\pi k/n)$ (cf., **227**).

232. *Hint.* If $z_1^n = z_0$ and $z_2^n = z_0$, then $(z_2/z_1)^n = z_2^n/z_1^n = 1$. *Answer.* z_1, $z_1\cdot\epsilon_n$, $z_1\cdot\epsilon_n^2, \dots$, $z_1\cdot\epsilon_n^{n-1}$ (cf., **231**).

233. By virtue of the result of **221** $z_A = z_{\overrightarrow{OA}}$, where O is the origin of the coordinates. *Answer.* a) the distance of the point z from the origin of the coordinates; b) the angle between the positive side of the axis Ox and the ray Oz; c) the distance between the points z_1 and z_2 (because $z_B - z_A = z_{\overrightarrow{AB}}$ (cf., **221**); d) the angle between rays Oz_2 and Oz_1 (cf., **226**).

234. See **233**. *Answer.* a), b) The circle of radius 1 (and R) with centre at the origin of the coordinates; c) the circle of radius R and centre at the point z_0; d) the disc of radius R with centre z_0 together with its bounding circle; e) the straight line perpendicular to the segment joining points z_1 and z_2, and passing through its middle point; f) the negative side of the real axis; g) the bisector of the first quadrant of the plane; h) the ray defining the angle φ with the positive side of the axis Ox.

235. See **229** and **232**. *Answer.* In the vertices of a regular n-gon with centre at the origin of the coordinates.

236. Let a point z_0 and an arbitrary real number $\varepsilon > 0$ be given. Choose $\delta = 1$ (independently of z_0 and of ε). Thus for every z, satisfying the condition $|z-z_0| < 1$, the inequality $|f(z) - f(z_0)| = |a-a| = 0 < \varepsilon$ is satisfied. Consequently the function $f(z) \equiv a$ is continuous for every value of the argument z. (For δ one may take an arbitrary positive real number).

237. Let a point z_0 and an arbitrary real number $\varepsilon > 0$ be given. Choose $\delta = \varepsilon$ (independently of z_0). Thus for every z satisfying the condition $|z-z_0| < \delta$ the inequality $|f(z) - f(z_0)| = |z - z_0| < \delta = \varepsilon$ is satisfied, i.e., we will have $|f(z) - f(z_0)| < \varepsilon$. Consequently the function of complex argument $f(z) = z$ is continuous for every value of the argument. Considering only the real values of the argument z we obtain that the function of real argument $f(x) = x$ is also continuous for all values of the argument.

238. Let a point z_0 and an arbitrary real number $\varepsilon > 0$ be chosen. If δ is an arbitrary real positive number and $|z-z_0| < \delta$, then $|f(z)-f(z_0)| = |z^2 - z_0^2| = |(z - z_0)(z + z_0)| =$ (cf., **226**) $= |z - z_0| \cdot |z + z_0| \leq \delta|z + z_0|$. But $|z + z_0| = |(z - z_0) + 2z_0| \leq$ (cf., **223**) $\leq |z - z_0| + |2z_0| < \delta + 2|z_0|$. So for $|z - z_0| < \delta$ we obtain

$$|f(z) - f(z_0)| < \delta \cdot (\delta + 2|z_0|).$$

Now choose a value δ satisfying the inequality

$$\delta \cdot (\delta + 2|z_0|) \leq \varepsilon.$$

If $z_0 = 0$ then put $\delta = \sqrt{\varepsilon}$ (taking the positive root of ε). If $z_0 \neq 0$ then consider the two real positive numbers $|z_0|$ and $\varepsilon/(3|z_0|)$ and choose as δ the smallest of these two numbers. Thus inequalities $\delta \leq |z_0|$ and $\delta \leq \varepsilon/(3|z_0|)$ will hold. It follows that

$$\delta \cdot (\delta + 2|z_0|) \leq \frac{\varepsilon}{3|z_0|} \cdot (|z_0| + 2|z_0|) = \varepsilon.$$

Hence if $|z - z_0| < \delta$, where δ takes the value we had chosen, then $|f(z) - f(z_0)| < \varepsilon$. Therefore the function of complex argument $f(z) = z^2$ is continuous for all values of the argument.

239. a) Let an arbitrary real number $\varepsilon > 0$ be given. We have

$$\begin{aligned} |h(z) - h(z_0)| &= |(f(z) + g(z)) - (f(z_0) + g(z_0))| \\ &= |(f(z) - f(z_0)) + (g(z) - g(z_0))| \\ \text{(cf., \textbf{223})} \quad &\leq |f(z) - f(z_0)| + |g(z) - g(z_0)|. \end{aligned}$$

Consider $\varepsilon/2$ instead of ε. Thus by virtue of the continuity of the function $f(z)$ at the point z_0, one can choose a real number $\delta_1 > 0$ such that for every z satisfying $|z - z_0| < \delta_1$, the inequality $|f(z) - f(z_0)| < \varepsilon/2$ holds. In the same way, by virtue of the continuity of the function $g(z)$ at the point z_0, we can choose a real number $\delta_2 > 0$ such that for every z satisfying $|z - z_0| < \delta_2$, the inequality $|g(z) - g(z_0)| < \frac{\varepsilon}{2}$ holds. Take as δ the smallest of δ_1 and δ_2. Thus for every z satisfying the condition $|z-z_0| < \delta$ both inequalities: $|f(z)-f(z_0)| < \varepsilon/2$ and $|g(z)-g(z_0)| < \varepsilon/2$

hold. So for every number z satisfying the condition $|z - z_0| < \delta$, we will have

$$|h(z) - h(z_0)| \leq |f(z) - f(z_0)| + |g(z) - g(z_0)| < \frac{\varepsilon}{2} + \frac{\varepsilon}{2} = \varepsilon,$$

i.e., $|h(z) - h(z_0)| \leq \varepsilon$. Hence the function $h(z) = f(z) + g(z)$ is continuous at the point z_0.

b) if $h(z) = f(z) - g(z)$ then

$$\begin{aligned} |h(z) - h(z_0)| &= |(f(z) - g(z)) - (f(z_0) - g(z_0))| \\ &= |(f(z) - f(z_0)) + (-(g(z) - g(z_0)))| \\ \text{(cf., } \mathbf{223}\text{)} \quad &\leq |f(z) - f(z_0)| + |-1| \cdot |g(z) - g(z_0)| \\ &= |f(z) - f(z_0)| + |g(z) - g(z_0)|. \end{aligned}$$

We thus obtained the same inequality

$$|h(z) - h(z_0)| \leq |f(z) - f(z_0)| + |g(z) - g(z_0)|$$

as in the case (a), and therefore the problem is solved as in the preceding case.

c) Let a real number $\varepsilon > 0$ be given. We have

$$\begin{aligned} |h(z) - h(z_0)| &= |f(z)g(z) - f(z_0)g(z_0)| \\ &= |f(z)g(z) - f(z)g(z_0) + f(z)g(z_0) - f(z_0)g(z_0)| \\ \text{(cf., } \mathbf{223}\text{)} \quad &\leq |f(z)(g(z) - g(z_0))| + |g(z_0)(f(z) - f(z_0))| \\ \text{(cf., } \mathbf{226}\text{)} \quad &= |f(z)| \cdot |g(z) - g(z_0)| + |g(z_0)| \cdot |f(z) - f(z_0)|. \end{aligned}$$

Now choose a real number $\delta > 0$ such that for every z satisfying the condition $|z - z_0| < \delta$ both terms of the sum obtained are smaller than $\varepsilon/2$.

1) If $f(z_0) \neq 0$, then consider the number $\varepsilon_1 = |f(z_0)| > 0$. Since the function $f(z)$ is continuous at the point z_0, for some real number $\delta' > 0$ the condition $|z - z_0| < \delta'$ involves $|f(z) - f(z_0)| < \varepsilon_1 = |f(z_0)|$. Thus for $|z - z_0| < \delta'$ we will have

$$\begin{aligned} |f(z)| &= |f(z) - f(z_0) + f(z_0)| \\ \text{(cf., } \mathbf{223}\text{)} \quad &\leq |f(z) - f(z_0)| + |f(z_0)| < |f(z_0)| + |f(z_0)| \\ &= 2|f(z_0)|, \end{aligned}$$

i.e.,

$$|f(z)| < 2|f(z_0)|.$$

Consider the real number $\varepsilon_2 = \varepsilon/(4 \cdot |f(z_0)|)$. Since the function $g(z)$ is continuous at the point z_0, for some real number $\delta'' > 0$ the condition $|z - z_0| < \delta''$ involves $|f(z) - f(z_0)| < \varepsilon_2 = \varepsilon/(4 \cdot |f(z_0)|)$. Choose as δ_1 the smallest of the numbers δ' and δ''. Thus for every z satisfying the condition $|z - z_0| < \delta_1$ both inequalities

$$|f(z)| < 2|f(z_0)| \quad \text{and} \quad |g(z) - g(z_0)| < \frac{\varepsilon}{4 \cdot |f(z_0)|}$$

will hold. Consequently we shall have

$$|f(z)| \cdot |g(z) - g(z_0)| < \frac{\varepsilon}{2}.$$

If $f(z_0) = 0$ then we follow another argument. Consider as ε_1 the number $\varepsilon_1 = 1$. Thus for some real number $\delta' > 0$ the condition $|z - z_0| < \delta'$ involves $|f(z) - f(z_0)| < \varepsilon_1 = 1$, and since $f(z_0) = 0$ we obtain $|f(z)| < 1$. Take as ε_2 the number $\varepsilon_2 = \varepsilon/2$. Thus for some real number $\delta'' > 0$ the condition $|z - z_0| < \delta''$ involves $|g(z) - g(z_0)| < \varepsilon_2 = \varepsilon/2$. If as δ_1 we take the smallest of the numbers δ' and δ'', then for every z satisfying the condition $|z - z_0| < \delta_1$ both inequalities

$$|f(z)| < 1 \quad \text{and} \quad |g(z) - g(z_0)| < \frac{\varepsilon}{2}$$

will hold, and consequently we shall have

$$|f(z)| \cdot |g(z) - g(z_0)| < \frac{\varepsilon}{2}.$$

2) if $g(z_0) \neq 0$ consider $\varepsilon_3 = \varepsilon/(2 \cdot |g(z_0)|)$. One thus finds a real number $\delta_2 > 0$ such that for every z the condition $|z - z_0| < \delta'$ will involve

$$|f(z) - f(z_0)| < \varepsilon_3 = \frac{\varepsilon}{2 \cdot |g(z_0)|},$$

and consequently the inequality

$$|g(z_0)| \cdot |f(z) - f(z_0)| < \frac{\varepsilon}{2}$$

will hold. If $g(z_0) = 0$ then as δ_2 one can take an arbitrary positive real number, because in this case for every z one has

$$|g(z_0)| \cdot |f(z) - f(z_0)| = 0 < \frac{\varepsilon}{2}.$$

Now choose as δ the minimum of the numbers δ_1 and δ_2. Thus for every z satisfying the condition $|z - z_0| < \delta$ both inequalities

$$|f(z)| \cdot |g(z) - g(z_0)| < \frac{\varepsilon}{2}$$

and

$$|g(z_0)| \cdot |f(z) - f(z_0)| < \frac{\varepsilon}{2}$$

will hold and consequently we shall have

$$|h(z) - h(z_0)| < \frac{\varepsilon}{2} + \frac{\varepsilon}{2} = \varepsilon.$$

Hence the function $h(z) = f(z) \cdot g(z)$ is continuous at the point z_0.

240. a) Let an arbitrary real number $\varepsilon > 0$ be given. One has

$$\begin{aligned} |h(z) - h(z_0)| &= \left|\frac{1}{g(z)} - \frac{1}{g(z_0)}\right| = \left|\frac{g(z_0) - g(z)}{g(z)g(z_0)}\right| \\ (\text{ cf.}, \mathbf{226}) &= \frac{|-(g(z) - g(z_0))|}{|g(z)g(z_0)|} = \frac{|g(z) - g(z_0)|}{|g(z)| \cdot |g(z_0)|}. \end{aligned}$$

Consider the number $\varepsilon_1 = |g(z_0)|/2 > 0$. Since the function $g(z)$ is continuous at the point z_0 then for some real number $\delta_1 > 0$ there will follow $|g(z) - g(z_0)| < \varepsilon_1 = |g(z_0)|/2$ from the condition $|z - z_0| < \delta_1$. We thus obtain that for every z satisfying the condition $|z - z_0| < \delta_1$ the following inequality holds:

$$\begin{aligned} |g(z)| &= |g(z_0) - (g(z_0) - g(z))| \\ (\text{cf.}, \mathbf{223}) &\geq |g(z_0)| - |g(z_0) - g(z)| \geq |g(z_0)| - \frac{|g(z_0)|}{2} \\ &= \frac{|g(z_0)|}{2}. \end{aligned}$$

Consider the number $\varepsilon_2 = \varepsilon \cdot |g(z_0)|^2/2 > 0$. There exists a real number $\delta_2 > 0$ such that the condition $|z - z_0| < \delta_2$ involves $|g(z) - g(z_0)| < \varepsilon_2 =$

$\varepsilon|g(z_0)|^2/2$. Choose as δ the smallest of δ_1 and δ_2. Thus for every z satisfying the condition $|z - z_0| < \delta$ the following inequalities will hold:

$$|g(z)| \geq \frac{|g(z_0)|}{2} \quad \text{and} \quad |g(z) - g(z_0)| < \frac{\varepsilon|g(z_0)|^2}{2},$$

and consequently we will have

$$|h(z) - h(z_0)| = \frac{|g(z) - g(z_0)|}{|g(z)| \cdot |g(z_0)|} < \frac{\varepsilon|g(z_0)|^2}{2} \cdot \frac{2}{|g(z_0)|^2} = \varepsilon.$$

In this way the function $h(z) = 1/g(z)$ is continuous at the point z_0.

b) Since the functions $f(z)$ and $g(z)$ are continuous at the point z_0, at the point z_0 the function $1/g(z)$ is continuous (cf., (a)), and consequently the function $h(z) = f(z)/g(z) = f(z) \cdot (1/g(z))$ is continuous as well (cf., **239**(b)).

241. Let an arbitrary real number $\varepsilon > 0$ be given. Since the function $f(z)$ is continuous at the point z_1, there exists a real number $\delta_1 > 0$ such that the condition $|z - z_0| < \delta_1$ involves $|f(z) - f(z_1)| < \varepsilon$. Now consider the number $\delta_1 > 0$. Since the function $g(x)$ is continuous at the point z_0, there exists a real number δ such that from $|z - z_0| < \delta$ it follows that $|g(z) - g(z_0)| < \delta_1$, i.e., $|g(z) - z_1| < \delta_1$. But thus the inequality $|f(g(z)) - f(z_1)| < \varepsilon$ does hold, i.e., we obtain $|f(g(z)) - f(g(z_0))| < \varepsilon$. Consequently for every z satisfying the condition $|z - z_0| < \delta$ we have $|h(z) - h(z_0)| < \varepsilon$, and therefore the function $h(z) = f(g(z))$ is continuous at the point z_0.

242. Let a point z_0 and a real number $\varepsilon > 0$ be given. For the function $f(x) = \sin(x)$ we find $|f(x) - f(x_0)| = |\sin(x) - \sin(x_0)| =$ (by a trigonometric formula) $|2\sin((x - x_0)/2)\cos((x + x_0)/2)| = 2 \cdot |\sin((x - x_0)/2)| \cdot |\cos((x + x_0)/2)| \leq |\sin((x - x_0)/2)|$, (since $|\cos((x + x_0)/2)| \leq 1$).

For $f(x) = \cos(x)$ we find $|f(x) - f(x_0)| = |\cos(x) - \cos(x_0)| =$ (by a trigonometric formula) $|-2\sin((x - x_0)/2)\sin((x + x_0)/2)| = 2 \cdot |\sin((x - x_0)/2)| \cdot |\sin((x + x_0)/2)| \leq 2|\sin(x - x_0)/2|$, (since $|\sin(x + x_0)/2| \leq 1$). In this way in both cases we obtain

$$|f(x) - f(x_0)| \leq \left|\sin\frac{x - x_0}{2}\right|.$$

Now choose $\delta > 0$ in such a way that for every x satisfying $|x - x_0| < \delta$ the inequality $|\sin((x - x_0)/2)| < \varepsilon/2$ be satisfied.

If $\varepsilon/2 > 1$ then the inequality $|\sin((x - x_0)/2)| < \varepsilon/2$ holds for all x and thus δ can be chosen arbitrarily. If $\varepsilon/2 \leq 1$ consider on the plane with coordinates x and y a circle of radius 1 with centre at the origin of the coordinates and draw the straight lines $y = \varepsilon/2$ and $y = -\varepsilon/2$ (Figure 49). For the angles α and β shown in the Figure we obtain $\sin\alpha = \varepsilon/2$, $\sin\beta = -\varepsilon/2$. Hence $\beta = -\alpha$.

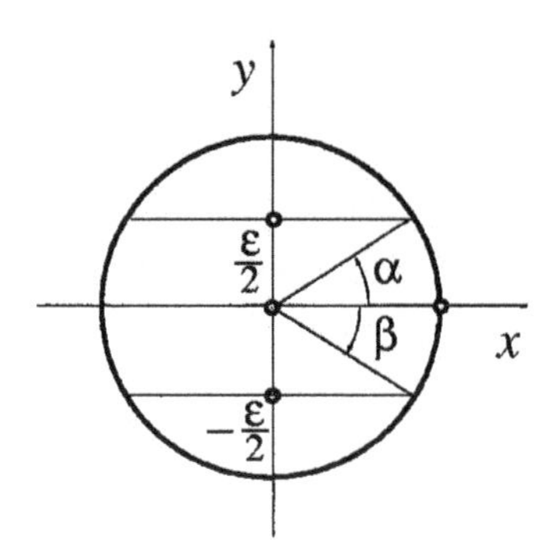

FIGURE 49

Choose $\delta = 2\alpha$. Thus from $|x - x_0| < \delta$ it follows that $|(x - x_0)/2| < \delta/2$, i.e., $-\alpha < (x - x_0)/2 < \alpha$. Hence $\sin(-\alpha) < \sin((x - x_0)/2) < \sin\alpha$, i.e., $-\varepsilon/2 < \sin((x - x_0)/2) < \varepsilon/2$ and $|\sin((x - x_0)/2)| < \varepsilon/2$. Thus

$$|f(x) - f(x_0)| \leq 2\left|\sin\frac{x - x_0}{2}\right| < \varepsilon.$$

Consequently the functions $f(x) = \sin(x)$ and $f(x) = \cos(x)$ are continuous for all the real values of the argument x.

243. Let a point x_0 and an arbitrary positive real number ε be given. We have to choose a real number $\delta > 0$ such that for every x satisfying the inequality $|x - x_0| < \delta$ (and, of course, $x \geq 0$) the inequality

$$|\sqrt[n]{x} - \sqrt[n]{x_0}| < \varepsilon \tag{3.6}$$

holds. This last inequality is equivalent to the inequalities

$$-\varepsilon < \sqrt[n]{x} - \sqrt[n]{x_0} < \varepsilon$$

and

$$\sqrt[n]{x_0} - \varepsilon < \sqrt[n]{x} < \sqrt[n]{x_0} + \varepsilon. \tag{3.7}$$

Since the function $\sqrt[n]{x}$ is strictly increasing (for $x \geq 0$) the inequalities (3.7) are equivalent when $\sqrt[n]{x_0} - \varepsilon \geq 0$ to the inequalities

$$(\sqrt[n]{x_0} - \varepsilon)^n < x < (\sqrt[n]{x_0} + \varepsilon)^n.$$

We therefore have

$$(\sqrt[n]{x_0} - \varepsilon)^n - x_0 < x - x_0 < (\sqrt[n]{x_0} + \varepsilon)^n - x_0. \tag{3.8}$$

In this case, taking as δ the smallest of numbers $x_0 - (\sqrt[n]{x_0} - \varepsilon)^n$ and $(\sqrt[n]{x_0} - \varepsilon)^n - x_0$ we obtain that (3.8) follows from the condition $|x - x_0| < \delta$, and these inequalities also involve the inequality (3.6). If in (3.7) $\sqrt[n]{x_0} - \varepsilon < 0$, then the inequality on the left is always satisfied (for $x \geq 0$), and (3.7) is equivalent to the inequality

$$x < (\sqrt[n]{x_0} + \varepsilon)^n,$$

which involves

$$x - x_0 < (\sqrt[n]{x_0} + \varepsilon)^n - x_0. \tag{3.9}$$

In this case it suffices to take

$$\delta = (\sqrt[n]{x_0} + \varepsilon)^n - x_0,$$

and for every x satisfying the condition $|x - x_0| < \delta$ the inequality (3.9) together with the inequalities (3.7) and (3.6) will be satisfied.

244. *Answer.* a) See Figure 50. *Hint.* $y(t) \equiv 0$; b), c) See Figure 50. *Hint.* $x(t) \equiv 0$; d) See Figure 50; $x(t) = t$, $y(t) = -t$, $y = -x$.

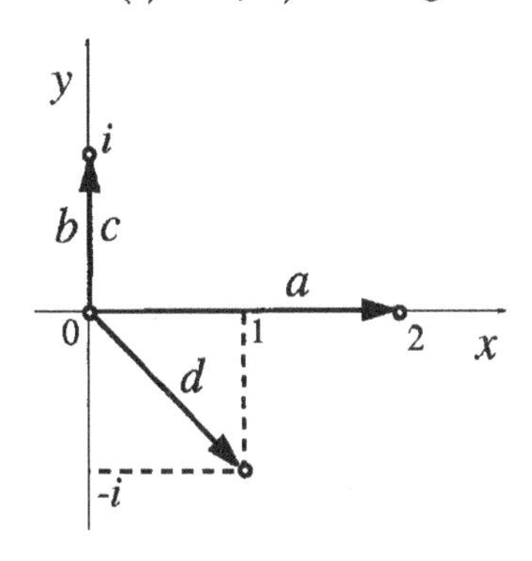

FIGURE 50

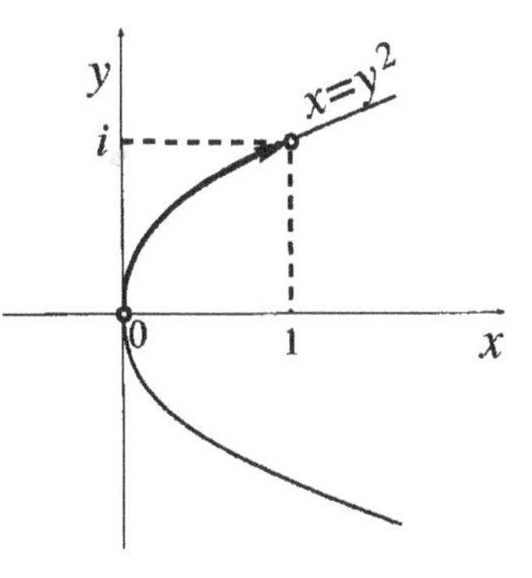

FIGURE 51

e) See Figure 51; $x(t) = t^2$, $y(t) = t$, $x = y^2$; f), g) a turn (the case (f)) and a double turn (the case (g)), both counterclockwise, along a circle of radius R, with initial point $z = R + 0i$ (Figure 52). *Hint.* (f) $|z(t)| = R$, $\arg z(t) = 2\pi t$, and (g) $\arg z(t) = 4\pi t$; h) the semi-circle of radius R (Figure 53); i) See Figure 54.

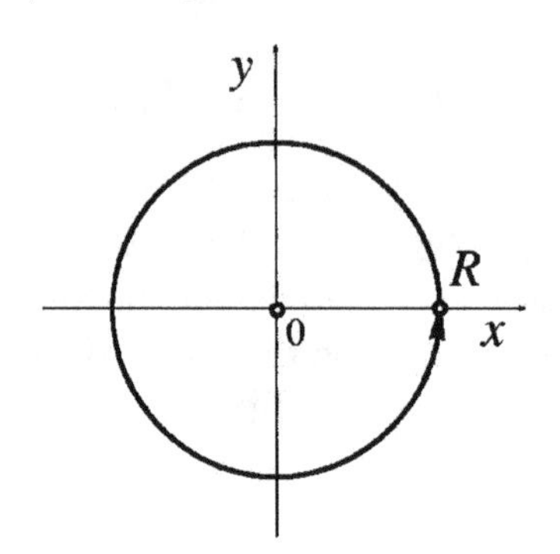

FIGURE 52

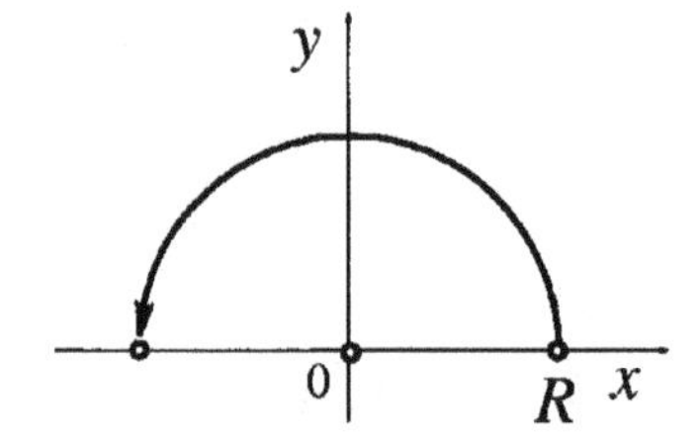

FIGURE 53

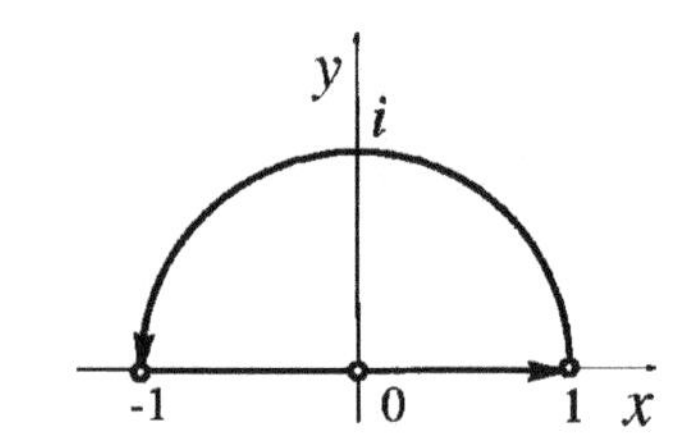

FIGURE 54

245. See Figure 55. By similarity one obtains $t = (x - a_0)/(a_1 - a_0) = (y - b_0)/(b_1 - b_0)$. When the point z moves along the segment from the position $z_0 = a_0 + b_0 i$ to the position $z_1 = a_1 + b_1 i$ the parameter varies from 0 to 1. Thus, for example, one can take $t = (x - a_0)/(a_1 - a_0) = (y - b_0)/(b_1 - b_0)$, obtaining $x = a_0 + (a_1 - a_0)t$, $y = b_0 + (b_1 - b_0)t$, $z(t) = (a_0 + (a_1 - a_0)t) + (b_0 + (b_1 - b_0)t)i$. It is easy to verify that this formula describes the initial segment for any position of the points z_0 and z_1, so, in particular, for $a_0 = a_1$ or for $b_0 = b_1$.

246. Since $z_A + z_{\overrightarrow{AB}} = z_B$ (cf., **221**) it follows that: the case (a) corresponds to the displacement of the curve by the vector corresponding to the complex number z_0 (Figure 56).

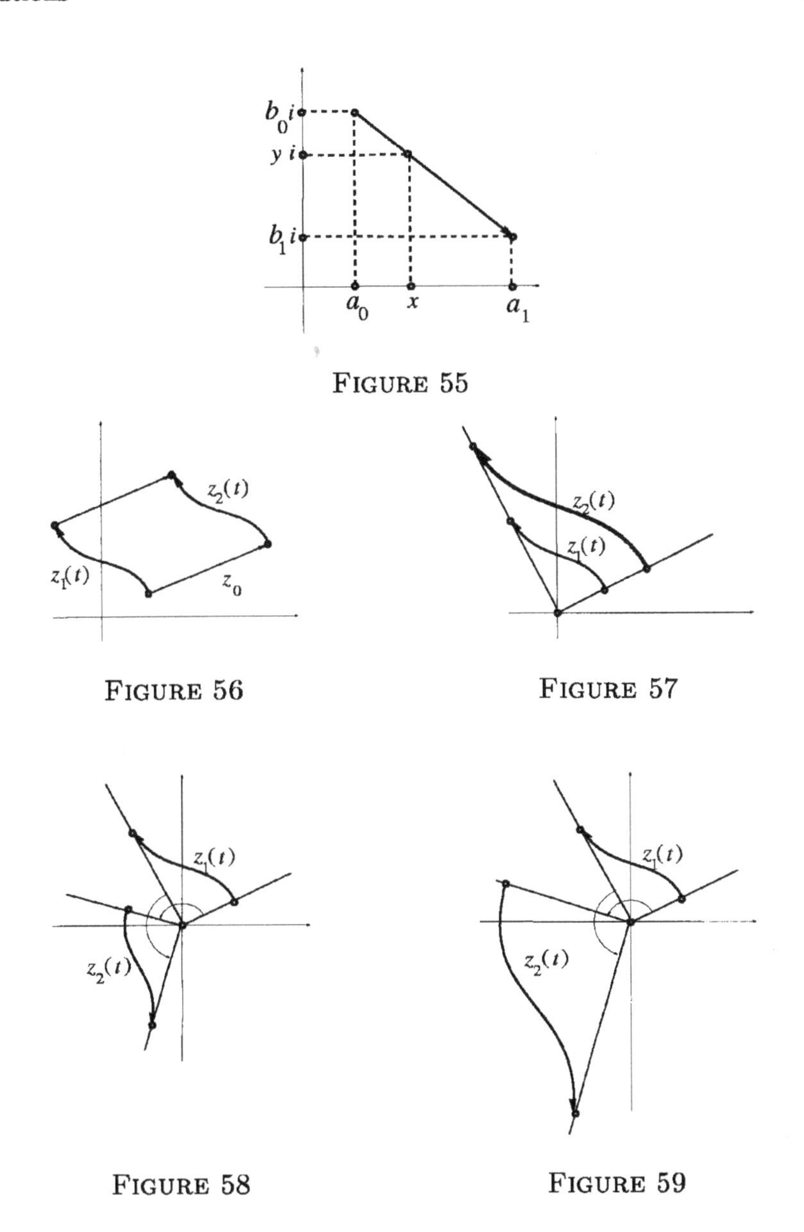

FIGURE 55

FIGURE 56

FIGURE 57

FIGURE 58

FIGURE 59

If $z_2 = z_0 \cdot z_1$ and $\arg z_0 = \varphi_0$, $\arg z_1 = \varphi_1$ then (cf., **226**) $|z_2| = |z_0| \cdot |z_1|$ and $\arg z_2 = \varphi_0 + \varphi_1$. Therefore the case (b) corresponds to a dilation of the curve C_1 by a factor a, fixing the origin of the coordinates (Figure 57); the case (c) corresponds to a rotation of the curve C_1 by the angle $\varphi_0 = \arg z_0$ around the origin of the coordinates (Figure 58); the case (d) corresponds to a dilation by z_0 simultaneous with a rotation by the angle $\varphi_0 = \arg z_0$ (Figure 59).

247. When t varies continuously from 0 to 1, $1-t$ varies continuously from 1 to 0. The function $z_2(t) = z_1(1-t)$ thus describes the same geometrical curve as C, but oriented in the opposite way.

248. When t varies continuously from 0 to $1/2$, $2t$ varies continuously from 0 to 1. When t varies continuously from $1/2$ to 1, $2t-1$ varies continuously from 0 to 1. Hence the function $z_3(t)$ given by the problem describes the curve which is obtained by drawing first the curve C_1 and then the curve C_2. The condition $z_1(1) = z_2(0)$ guarantees the continuity of the obtained curve.

249. *Answer.* $\pi t + 2\pi k$, $k = 0, \pm 1, \pm 2, \dots$.

250. *Answer.* a) πt; b) $2\pi + \pi t$; c) $-4\pi + \pi t$; d) $2\pi k + \pi t$.

251. For every t we have $\varphi(t) - \varphi'(t) = 2\pi k$, where the values of k may be distinct for all t. We thus write $\varphi(t) - \varphi'(t) = 2\pi k(t)$. It follows that $k(t) = (\varphi(t) - \varphi'(t))/(2\pi)$. Since the functions $\varphi(t)$ and $\varphi'(t)$ are continuous for $0 \le t \le 1$, the function $k(t)$ is also continuous for $0 \le t \le 1$ (cf., **239**, **240**). But since the function $k(t)$ takes only integer values it is continuous only if it is a constant, i.e., if $k(t) = k$, where k is some given integer which does not depend on t. Therefore $(\varphi(t) - \varphi'(t))/(2\pi) = k$ and $\varphi(t) - \varphi'(t) = 2\pi k$.

252. Let $\varphi(t)$ and $\varphi'(t)$ be two functions describing the continuous variation of $\arg z(t)$ and $\varphi(0) = \varphi'(0) = \varphi_0$. Thus (cf., **251**) $\varphi(t) - \varphi'(t) = 2\pi k$, where k is a given integer. But $\varphi(0) - \varphi'(0) = 0$, hence $k = 0$. Consequently $\varphi(t) - \varphi'(t) \equiv 0$ and $\varphi(t) \equiv \varphi'(t)$.

253. Let $\varphi(t)$ and $\varphi'(t)$ be two functions describing the continuous variation of $\arg z(t)$. Thus (cf., **251**) $\varphi(t) - \varphi'(t) = 2\pi k$, where k is a given integer. In particular, $\varphi(0) - \varphi'(0) = 2\pi k$. Consequently $\varphi(t) - \varphi'(t) = \varphi(0) - \varphi'(0)$ and $\varphi(t) - \varphi(0) = \varphi'(t) - \varphi'(0)$.

254. a) One can take $\varphi(t) = \pi t$. *Answer.* $\varphi(1) - \varphi(0) = \pi$ (see Figure 53); b) $\varphi(t) = 2\pi t$. *Answer.* 2π (see Figure 52); c) $\varphi(t) = 4\pi t$. *Answer.* 4π; d) *Answer.* $\pi/2$ (see Figure 60)

255. *Answer.* a) $3\pi/2$; b) $-\pi/2$.

256. *Answer.* a) 1; b) -2; c) 2; d) 0.

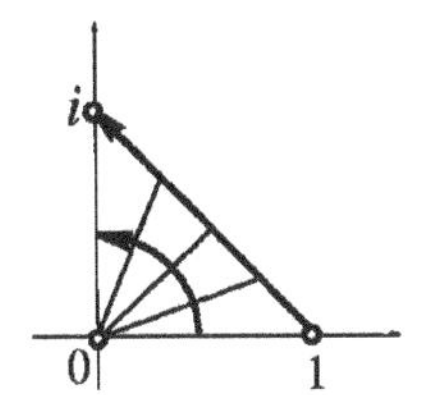

FIGURE 60

257. Suppose one has chosen as the initial point on an oriented continuous closed curve C the point A in one case and the point B in another case (Figure 61). If the variation of the argument along C in the segment AB (according to the orientation of the curve C) is equal to φ_1 and in the segment BA it is equal to φ_2, then the variation of the argument along the entire curve C is evidently equal to $\varphi_1 + \varphi_2$.

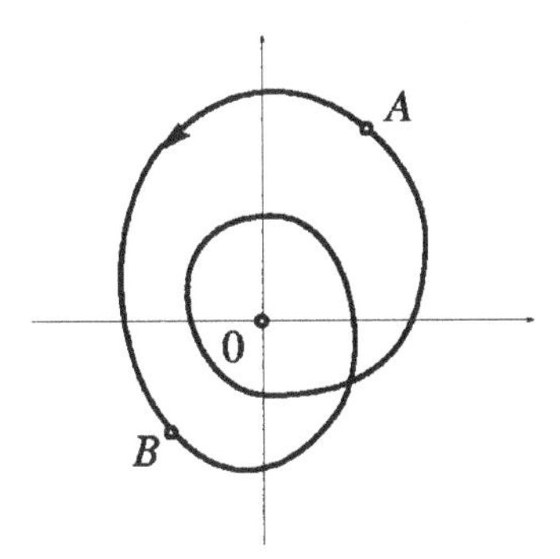

FIGURE 61

258. a),b),c) If $z_2(t) = z_0 \cdot z_1(t)$ and $\varphi_1(t)$ is a function describing the continuous variation of $\arg z_1(t)$, then the function $\varphi_2(t) = \varphi_1(t) + \varphi_0$, where $\varphi_0 = \arg z_0$, describes the continuous variation of $\arg z_2(t)$ (cf., **226**, **239**). It follows that $\varphi_2(1) - \varphi_2(0) = \varphi_1(1) - \varphi_1(0)$.

Answer. k times. This result obviously also depends on the result of Problem **246**.

d) If $\arg z = \varphi$ then $\arg \overline{z} = -\varphi$ (cf., **218**(b)). Therefore if $\varphi_1(t)$ is a function which describes the continuous variation of $\arg z_1(t)$, then the function $\varphi_2(t) = -\varphi_1(t)$ describes the continuous variation of $\arg z_2(t)$. It follows that $\varphi_2(1) - \varphi_2(0) = -(\varphi_1(1) - \varphi_1(0))$.

Answer. $-k$ times.

259. *Answer.* a) 1; b) 0; c) 1; d) 2.

260. *Solution.* If $\varphi_1(t)$ and $\varphi_2(t)$ are two functions which describe the continuous variation of the argument along the curves C_1 and C_2, then as a function $\varphi(t)$ describing the continuous variation of the argument along C on can take: in the case (a) $\varphi(t) = \varphi_1(t) + \varphi_2(t)$; in the case (b) $\varphi(t) = \varphi_1(t) - \varphi_2(t)$ (cf., **226, 239**). It follows that $\varphi(1) - \varphi(0) = (\varphi_1(1) \pm \varphi_2(1)) - (\varphi_1(0) \pm \varphi_2(0)) = (\varphi_1(1) - \varphi_1(0)) \pm (\varphi_2(1) - \varphi_2(0)) = \varphi_1 \pm \varphi_2$.

Answer. a) $\varphi_1 + \varphi_2$; b) $\varphi_1 - \varphi_2$.

261. *Hint.* $w_0(t) = [z(t)]^2$ (use de Moivre's formula; cf., **227**):

a) $w_0(t) = R^2(\cos \pi t + i \sin \pi t)$ (a semi-circle of radius R^2);

b) $w_0(t) = R^2(\cos 2\pi t + i \sin 2\pi t)$ (a circle of radius R^2);

c) $w_0(t) = R^2(\cos 4\pi t + i \sin 4\pi t)$ (a circle of radius R^2, twice covered).

262. *Hint.* Use the result of Problem **260**(a). *Answer.* a) 2φ; b) 3φ; c) $n\varphi$.

263. Let $z(t)$ be the parametric equation of the curve C and $z_1(t) = z(t) - z_0$. By hypothesis the variation of $\arg z_1(t)$ is equal to $2\pi k$. The parametric equation of the curve $f(C)$ is $w_0(t) = (z(t) - z_0)^n = [z_1(t)]^n$. The variation of the argument of $w_0(t)$ is thus equal to $2\pi kn$ (cf., **262**(c)). Hence the curve $f(C)$ turns around the point $w = 0$ kn times. *Answer.* kn times.

264. By hypothesis the variation of the argument of $z(t)$ is equal to $2\pi k_1$, of the argument of $z(t) - 1$ to $2\pi k_2$, of the argument of $z(t) - i$ to $2\pi k_3$ and of the argument of $z(t) + i$ to $2\pi k_4$. By the result of Problem **260**(a) one obtains:

a) $w_0(t) = z(t) \cdot (z(t) - 1)$. The variation of the argument of $w_0(t)$ is equal to $2\pi k_1 + 2\pi k2 = 2\pi(k_1 + k_2)$. *Answer.* The curve $f(C)$ turns around the point $w = 0$ $k_1 + k_2$ times.

b) $w_0(t) = (z(t) - i)(z(t) + i)$. *Answer.* $k_3 + k_4$ times.

c) $w_0(t) = [(z(t)]^4 \cdot [(z(t) + i]^4$. *Answer.* $4(k_1 + k_4)$ times.

d) $w = (z - 1)(z^2 + 1) = (z - 1)(z + i)(z - i)$. *Answer.* $(k_1 + k_3 + k_4)$ times.

265.

$$\begin{aligned} |a_0 z^n + \ldots + a_{n-1} z| &\leq |a_0 z^n| + \ldots + |a_{n-1} z| \quad (\text{cf., } \mathbf{223}(\text{a})) \\ &= |a_0| \cdot |z|^n + \ldots + |a_{n-1}| \cdot |z| \quad (\text{cf., } \mathbf{226}) \\ &= |a_0| \cdot R_1^n + \ldots + |a_{n-1}| \cdot R_1 \\ (\text{since } R_1 < 1) \quad &\leq |a_0| \cdot R_1 + \ldots + |a_{n-1}| \cdot R_1 \leq nAR_1 \\ (\text{since } |a_i| \leq A) \quad &< nA \frac{|a_n|}{10nA} = \frac{|a_n|}{10} \quad \left(\text{since} \quad R_1 < \tfrac{|a_n|}{10nA}\right). \end{aligned}$$

266.

$$\begin{aligned} \left|\frac{a_1}{z} + \ldots + \frac{a_n}{z^n}\right| &\leq \left|\frac{a_1}{z}\right| + \ldots + \left|\frac{a_n}{z^n}\right| \quad (\text{cf., } \mathbf{223}(\text{a})) \\ (\text{cf., } \mathbf{226}) \quad &= \frac{|a_1|}{|z|} + \ldots + \frac{|a_n|}{|z^n|} = \frac{|a_1|}{R_2} + \ldots + \frac{|a_n|}{R_2^n} \\ (\text{since } R_2 > 1) \quad &\leq \frac{|a_1|}{R_2} + \ldots + \frac{|a_n|}{R_2} \\ (\text{since } |a_i| \leq A) \quad &\leq \frac{nA}{R_2} \\ \left(\text{since} \quad R_2 > \tfrac{10nA}{|a_0|}\right) \quad &< \frac{nA|a_0|}{10nA} = \frac{|a_0|}{10}. \end{aligned}$$

267. If $|z| = R_1$ then (cf., **265**) $|f(z) - a_n| < |a_n|/10$. Consequently the curve $f(C_{R_1})$ lies entirely in the disc of radius $|a_n|/10$ with centre at the point $w = a_n$ (Figure 62). It is evident that the curve does not turn at all around the point $w = 0$. Therefore $\nu(R_1) = 0$.

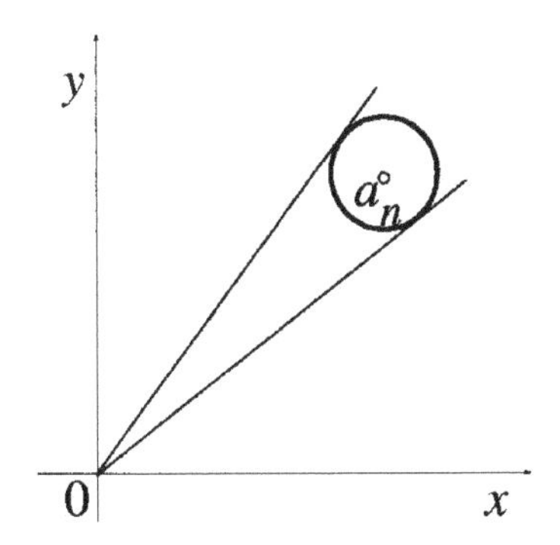

FIGURE 62

Let us now prove that $\nu(R_2) = n$. We give first a not strictly exact, but nice, proof called 'the lady with her little dog'. From the result of Problem **266** we obtain that if $|z| = R_2$, then

$$\begin{aligned}
|f(z) - a_0 z^n| &= |a_1 z^{n-1} + \ldots + a_n| = \left|z^n \left(\frac{a_1}{z} + \ldots + \frac{a_n}{z^n}\right)\right| \\
&= |z^n| \cdot \left|\frac{a_1}{z} + \ldots + \frac{a_n}{z^n}\right| \qquad (\text{cf., } \mathbf{226}) \\
&< \frac{|a_0| \cdot |z^n|}{10} \\
&= \frac{|a_0| \cdot |z|^n}{10} = \frac{|a_0| \cdot R_2^n}{10}.
\end{aligned}$$

Note that $|a_0| \cdot R_2^n = R$. Thus $|f(z) - a_0 z^n| < R/10$ for $|z| = R_2$. When z covers once the circle of radius R_2 the point $w = a_0 z^n$ (the 'lady') covers the circle of radius R n times. Since $|f(z) - a_0 z^n| < R/10$ the point $w = f(z)$ ('the little dog') cannot remain farther than $R/10$ from the lady. But thus if the lady turns n times around the point $w = 0$ along a circle of radius R, her dog is also obliged to go n times around the point $w = 0$ (Figure 63). It follows that $\nu(R_2) = n$.

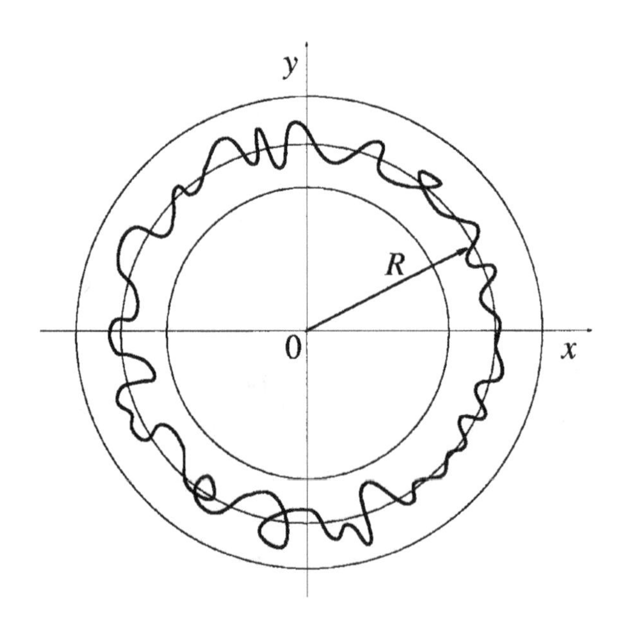

FIGURE 63

A more exact proof of the equality $\nu(R_2) = n$ can be obtained in the following way. If $|z| = R_2$ then (cf., **266**) $|(f(z)/z^n) - a_0| < |a_0|/10$. So we obtain, as well as in the proof of the equality $\nu(R_1) = 0$, that when the point $z(t)$ covers the circle C_{R_2} the variation of the argument of $f(z)/z^n$ vanishes. The variation of the argument of z is equal to 2π, and consequently the variation of the argument of z^n is equal to $2\pi n$ (cf., **262**). Since $f(z) = (f(z)/z^n) \cdot z^n$ (cf., **260**) the variation of the argument of $f(z)$ is equal to $2\pi n$, i.e., the curve $f(C_{R_2})$ turns n times around the point $w = 0$. Hence $\nu(R_2) = n$.

Answer $\nu(R_1) = 0$ and $\nu(R_2) = n$.

268. Divide the polynomial $P(z) = a_0 z^n + \ldots + a_{n-1} z + a_n = 0$ by the binomial $z - z_0$ with remainder (for example, by the Euclidean algorithm; cf., §2.1). The remainder will be some complex number r, and the quotient will be some polynomial $Q(z)$. We thus have the equation

$$P(z) = Q(z) \cdot (z - z_0) + r.$$

Replacing z by z_0 we obtain

$$P(z_0) = Q(z_0) \cdot (z_0 - z_0) + r = 0 + r = r.$$

But $P(z_0) = 0$, since by hypothesis z_0 is a root of the polynomial $P(z)$. So from the equality $P(z_0) = r$ we obtain that $r = 0$. Hence $P(z) = Q(z) \cdot (z - z_0)$.

269. Let $P(z) = a_0 z^n + \ldots + a_{n-1} z + a_n = 0$. By the fundamental theorem of algebra of complex numbers the equation $P(z) = 0$ has a root z_1. By Bézout's theorem (cf., **268**) $P(z) = (z - z_1)Q(z)$, from which it is easy to see that $Q(z)$ has the form $Q(z) = a_0 z^{n-1} + b_1 z^{n-2} + \ldots + b_{n-1}$. If $n - 1 \geq 1$ then the equation $Q(z) = 0$ has a root z_2 from which $Q(z) = (z - z_2)R(z)$ and $R(z) = a_0 z^{n-2} + \ldots$. Therefore $P(z) = (z - z_1)(z - z_2)R(z)$. Continuing this procedure we obtain at the nth step a quotient which is a complex constant, obviously equal to a_0. The final result is thus the required decomposition.

270. Denoting by $P(z)$ the polynomial $a_0 z^n + \ldots + a_{n-1} z + a_n$ we obtain (cf., **212**) $\overline{P(z)} = P(\overline{z})$. By hypothesis $P(z_0) = 0$. It follows that $P(\overline{z_0}) = \overline{P(z_0)} = \overline{0} = 0$, i.e., $\overline{z_0}$ is a root of the equation $P(z) = 0$.

271. Let $P(z) = a_0 z^n + \ldots + a_{n-1} z + a_n = 0$. Since all the a_is are real numbers and z_0 is a root of the equation $P(z) = 0$, $\overline{z_0}$ is also a root of the equation $P(z) = 0$ (cf., **270**).

Since z_0 is not a real number, $z_0 \neq \overline{z_0}$. Since z_0 and $\overline{z_0}$ are roots of the equation $P(z) = 0$, in the decomposition (cf., **269**) $P(z) = a_0(z - z_1)(z-z_2)\cdot\ldots\cdot(z-z_n)$ we must find the factors $(z-z_0)$ and $(z-\overline{z_0})$. We can write: $P(z) = (z-z_0)(z-\overline{z_0})Q(z) = (z^2-(z_0+\overline{z_0})z+z_0\overline{z_0})\cdot Q(z)$. In this way the polynomial $P(z)$ is divisible by the polynomial of second degree $z^2-(z_0+\overline{z_0})z+z_0\overline{z_0}$ whose coefficients are real (cf., the solution of Problem **211**).

272. Let $P(z)$ be the given polynomial. If the degree of the polynomial $P(z)$ is higher than 2 then the equation $P(z) = 0$ has, by the fundamental Theorem of algebra of complex numbers, a root z_0. If z_0 is a real number we divide $P(z)$ by $z - z_0$. We obtain $P(z) = (z-z_0)Q(z)$ (cf., **268**). If z_0 is not a real number the polynomial $P(z)$ is divisible by a polynomial of second degree with real coefficients (cf., **271**). In both cases the quotient is a polynomial, with real coefficients, which results, for example, from the Euclidean algorithm (cf., §2.1). This quotient is again divisible by some polynomial of the first or second degree with real coefficients, etc.. This procedure ends when the quotient obtained has first or second degree. So we have obtained the required decomposition.

273. $z^5-z^4-2z^3+2z^2+z-1 = (z-1)(z^4-2z^2+1) = (z-1)(z^2-1)^2 = (z-1)^3(z+1)^2$. *Answer.* 1 is a root of order 3, -1 is a root of order 2.

274. Compare the coefficients of the terms of the same degree in the two members of the given equalities. Let

$$P(z) = a_0z^n + a_1z^{n-1} + \ldots + a_{n-1}z + a_n,$$
$$Q(z) = b_0z^m + b_1z^{m-1} + \ldots + b_{m-1}z + b_m.$$

Thus

$$P'(z) = a_0nz^{n-1} + a_1(n-1)z^{n-2} + \ldots + a_{n-1},$$
$$Q'(z) = b_0mz^{m-1} + b_1(m-1)z^{m-2} + \ldots + b_{m-1}.$$

it is easy to see that in the case (a), for every k the coefficients of z^k in the two members of the equality are equal to $(a_{n-1-k}+b_{m-1-k})(k+1)$, and that in the case (b) the coefficients of z^k in the two members are equal to $ca_{n-1-k}(k+1)$.

c) For brevity let us use the summation symbol $\sum$. The symbol $\sum_{i=0}^{n}$ (respectively, $\sum_{i+j=r}$) means that one has to consider the expression which lies on the right of this symbol for $i = 0,1,2,\ldots,n$ (respectively, for all i, j such that $i+j=r$) and to add all the expressions so obtained. By this notation we write

$$P(z) = \sum_{i=0}^{n} a_i z^{n-i}, \qquad Q(z) = \sum_{j=0}^{m} b_j z^{m-j},$$
$$P'(z) = \sum_{i=0}^{n-1} a_i(n-i)z^{n-i-1}, \quad Q'(z) = \sum_{j=0}^{m-1} b_j(m-j)z^{m-j-1}.$$

Let k be any integer such that $0 \leq k \leq n+m-1$. We are looking for the coefficient of z^{k+1} in the product $P(z) \cdot Q(z)$. Since $(n-i)+(m-j) = k+1$ if and only if $i+j = n+m-k-1$, the coefficient of z^{k+1}, expanding and collecting the terms of degree $k+1$ in the polynomial $P(z) \cdot Q(z)$, is equal to $\sum_{i+j=n+m-k-1} a_i b_j$. Consequently the coefficient of z^k in the polynomial $(P(z) \cdot Q(z))'$ is equal to $\sum_{i+j=n+m-k-1} a_i b_j (k+1)$. Exactly in the same way we obtain that the coefficient of z^k in the polynomial $P'(z) \cdot Q(z)$ is equal to $\sum_{i+j=n+m-k-1} a_i b_j (n-i)$, and in the polynomial $P(z) \cdot Q'(z)$ is equal to $\sum_{i+j=n+m-k-1} a_i b_j (m-j)$. Hence the coefficient of z^k in the polynomial $P'(z) \cdot Q(z) + P(z) \cdot Q'(z)$ is equal to

$$\sum_{i+j=n+m-k-1} a_i b_j((n-i)+(m-j)) = \sum_{i+j=n+m-k-1} a_i b_j (k+1),$$

because $(n-i)+(m-j) = k+1$. The expression obtained coincides with the coefficient of z^k in the polynomial $(P(z) \cdot Q(z))'$.

275. For $n=1$ the claim is true because $(z-z_0)' = 1 = 1 \cdot (z-z_0)^0$. It also holds for $n=2$: $((z-z_0)^2)' = (z^2 - 2z_0 z + z_0^2)' = 2z - 2z_0 = 2(z-z_0) = 2(z-z_0)^1$. Suppose that it holds for $n=k$, i.e., that $((z-z_0)^k)' = k(z-z_0)^{k-1}$. We prove that it also holds for $n=k+1$. We obtain $((z-z_0)^{k+1})' = ((z-z_0)^k \cdot (z-z_0))' =$ (cf., **274**(b)) $= ((z-z_0)^k)' \cdot (z-z_0) + (z-z_0)^k \cdot (z-z_0)' = k(z-z_0)^{k-1} \cdot (z-z_0) + (z-z_0)^k = (k+1)(z-z_0)^k$. So if our claim is true for $n=k$ then it is also true for $n=k+1$. Since it holds for $n=1$ and $n=2$ it holds for all integers $n \geq 1$.

276. By hypothesis $P(z) = (z-z_0)^k \cdot Q(z)$, where the polynomial $Q(z)$ is not divisible by $z - z_0$. It follows that $P'(z) =$ (cf., **274**(b)) $= ((z-z_0)^k)' \cdot Q(z) + (z-z_0)^k \cdot Q'(z) =$ (cf., **275**) $= k(z-z_0)^{k-1}Q(z) + (z-z_0)^k Q'(z) = (z-z_0)^{k-1}(kQ(z) + (z-z_0)Q'(z))$. The polynomial within the last brackets is not divisible by $z - z_0$, because otherwise the polynomial $Q(z)$ should have been divisible by $z - z_0$. Consequently the polynomial $P'(z)$ is divisible by $(z-z_0)^{k-1}$ and is not divisible by $(z-z_0)^k$.

277. *Answer.* a) ± 1; b) $\pm i$; c) $\pm(\sqrt{2}/2 + i\sqrt{2}/2)$; d) $\pm(\sqrt{6}/2 + i\sqrt{6}/2)$ (here $\sqrt{2}$ and $\sqrt{6}$ are the positive values of the square roots).

278. The continuous image under the mapping $w = \sqrt{z}$ of the upper (respectively, lower) semi-circle starting at the point $w = 1$ is the arc AB (respectively, the arc AC) (Figure 64). The curve AB ends at point i, curve AC at point $-i$.

Answer. a) $w(-1) = i$; b) $w(-1) = -i$.

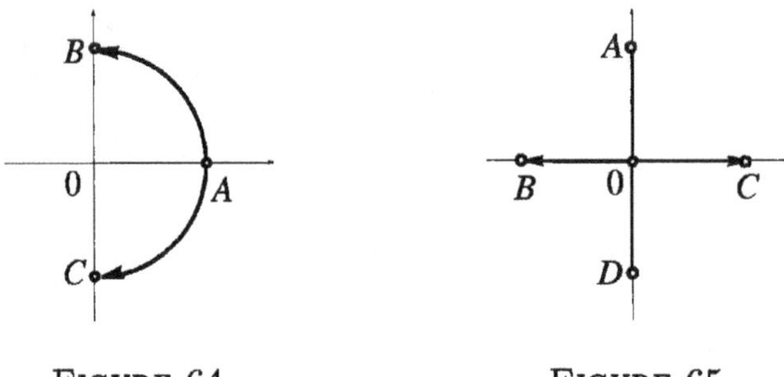

FIGURE 64 FIGURE 65

279. For $0 \le t \le \frac{1}{2}$, $w_0(t)$ may take two values: $w_0(t) = \pm i\sqrt{1-2t}$ (the value of the square root is considered positive). Also for $\frac{1}{2} < t \le 1$ $w_0(t)$ may take two values: $w_0(t) = \pm\sqrt{2t-1}$. For $t = \frac{1}{2}$, $w_0(t)$ takes a unique value: $w(\frac{1}{2}) = 0$.

Answer (see Figure 65). a) The continuous images are represented by the broken lines AOB and AOC; b) the continuous images are represented by the broken lines DOB and DOC.

280. Let C_1 be the continuous image of the curve C under the mapping $w = \sqrt{z}$, and let the variation of the argument along the curve C_1 be equal to φ_1. Thus the curve C is the image of the curve C_1 under the mapping $z = w^2$, and (cf., **262**(a)) $\varphi = 2\varphi_1$. Therefore $\varphi_1 = \varphi/2$.

Answer. $\varphi/2$.

281. Let $w = \sqrt{z}$ and $w_0(t)$ the continuous image of the given curve. Since $i = \cos(\pi/2) + i\sin(\pi/2)$, $\sqrt{i}$ (and $w_0(1)$) may take two values: $\cos(\pi/4) + i\sin(\pi/4) = \sqrt{2}/2 + i\sqrt{2}/2$ and $\cos(5\pi/4) + i\sin(5\pi/4) = -\sqrt{2}/2 - i\sqrt{2}/2$. In accord with the condition $w_0(0) = -1$ one may take $\arg w_0(0) = \pi$.

a) the variation of the argument along the given segment is, evidently, $\pi/2$. Consequently (cf., **280**), the variation of the argument along the curve $w_0(t)$ will be equal to $\pi/4$ and of the argument of $w_0(1)$ equal to $\pi + \pi/4 = 5\pi/4$.

Answer. $\sqrt{i} = -\sqrt{2}/2 - i\sqrt{2}/2$.

b) $z(t) = \cos(-3\pi t/2)) + i\sin(-3\pi t/2))$. In order for $\arg z(t)$ to vary continuously one may choose $\varphi(t) = -3\pi t/2$. The variation of the argument along the curve will thus be $\varphi(1) - \varphi(0) = -3\pi/2$. Hence the variation of the argument along curve $w_0(t)$ (cf., **280**) is equal to $-3\pi/4$ and the argument of $w_0(1)$ is equal to $\pi + (-3\pi/4) = \pi/4$.

Answer. $\sqrt{i} = \sqrt{2}/2 + i\sqrt{2}/2$.

c) $\varphi(t) = 5\pi t/2$. The variation of the argument along the given curve is: $\varphi(1) - \varphi(0) = 5\pi/2$. The variation of the argument along curve $w_0(t)$ is equal to $5\pi/4$. The argument of w_0 is equal to $\pi + (5\pi/4) = 9\pi/4$.

Answer. $\sqrt{i} = \sqrt{2}/2 + i\sqrt{2}/2$.

282. Let $w = \sqrt{z}$, let $w_0(t)$ be the continuous image of the given curve with $w_0(0) = 1$. We have to define $w_0(1)$. We may take $\arg w_0(0) = 0$.

a) the variation of $\arg z(t)$ is equal to 2π. Therefore the variation of $\arg w_0(t)$ is equal to π (cf., **280**), and $\arg w_0(1) = 0 + \pi = \pi$.

Answer. $\sqrt{1} = -1$.

b) $z(t) = \cos(-4\pi t) + i\sin(-4\pi t)$. The variation of $\arg z(t)$ is equal to -4π. The variation of $\arg w_0(t)$ is equal to -2π, and $\arg w_0(1) = 0 - 2\pi = -2\pi$.

Answer. $\sqrt{1} = 1$.

c) the given curve is a circle of unit radius, whose centre is moved to the point $z = 2$ (cf., **246**(a)). This curve does not turn at all around the point $z = 0$, therefore the variation of $\arg z(t)$ vanishes. It follows that the variation of $\arg w_0(t)$ is also equal to zero.

Answer. $\sqrt{1} = 1$.

283. Let $w_0(t)$ be the continuous image of the curve C under the mapping $w = \sqrt{z}$. Since $z(1) = z(0)$, either $w_0(1) = w_0(0)$ or $w_0(1) = -w_0(0)$. In order to have $w_0(1) = w_0(0)$, it is necessary and sufficient that $\arg w_0(t)$ be equal to $2\pi k$, where k is any integer. To obtain this the variation of $\arg z(t)$ must be equal to $4\pi k$ (cf., **280**), i.e., the curve C must turn $2k$ times around the point $z = 0$.

284. Let L_1 and L_2 be the continuous images of the curves C_1 and C_2 under the mapping $w = \sqrt{z}$. If the curves L_1 and L_2 start from the same point w_0 (Figure 66) then the curve $L_1^{-1}L_2$ is a continuous image of the curve $C_1^{-1}C_2$. The ends of the curve $L_1^{-1}L_2$ (points A and B) will coincide if and only if the curve $C_1^{-1}C_2$ turns around the point $z = 0$ an even number of times (cf., **283**).

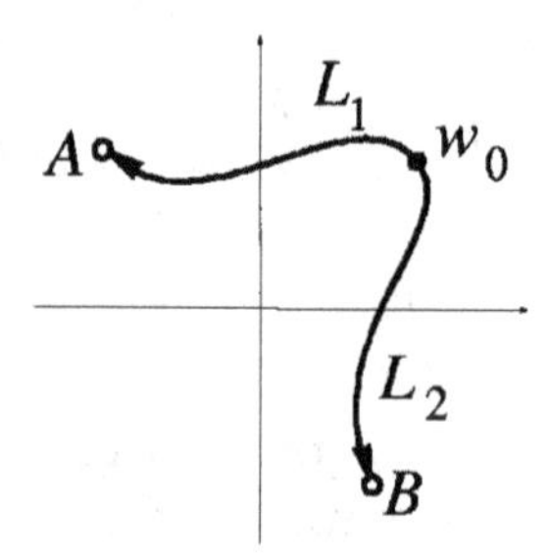

FIGURE 66

285. Let C be a curve not traversing the cut and joining the points z_0 and z_1. Suppose that by choosing two distinct values at the point z_0 and defining the function $\sqrt{z}$ by continuity along the curve C we obtain the same value at the point z. Consider thus the curve C^{-1}, i.e., the curve C oriented in the opposite way. We obtain that the value at the initial point of C^{-1} (z) is the same in both cases, but the values at the final point (z_0), defined by continuity, are different. This is not possible by virtue of the uniqueness of the continuous image, because the curve C does not pass through the point $z = 0$. It follows that our claim that ${}_1\sqrt{z} ={}_2\sqrt{z}$ is not true.

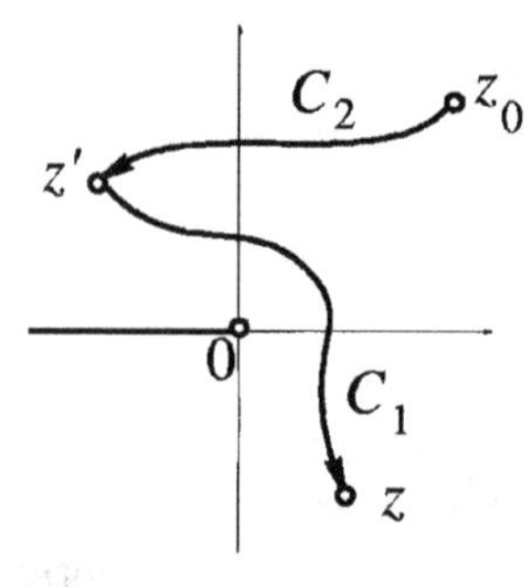

FIGURE 67

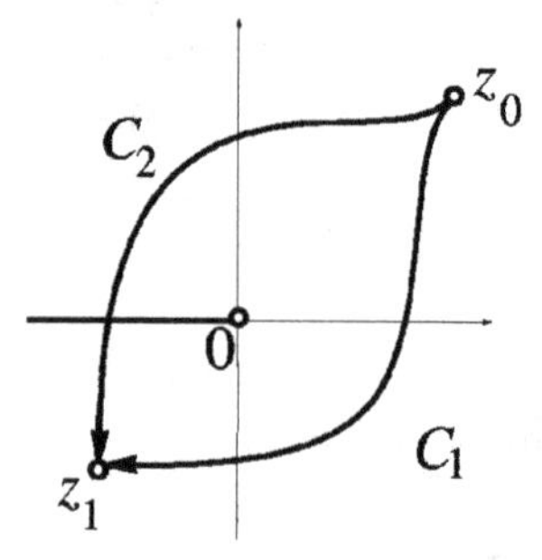

FIGURE 68

286. Let z be an arbitrary point outside the cut, and let C_1 be a continuous curve starting from z' and ending at z without crossing the cut. Let us draw another curve C_2, not crossing the cut, going from the point z_0 to the point z' (Figure 67). By hypothesis we have chosen the value $w' ={}_1\sqrt{z'}$. This means that if we choose $\sqrt{z_0} = w'_0$ and we define $\sqrt{z'}$ by continuity along the curve C_2 we shall obtain exactly w'. But thus the value of $\sqrt{z}$, defined by continuity along the curve C_1 according to the condition $\sqrt{z'} = w'$ coincides, as we easily see, with the value of $\sqrt{z}$ defined by continuity along the curve C_2C_1 according to the condition $\sqrt{z_0} = w'_0$. Hence the value of $\sqrt{z}$ for every z outside the cut is equal to ${}_1\sqrt{z}$.

287. Let C_1 be a continuous curve joining the points z_0 and z_1 and not crossing the cut (Figure 68). Let $w'_1 = \sqrt{z_1}$ be the value of the function $\sqrt{z}$ defined by continuity along the curve C_1 according to the condition $\sqrt{z_0} = w_0$. Since the curve C_1 does not pass through the cut the values w_0 and w'_1 correspond to the same branch of the function $\sqrt{z}$. The curve $C_1^{-1}C_2$ turns once around the point $z = 0$. The values w_1 and w'_1 are thus different (cf., **283**). Since w_0 and w'_1 correspond to the same branch of the function $\sqrt{z}$, w_0 and w_1 correspond to different branches.

288. If $z_0 \neq 0$, then the circle with centre at the point z_0 and with a sufficiently small radius does not turn at all around the point $z = 0$. Therefore the variation of $\arg w_0(t)$ vanishes, and consequently the variation of $\arg w_0(t)$ is zero, i.e., the value of $\sqrt{z}$ does not change. The variation of $\arg z(t)$ along a circle with centre at the point $z = 0$ is equal to 2π. In this case the variation of $\arg w_0(t)$ is equal to π. The value of $\sqrt{z}$ during a turn around the point $z = 0$ thus changes into the opposite value.

289. The curve $z(t)$ is the image of the curve $w_0(t)$ under the mapping $z(w) = w^3$. Consequently if φ_1 is the variation of the argument along the curve $w_0(t)$ then $\varphi = 3\varphi_1$ (cf., **262** (b)), from which $\varphi_1 = \varphi/3$.

290. If $z_0 \neq 0$, then the circle with centre at the point z_0 and with a sufficiently small radius does not turn at all around the point $z = 0$. Consequently the variation of $\arg z(t)$ along this circle vanishes. But thus the variation of $\arg w_0(t)$ also vanishes (cf., **289**), i.e., the value of function $w = \sqrt[3]{z}$ does not vary. This means that none of the points $z \neq 0$ is a branch point.

The variation of $\arg z(t)$ along a circle with centre at the point $z = 0$ is equal to 2π. Thus the variation of $\arg w_0(t)$ is equal to $2\pi/3$. The value of the function $w = \sqrt[3]{z}$ after a simple turn around the point $z = 0$ turns out to be multiplied by $\epsilon_3 = \cos(2\pi/3) + i\sin(2\pi/3)$, i.e., the point $z = 0$ is a branch point of the function $\sqrt[3]{z}$.

Answer. $z = 0$.

291. Let $z(t)$ be a continuous curve, not crossing the cut and joining the point $z = 1$ to the given point. Let $w_0(t)$ be the continuous image of this curve under the mapping $w = \sqrt[3]{z}$ and let $\arg w_0 = \varphi_0$ be chosen. If the variation of $\arg z(t)$ is equal to φ, then the variation of $\arg w_0(t)$ is equal to $\varphi/3$ (cf., **289**). Consequently $\arg w_0(1) = \varphi_0 + \varphi/3$. One may choose $\varphi_0 = 0$ for the branch $f_1(z)$, $\varphi_0 = 2\pi/3$ for the branch $f_2(z)$, and $\varphi_0 = -2\pi/3$ for the branch $f_3(z)$.

a) $\varphi_0 = 0$, $\varphi = \pi/2$, $\arg w_0(1) = \pi/6$.
Answer. $f_1(i) = \cos(\pi/6) + i\sin(\pi/6) = \sqrt{3}/2 + i/2$.
b) $\varphi_0 = 2\pi/3$, $\varphi = \pi/2$, $\arg w_0(1) = 5\pi/6$.
Answer. $f_2(i) = \cos(5\pi/6) + i\sin(5\pi/6) = -\sqrt{3}/2 + i/2$.
c) $\varphi_0 = 0$, $\varphi = 0$, $\arg w_0(1) = 0$.
Answer. $f_1(8) = 2$.
d) $\varphi_0 = -2\pi/3 =$, $\varphi = 0$, $\arg w_0(1) = -2\pi/3$.
Answer. $f_3(8) = 2(\cos(2\pi/3) - i\sin(2\pi/3)) = -1 - i\sqrt{3}$.
e) $\varphi_0 = -2\pi/3$, $\varphi = -\pi/2$, $\arg w_0(1) = -5\pi/6$.
Answer. $f_3(-i) = \cos(5\pi/6) - i\sin(5\pi/6) = -\sqrt{3}/2 - i/2$.

292. As for the function $w = \sqrt{z}$, one proves that after making the cut from the point $z = 0$ to infinity, for example along the negative side of the real axis, the function $\sqrt[3]{z}$ turns out to be decomposed into three single-valued continuous branches. During a simple counterclockwise turn around the point $z = 0$, $\arg w$ varies by $2\pi/3$, during a double turn by $4\pi/3$, and only after a triple turn around the point $z = 0$ does the value of the function $\sqrt[3]{z}$ come back to its initial value. The scheme of the Riemann surface of the function $\sqrt[3]{z}$ has thus the form shown in Figure 69. The Riemann surface is represented in Figure 70[2].

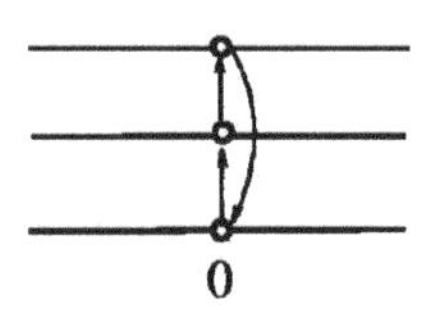

FIGURE 69

293. Suppose first that the curve C does not pass through the point $z = 0$. Let $\varphi(t)$ be a function which describes the continuous variation of $\arg z(t)$ (cf., Theorem 6, §2.7), and let $r(t) = |z(t)|$. Thus $\varphi(t)$ and $r(t)$

[2]To know the meaning of the details of this figure see the section: Drawings of Riemann surfaces.

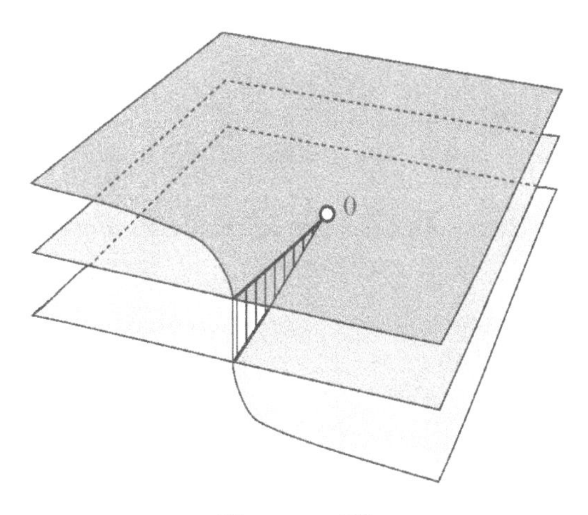

FIGURE 70

are continuous functions and

$$z(t) = r(t)(\cos\varphi(t) + i\sin\varphi(t)).$$

Let $\rho(t)$ be the positive real value of $\sqrt[n]{r(t)}$. Thus $\rho(t)$ is a continuous function (cf., **243.**) and the n continuous curves with parametric equations

$$w_k(t) = \rho(t) = \left(\cos\left(\frac{\varphi(t)}{n} + \frac{2\pi k}{n}\right) + i\sin\left(\frac{\varphi(t)}{n} + \frac{2\pi k}{n}\right)\right),$$
$$k = 0, 1, \ldots, n-1,$$

are the continuous images of the curve $z(t)$ under the mapping $w(z) = \sqrt[n]{z}$ (cf., **229**). Since $w_k(0)$ on these curves takes all values of $\sqrt[n]{z(0)}$, one of these curves will begin at the point w_0.

If the curve C passes through the point $z = 0$, then the points at which $z(t) = 0$ divide the curve C into segments. In this case we have, as before, a continuous image for every segment of the curve, and we take thus as the initial segment the image which starts from the point w_0. If $z(t) = 0$ then $w(z(t)) = 0$ also. Hence the images obtained can be joined in one unique continuous curve, which is the required curve.

294. See solution **280** and **289**. *Answer.* φ/n.

295. See solution **290**. *Answer.* $z = 0$.

296. $w_0(t)$, for every t, is one of the values of $\sqrt[n]{z(t)}$. All the values of $\sqrt[n]{z(t)}$ for a given t are $w_0(t)$, $w_0(t)\cdot\epsilon_n$, $w_0(t)\cdot\epsilon_n^2, \ldots, w_0(t)\cdot\epsilon_n^{n-1}$ (cf.,

232). Since the value of $\sqrt[n]{z}$ at the initial point of the curve $z(t)$ can be chosen in n different ways, we have exactly n continuous curves, images of the curve $z(t)$ under the mapping $w = \sqrt[n]{z}$ (the uniqueness may be lost only at the point $z = 0$, but the curve $z(t)$ does not pas through this point).

Answer. The n continuous images are the curves $w_0(t)$, $w_1(t) = w_0(t) \cdot \epsilon_n$, $w_2(t) = w_0(t) \cdot \epsilon_n^2, \dots, w_{n-1}(t) = w_0(t) \cdot \epsilon_n^{n-1}$.

297. Let $z(t)$ be an arbitrary curve joining the point 1 to an arbitrary point without crossing the cut. From the solution of Problem **296**, we obtain that if the value of the function at the initial point of this curve is multiplied by ϵ_n^k then the values at the final point, defined by continuity, turns out to be multiplied by ϵ_n^k. Hence $f_i(z) = f_0(z) \cdot \epsilon_n^i$.

Answer. $f_i(z) = f_0(z) \cdot \epsilon_n^i$.

298. Solving Problem **297** we have found that the n branches of the function $\sqrt[n]{z}$ are related each other in this way: $f_i(z) = f_0(z) \cdot \epsilon_n^i$ $(i = 0, 1, \dots, n-1)$. The unique branch point of the function $\sqrt[n]{z}$ is the point $z = 0$. During a simple turn around this point the argument of the function $\sqrt[n]{z}$ changes by $2\pi/n$ (cf., **294**), i.e., the value of the function $\sqrt[n]{z}$ varies by ϵ_n. Consequently the scheme of the Riemann surface of the function $\sqrt[n]{z}$ has the form shown in Figure 71 (the Riemann surface for $n = 6$ is shown in Figure 119).

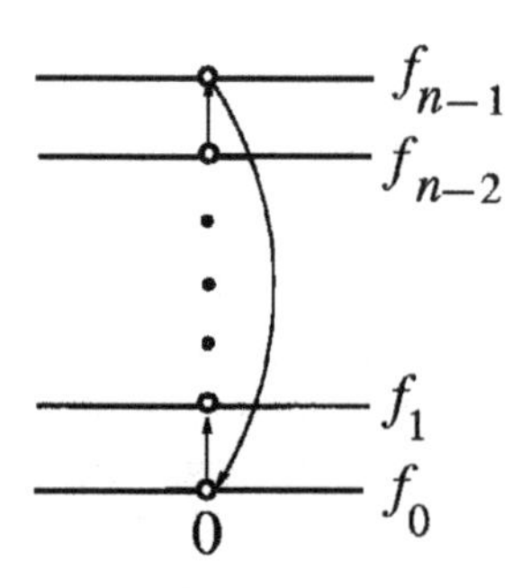

FIGURE 71

299. During a turn around the point $z = 1$, $\arg(z-1)$ varies by 2π and it does not change during any other turn around the other points (along sufficiently small circles). During a turn around the point $z = 1$ $\arg\sqrt{z-1}$ thus changes by π and it remains constant during any other turn around the other points. Consequently the sole branch point is $z = 1$, turning around which the value of the function $\sqrt{z-1}$ turns out to be multiplied by -1. As for the function $\sqrt{z}$, one proves that after have made an arbitrary cut from the point $z = 1$ to infinity the image plane turns out to be decomposed into two single-valued continuous branches of the function $\sqrt{z-1}$. The scheme of the Riemann surface of the function $\sqrt{z-1}$ is shown in Figure 72.

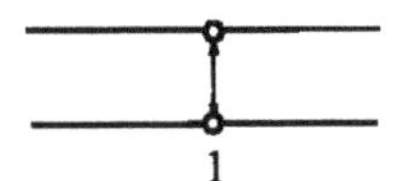

FIGURE 72

300. See the solution **299**. The sole branch point is the point $-i$ (because $z + i = z - (-i)$), turning round which the function $\sqrt[n]{z+i}$ turns out to be multiplied by ϵ_n. The scheme of the Riemann surface of the function $\sqrt[n]{z+i}$ is shown in Figure 73.

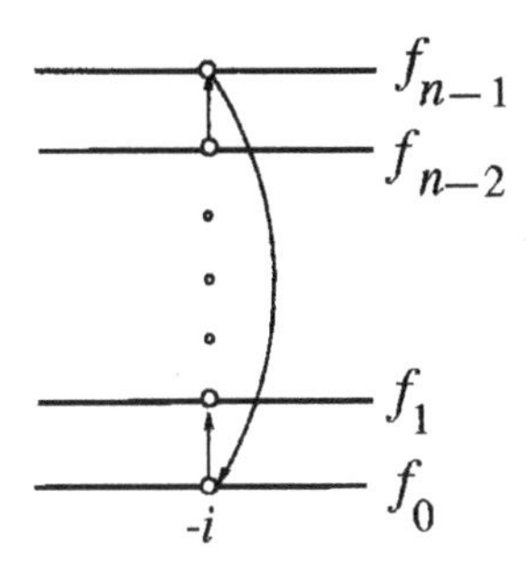

FIGURE 73

301. *Hint.* Consider the mapping $w = \sqrt[n]{f(z)}$ as the composition of two mappings: $\tau = f(z)$ and $w = \sqrt[n]{\tau}$ (cf., **293**).

302. The mapping $w(z) = \sqrt[n]{f(z)}$ can be imagined as the composition of two mappings:

$$\tau = f(z) \quad \text{and} \quad w = \sqrt[n]{\tau}.$$

If C is a continuous curve on the z plane then on the τ plane there is only one image $C' = f(C)$ of this curve. Since $f(z)$ is a continuous function,

C' is a continuous curve. If the curve C'' with equation $w_0(t)$ is one of the continuous images of the curve C' under the mapping $w(\tau) = \sqrt[n]{\tau}$, then the curves with the equations $w_i(t) = w_0(t)\epsilon_n^i$ $(i = 1, 2, \dots, n - 1)$ are, as well, continuous images of the curve C' under the mapping $w(\tau) = \sqrt[n]{\tau}$ (cf., **296**), and, consequently, they are continuous images of the curve C under the mapping $w(z) = \sqrt[n]{f(z)}$. Thus if the value of the function $w(z) = \sqrt[n]{f(z)}$ at the initial point is multiplied by ϵ_n^i the value at the final point of the curve C, defined by continuity, will be multiplied by ϵ_n^i. Therefore if $w_0(z)$ is a continuous single-valued branch of the function $\sqrt[n]{f(z)}$, then all continuous single-valued branches are obtained by multiplying $w_0(z)$ by ϵ_n^i $(i = 1, 2, \dots, n - 1)$.

Answer. $w_0(z)$, $w_1(z) = w_0(z) \cdot \epsilon_n$, $w_2(z) = w_0(z) \cdot \epsilon_n^2, \dots, w_{n-1}(z) = w_0(z) \cdot \epsilon_n^{n-1}$.

303. a) During a turn around the point $z = 0$ or $z = i$, $\arg z(z - i)$ varies by 2π (cf., **260**), and $\arg \sqrt{z(z - i)}$ varies by π (cf., **280**), i.e., the value of the function $\sqrt{z(z - i)}$ is multiplied by -1. To separate the single-valued continuous branches of the function $\sqrt{z(z - i)}$ it suffices to make two cuts respectively from the point $z = 0$ and from the point $z = i$ to infinity (the proof is the same as for the function $w = \sqrt{z}$). The scheme of the function $\sqrt{z(z - i)}$ is shown in Figure 74 (the Riemann surface is shown in Figure 120a).

b) See Figure 75 (the Riemann surface is shown in Figure 120b). *Hint.* $z^2 + 1 = (z - i)(z + i)$.

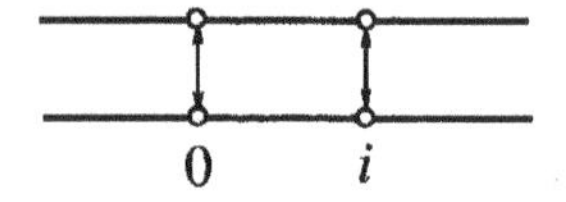

FIGURE 74

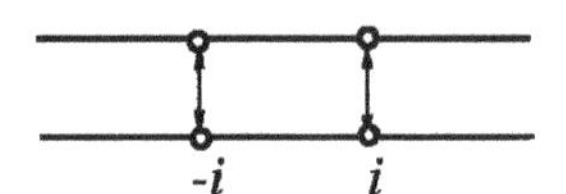

FIGURE 75

304. See **303**. a) Since $z^2 - 1 = (z - 1)(z + 1)$, during a turn around each of the points $z = 1$ and $z = -1$, the value of the function $\sqrt[3]{z^2 - 1}$ is multiplied by $\epsilon_3 = \cos 2\pi/3 + i \sin 2\pi/3$. The scheme sought is shown in Figure 76 (the Riemann surface is shown in Figure 121).

b) After a turn around the point $z = 0$ the value of the function $\sqrt[3]{(z - 1)^2 z}$ turns out to be multiplied by ϵ_3. During a turn around the point $z = 1$ $\arg (z - 1)^2 z$ varies by 4π and $\arg \sqrt[3]{(z - 1)^2 z}$ varies by $4\pi/3$, i.e., the value of the function $\sqrt[3]{(z - 1)^2 z}$ is multiplied by ϵ_3^2. The scheme sought is shown in Figure 77 (the Riemann surface is shown in Figure 122).

c) See Figure 78. *Hint.* $z^2 + 1 = (z - i)(z + i)$.

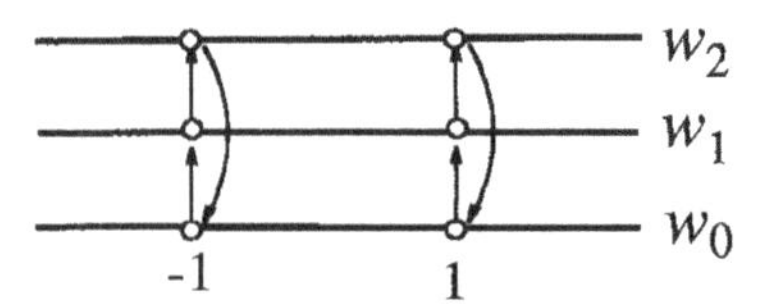

FIGURE 76

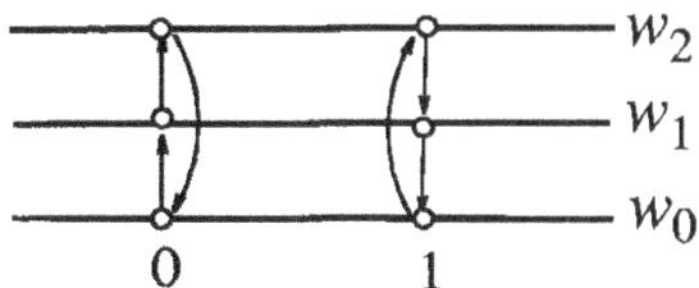

FIGURE 77

FIGURE 78

305. The single-valued continuous branches of the function $\sqrt{z^2}$, defined on the entire z plane, are $w_0(z) = z$, $w_1(z) = -z$. When one passes through the point $z = 0$, $\arg z^2$ changes by 4π and $\arg \sqrt{z^2}$ by 2π, i.e., the value of the function $\sqrt{z^2}$ does not change. The scheme sought consists of two disjoints sheets.

306. The problem is solved in the same way as Problem **304**. a) See Figure 79. *Hint.* $z^2 + 2 = (z - i\sqrt{2})(z + i\sqrt{2})$. b) See Figure 80 (the Riemann surface is shown in Figure 123). c) See Figure 81. d) See Figure 82 (the Riemann surface is shown in Figure 124). *Hint.* $(z^2 - 1)^3(z + 1)^3 = (z - 1)^3(z + 1)^6$. e) See Figure 83 (the Riemann surface is shown in Figure 125). *Hint.* $z^3 - 1 = (z - 1)(z - \epsilon_3)(z - \epsilon_3^2)$.

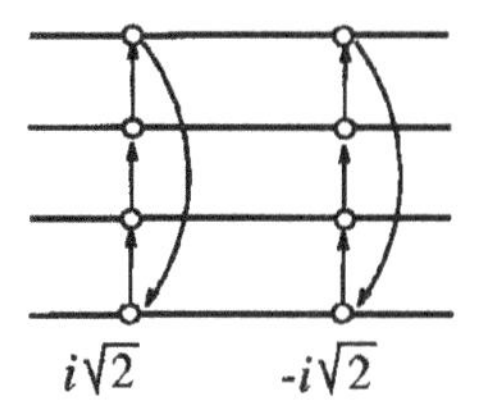

FIGURE 79

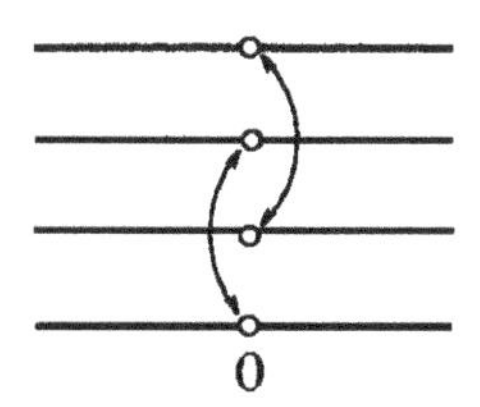

FIGURE 80

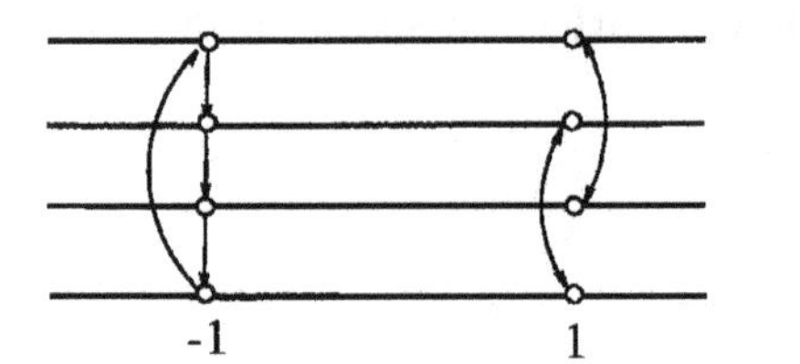

FIGURE 81

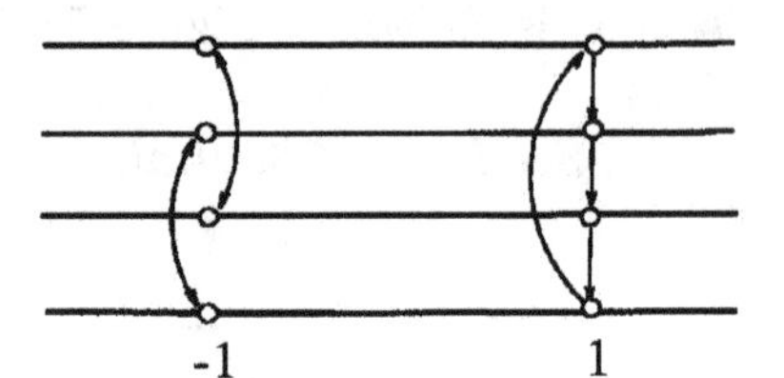

FIGURE 82

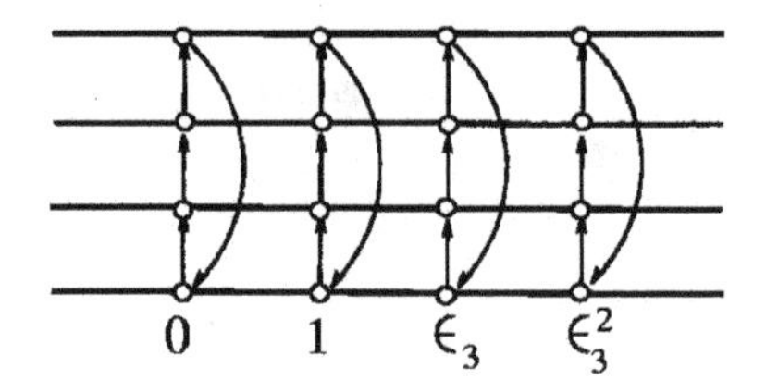

FIGURE 83

307. See Figure 84. *Hint.* During a turn around the point $z = 0$, $\arg z$ varies by 2π, $\arg 1/z$ by -2π (cf., **260**(b)) and $\arg\sqrt{1/z}$ by $-\pi$, i.e., the value of the function $\sqrt{1/z}$ is multiplied by -1.

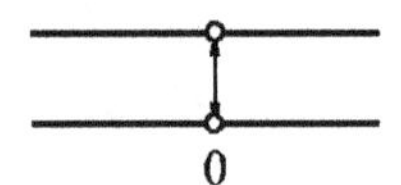

FIGURE 84

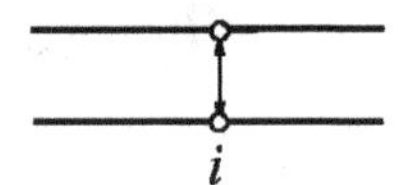

FIGURE 85

308. a) See Figure 85. b) During a turn around the point $z = 1$, $\arg((z-1)/(z+1))$ varies by 2π, and, during a turn around the point $z = -1$ by -2π (cf., **260**). Consequently around the point $z = 1$ the value of the function $\sqrt[3]{(z-1)/(z+1)}$ is multiplied by ϵ_3, and around the point $z = -1$ by $\epsilon_3^{-1} = \epsilon_3^2$. The required scheme is shown in Figure 86. c) See Figure 87 (the Riemann surface is shown in Figure 126).

309. Let a value $w_0 = w(z_0)$ be chosen at the point z_0 and z_1 be another point. If C_1 and C_2 are two arbitrary continuous curves joining z_0 and z_1 without crossing the cuts (Figure 88), then evidently the curve C_1 can be continuously deformed into the curve C_2 without passing through the branch points. Since the function $w(z)$ possesses the monodromy

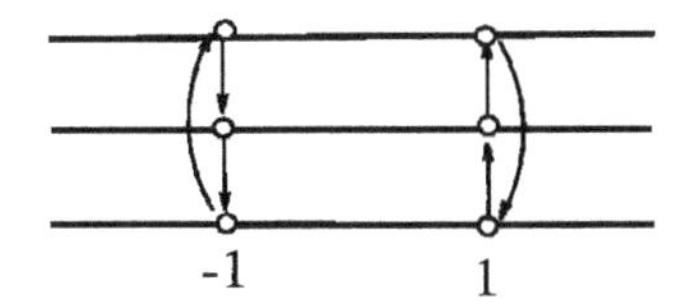

FIGURE 86

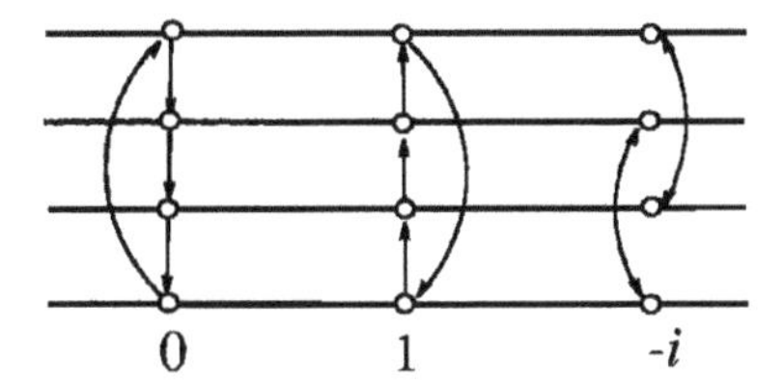

FIGURE 87

property, the values of $w(z_1)$, defined by continuity along the curves C_1 and C_2, coincide. Consequently the value $w(z_1)$ is uniquely defined by continuity along an arbitrary curve joining z_0 and z_1 without crossing the cut.

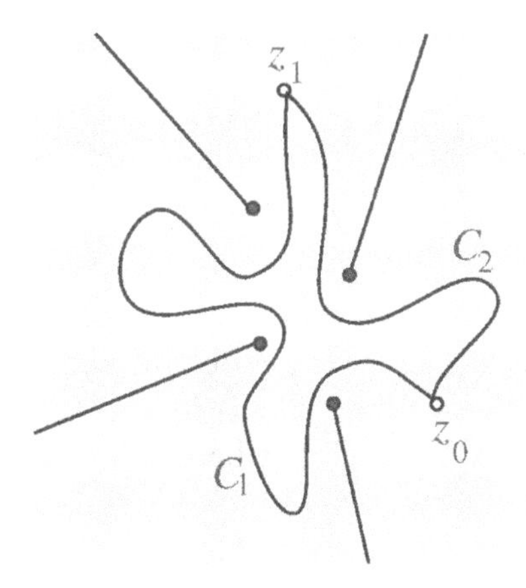

FIGURE 88

310. See Figure 89. Suppose that on moving along AB one moves from the ith branch to the jth branch. We want to know at which branch we will arrive starting from the ith branch and traversing the cut along

CD.

Since the function $w(z)$ has a finite number of branch points then one can choose the curves AB and CD sufficiently short and the curves CA and BD sufficiently close to the cut, in such a way that in the region bounded by the curve $CABDC$ there are no branch points of the function $w(z)$. In this case one can evidently transform the curve $CABD$ into the curve CD without passing through any branch point. Since the function $w(z)$ possesses the monodromy property, the function $w(z)$ at the point D is uniquely defined by continuity along the curves CD and $CABD$. Starting from the ith branch and covering the curve $CABD$, moving first on the ith branch, one passes later to the jth branch, moving finally on it. In this way, along the curve $CABD$ and therefore also along the curve CD, one moves from the ith branch to the jth branch, exactly as along the curve AB.

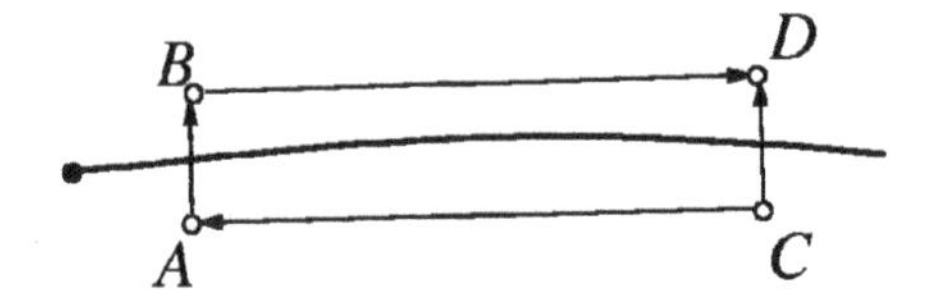

FIGURE 89

311. a) *Hint.* $\sqrt[3]{-8} = -2\epsilon_3^i$ $(i = 0, 1, 2)$; $\sqrt{2i} = \pm(1 + i)$. *Answer.* (Here $\sqrt{3}$ is the positive value of the square root) $-1 + i$, $-3 - i$, $2 + (\sqrt{3} + 1)i$, $(\sqrt{3} - 1)i$, $2 + (1 - \sqrt{3})i$, $-(\sqrt{3} + 1)$; b) $\pm(1/2 + i), \pm 1/2$; c) $\pm(1 + i)$, 0; d) $\pm(1 + i)$; e) $4i$, 0.

312. *Hint.* It suffices to prove that the functions $h(z) = z$ and $h(z) = a$ possess the properties stated by the problem, and that if the functions $f(z)$ and $g(z)$ possess these properties then the functions $f(z) - g(z)$, $f(z) \cdot g(z)$, $f(z)/g(z)$, $[f(z)]^n$, $\sqrt[n]{f(z)}$ (where n is an integer) also possess these properties.

Solution. 1) If $h(z) = z$ then $w_0 = h(z_0) = z_0$. The curve sought is the curve with the parametric equation $w_0(t) = z(t)$, where $z(t)$ is the parametric equation of the curve C.

2) If $h(z) = a$ then $w_0 = a$ and the required curve is the curve with the equation $w_0(t) = a$ (degenerated to a point).

3) Suppose that $h(z) = f(z) + g(z)$ and that the statement of the problem is true for the functions $f(z)$ and $g(z)$. By the definition of the

sum of multi-valued functions we have $w_0 = w_0' + w_0''$, where w_0' is one of the values of $f(z_0)$ and $w''(0)$ is one of the values of $g(z_0)$. Since for the functions $f(z)$ and $g(z)$ the statement of the problem holds, there exist two continuous images $C' = f(C)$ and $C'' = g(C)$, starting respectively from the points w_0' and w_0''. If $w'(t)$ and $w''(t)$ are the parametric equations of the curves C' and C'' then the function $w_0(t) = w'(t) + w''(t)$ (which is continuous, being the sum of continuous functions) is the parametric equation of the required curve because $w_0(0) = w'(0)+w''(0) = w_0'+w_0'' = w_0$. In an identical way one considers the case in which $h(z) = f(z)-g(z)$, $h(z) = f(z) \cdot g(z)$, $h(z) = f(z)/g(z)$ (in this last case the continuous function sought is $w_0(t) = w'(t)/w''(t)$, because by hypothesis the curve C does not pass through the points at which the function $h(z)$ is not defined, and consequently $w''(t) \neq 0$.

4) Suppose that $h(z) = \sqrt[n]{f(z)}$ and that for $f(z)$ the statement of the problem is true. By the definition of the function $\sqrt[n]{f(z)}$ we have $w_0^n = \tau_0$, where τ_0 is one of the values of $f(z_0)$. The mapping $h(z) = \sqrt[n]{f(z)}$ can be considered as the composition of two mappings, $\tau = f(z)$ and $w = \sqrt[n]{\tau}$. Since for the function $f(z)$ the statement of the problem holds, there exists at least one continuous image C' of the curve C under the mapping $\tau = f(z)$ beginning at the point τ_0. By virtue of the result of Problem **293**, there exists at least one continuous image C'' of the curve C' under the mapping $w = \sqrt[n]{\tau}$, beginning at the point w_0. The curve C'' is the curve required.

313. At the point z_0, chosen arbitrarily, the function $h(z)$ takes nm values: $h_{i,j}(z_0) = f_i(z_0) + g_j(z_0)$, where $i = 1, \ldots, n$; $j = 1, \ldots, m$. Since the sum of continuous functions is a continuous function, the single-valued continuous branches of the function $h(z)$ are the nm following functions: $h_{i,j}(z) = f_i(z) + g_j(z)$, where $i = 1, \ldots, n$; $j = 1, \ldots, m$.

314. a) See Figure 90. *Hint.* Use the schemes of the Riemann surfaces of the functions $\sqrt{z}$ and $\sqrt{z-1}$ (cf., **288**, **299**). b) See Figure 91. *Hint.* Cf., **304**, **307**. c) See Figure 92. *Hint.* Cf., **288**, **292**. d) See Figure 93 (the Riemann surface is shown in Figure 127). *Hint.* Draw first the schemes of the Riemann surfaces of the functions $\sqrt{z^2-1}$ and $\sqrt[4]{z-1}$.

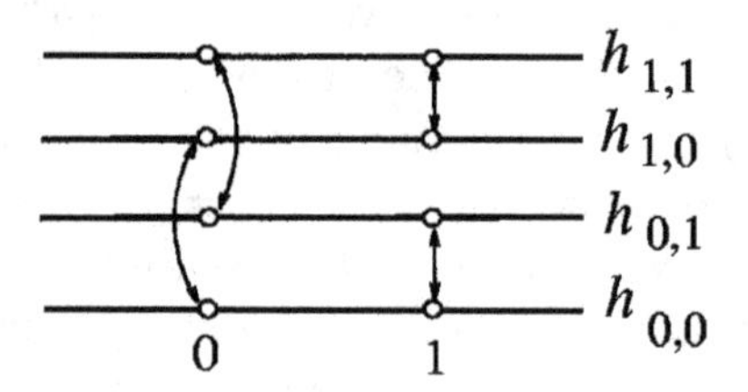

FIGURE 90

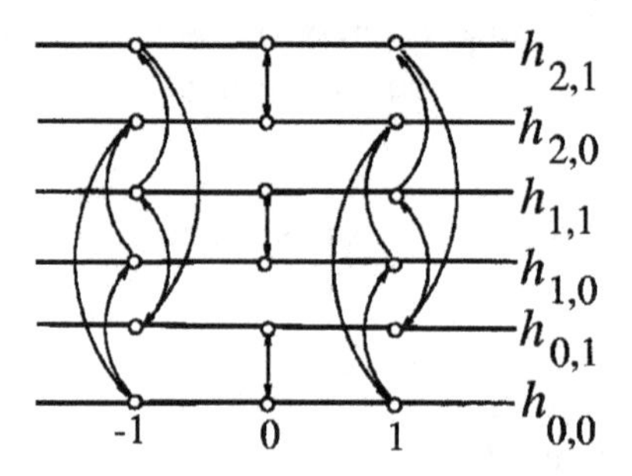

FIGURE 91

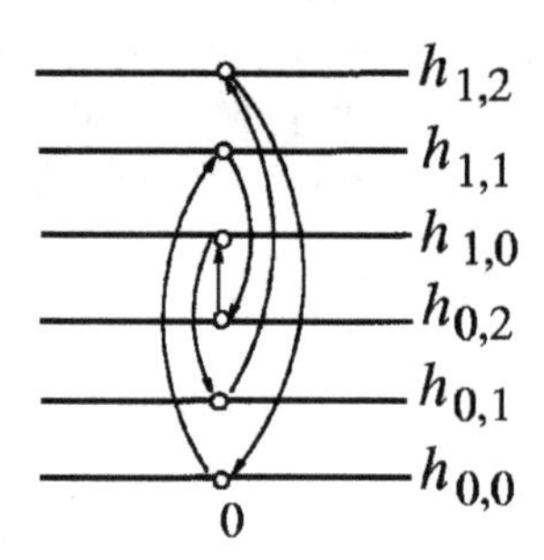

FIGURE 92

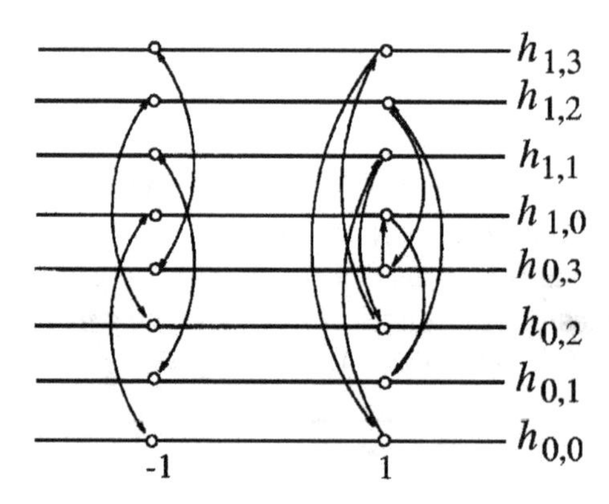

FIGURE 93

315. *Answer.* a) Three values: $2, -2, 0$. b) Seven values: $2, 0, 1+i, -1+i, -2, 1-i, -1-i$. c) Six values: $2, 2\epsilon_3, 2\epsilon_3^2, 1+\epsilon_3, 1+\epsilon_3^2, -1$.

316. a) Let $f_0(z)$ and $f_1(z) = -f_0(z)$ be the single-valued continuous branches of the function $\sqrt{z}$. The scheme of the Riemann surface of the function $h(z) = \sqrt{z}+\sqrt{z}$, built by the formal method, is shown in Figure 94. The branches $h_{0,1}(z) = f_0(z)+f_1(z) \equiv 0$ and $h_{1,0}(z) = f_1(z)+f_0(z) \equiv 0$ coincide. To obtain the correct scheme of the Riemann surface of the function $h(z) = \sqrt{z} + \sqrt{z}$ we therefore have to identify the branches $h_{0,1}$ and $h_{1,0}$. This scheme is shown in Figure 95.

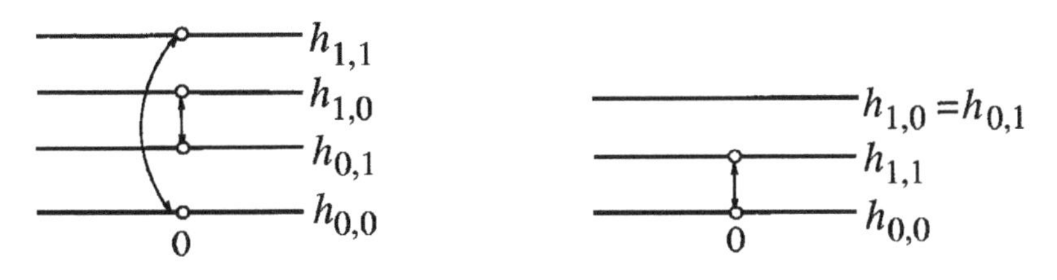

FIGURE 94 FIGURE 95

b) Let $f_0(z)$ and $f_1(z) = -f_0(z)$ be the single-valued continuous branches of the function $\sqrt{z}$. Thus $[f_0(z)]^2 = z$ and $[f_0(z)]^4 = z^2$. Consequently $f_0(z)$ is one of the single-valued continuous branches of the function $\sqrt[4]{z^2}$. The branches of this function are: $g_0(z) = f_0(z)$, $g_1(z) = i \cdot f_0(z)$, $g_2(z) = -f_0(z)$, $g_3(z) = -i \cdot f_0(z)$. The scheme of the Riemann surface of the function $h(z) = \sqrt{z} + \sqrt[4]{z^2}$, built by the formal method, is shown in Figure 96. The correct scheme (Figure 97) is obtained by identifying the coincident branches $h_{0,2}(z) \equiv 0$ and $h_{1,0}(z) \equiv 0$.

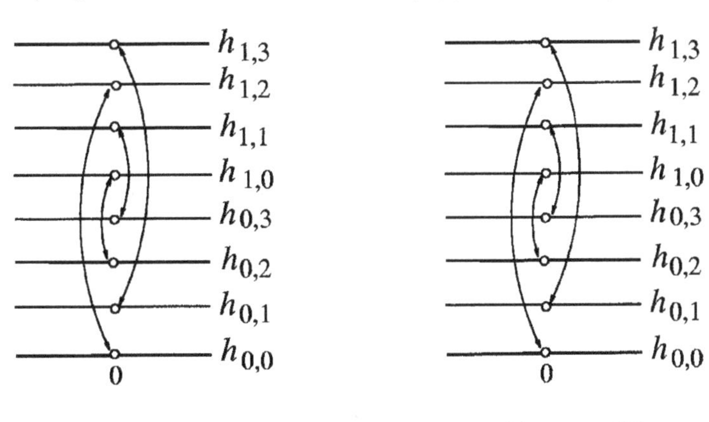

FIGURE 96 FIGURE 97

c) Let $f_0(z)$ be one of the single-valued continuous branches of the function $\sqrt[3]{z}$. Thus the branches are $f_0(z)$, $f_1(z) = f_0(z) \cdot \epsilon_3$, $f_2(z) = f_0(z) \cdot \epsilon_3^2$. The scheme of the Riemann surface of the function $h(z) = \sqrt[3]{z} + \sqrt[3]{z}$, built by the formal method, is shown in Figure 98. To obtain the correct scheme (Figure 99) one has to identify the following coincident branches: $h_{0,1}(z)$ and $h_{1,0}(z)$, $h_{0,2}(z)$ and $h_{2,0}(z)$, $h_{1,2}(z)$ and $h_{2,1}(z)$.

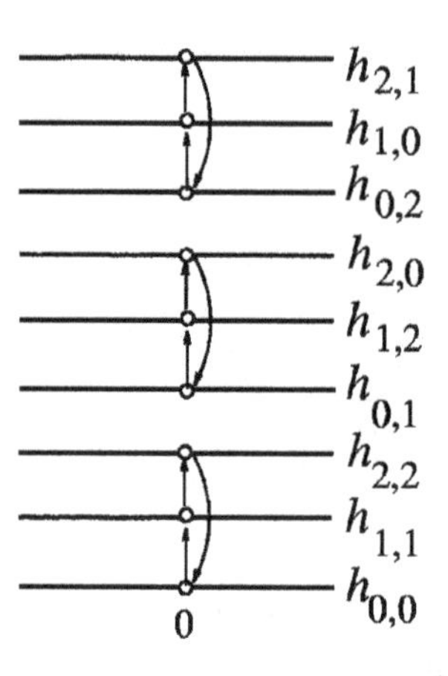

FIGURE 98

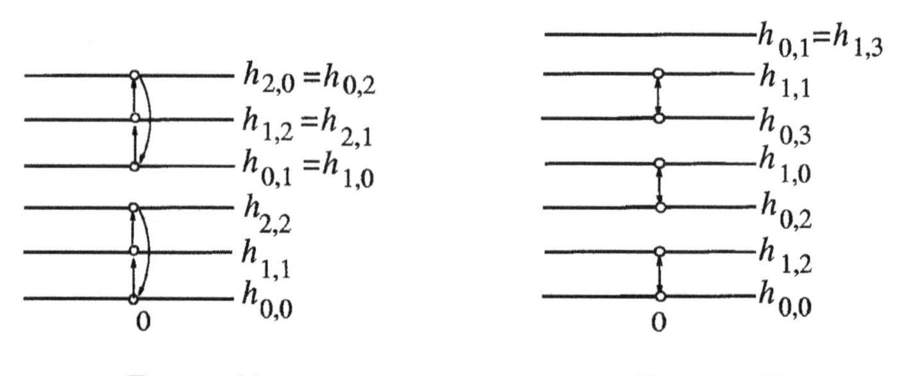

FIGURE 99

FIGURE 100

317. a) Let $f_0(z)$ and $f_1(z) = -f_0(z)$ be the single-valued continuous branches of the function $\sqrt{z}$. Thus $[f_0(z)]^2 = z$ and $[f_0(z)]^4 = z^2$. The single-valued continuous branches of the function $g(z) = \sqrt[4]{z^2}$ are therefore $g_0(z) = f_0(z)$, $g_1(z) = i \cdot f_0(z)$, $g_2(z) = -f_0(z)$, $g_3(z) = -i \cdot f_0(z)$. One builds the scheme of the Riemann surface of the function $h(z) =$

$i\sqrt{z} - \sqrt[4]{z^2}$ by the formal method, afterwards one identifies the following coincident branches: $h_{0,1}(z) = h_{1,3}(z) \equiv 0$. The remaining branches are distinct: it suffices to calculate their values at the point $z = 1$. The correct scheme is shown in Figure 100.

b) Let $f_0(z)$ and $f_1(z) = -f_0(z)$ be the single-valued continuous branches of the function $\sqrt{z-1}$, and let $g_0(z)$, $g_1(z) = i \cdot g_0(z)$, $g_2(z) = -g_0(z)$, $g_3(z) = -i \cdot g_0(z)$ be the single-valued continuous branches of the function $g(z) = \sqrt[4]{z}$. We build the scheme of the Riemann surface of the function $h(z) = \sqrt{z-1} \cdot \sqrt[4]{z}$ by the formal method (Figure 101) and we identify the coincident branches: $h_{0,0}(z) = h_{1,2}(z)$, $h_{0,1}(z) = h_{1,3}(z)$, $h_{0,2}(z) = h_{1,0}(z)$, $h_{0,3}(z) = h_{1,1}(z)$. The remaining branches are all distinct: it suffices to calculate their values at the point $z = 2$. The correct scheme shown in Figure 102.

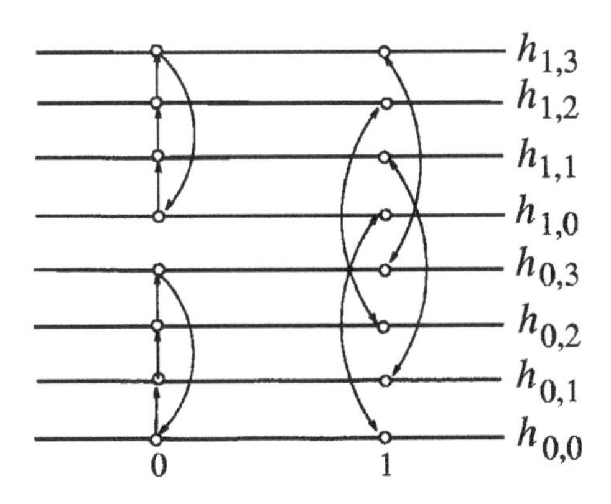

FIGURE 101

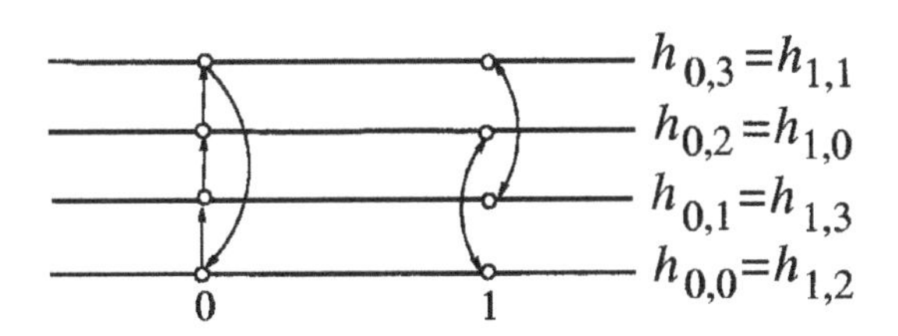

FIGURE 102

c) See Figure 103. The solution is similar to the solution of the case (b).

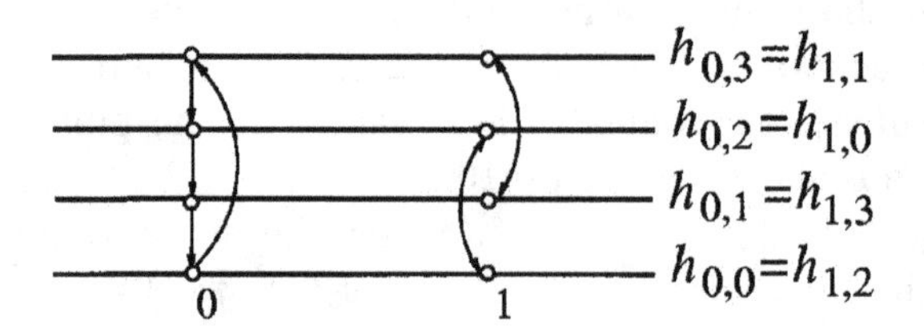

FIGURE 103

d) The function $f(z) = \sqrt{z} + \sqrt{z}$ has 3 single-valued continuous branches: $f_0(z) \equiv 0$, $f_1(z)$ and $f_2(z) = -f_1(z)$ (cf., solution **316**(a)). The function $g(z) = \sqrt[3]{z(z-1)}$ has 3 single-valued continuous branches as well: $g_0(z)$, $g_1(z) = \epsilon_3 \cdot g_0(z)$, $g_2(z) = \epsilon_3^2 \cdot g_0(z)$. The branches $h_{0,0}(z)$, $h_{0,1}(z)$ and $h_{0,2}(z)$ coincide: $h_{0,0}(z) = h_{0,1}(z) = h_{0,2}(z) \equiv 0$. The remaining branches are all distinct: it suffices to calculate their values at the point $z = 2$. The correct scheme is shown in Figure 104 (the Riemann surface is shown in Figure 128).

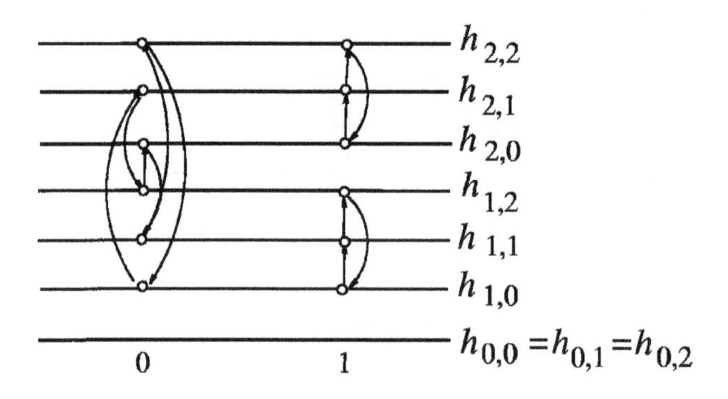

FIGURE 104

318. *Answer.* The single-valued continuous branches are the functions $h_i(z) = [f_i(z)]^n$, where $i = 1, 2, \dots, m$.

319. a) If $f_0(z)$, $f_1(z) = i \cdot f_0(z)$, $f_2(z) = -f_0(z)$, $f_3(z) = -i \cdot f_0(z)$ are the single-valued continuous branches of the function $f(z) = \sqrt[4]{z}$, then $[f_0(z)]^2 = [f_2(z)]^2$ and $[f_1(z)]^2 = [f_3(z)]^2$. The correct scheme is shown in Figure 32 (§2.9).

b) Cf., **316**(a). If $f_0 \equiv 0$, $f_1(z)$ and $f_2(z) = -f_1(z)$ are the single-valued continuous branches of the function $f(z) \equiv \sqrt{z}+\sqrt{z}$, then $[f_1(z)]^2 =$

$[f_2(z)]^2$. The correct scheme is shown in Figure 105.

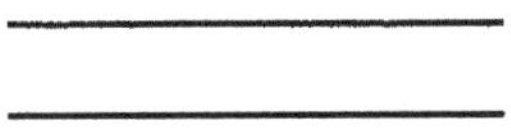

FIGURE 105

c) If $f_0(z)$ is one of the single-valued continuous branches of the function $f(z) = \sqrt{z} \cdot \sqrt[3]{z-1}$ then the branches are: $f_0(z)$, $f_1(z) = f_0(z) \cdot \epsilon_3$, $f_2(z) = f_0(z) \cdot \epsilon_3^2$, $f_3(z) = -f_0(z)$, $f_4(z) = -f_0(z) \cdot \epsilon_3$, $f_5(z) = -f_0(z) \cdot \epsilon_3^2$. Therefore $[f_0(z)]^3 = [f_1(z)]^3 = [f_2(z)]^3$ and $[f_3(z)]^3 = [f_4(z)]^3 = [f_5(z)]^3$. The required scheme is shown in Figure 32 (§2.9).

320. Suppose that the point z_0 is not a branch point of the function $f(z)$. Thus, by one turn around the point z_0 along a circle of radius sufficiently small, the value of $f(z)$ does not change.

Suppose all values of $f(z_0)$ be different from 0. Thus the continuous images under the mapping $w = f(z)$ of circles with centre at z_0 and with radii sufficiently small are closed continuous curves lying close to the point $w = f(z_0)$. Since all values of $f(z_0)$ are different from 0, all these curves–images avoid the point $w = 0$ for circles with sufficiently small radii. It follows that $\arg f(z)$ does not change. But thus also the value of the function $\sqrt[n]{f(z)}$ does not change. Hence the only possible branch points of the function $\sqrt[n]{f(z)}$ are the branch points of the function $f(z)$ and the points where one of the values of $f(z)$ vanishes.

Answer. The branch points of $f(z)$ and the points where one of the values of $f(z)$ vanishes.

321. Since $g(z)$ is a continuous function on the plane with the cuts described, $[g(z)]^n$ is also a continuous function. Since for every z $g(z)$ is one of the values of $\sqrt[n]{f(z)}$, $[g(z)]^n$ is, for every z, one of the values of the function $f(z)$. Consequently $[g(z)]^n$ is a single-valued continuous branch of the function $f(z)$ according to the chosen cuts.

322. Cf., **302**. *Answer.* $g(z)$, $g(z) \cdot \epsilon_n$, $g(z) \cdot \epsilon_n^2, \ldots, g(z) \cdot \epsilon_n^{n-1}$.

323. *Hint.* Since $w_0(t)$ is a continuous curve, $w_k(t)$ is also a continuous curve; moreover, $[w_k(t)]^n = [w_0(t)]^n \cdot \epsilon_n^{kn} = [w_0(t)]^n$, and consequently $[w_k(t)]^n$ is equal to one of the values of $f(z(t))$.

324. *Hint.* From the result of Problem **323** it follows that if the value of the function $\sqrt[n]{f(z)}$ at the initial point of the curve C is multiplied by

ϵ_n^k, then the value of the function $\sqrt[n]{f(z)}$ at the final point of the curve C is also multiplied by ϵ_n^k.

325. Let $f_0(z)$ and $f_1(z) = -f_0(z)$ be the single-valued continuous branches of the function $\sqrt{z}$ such that $f_0(1) = 1$ and $f_1(1) = -1$. Thus $f_0(z) - 1$ and $f_1(z) - 1$ are the single-valued continuous branches of the function $\sqrt{z} - 1$. Each one of these values corresponds to two branches of the function $\sqrt{\sqrt{z} - 1}$. If $\sqrt{z} - 1 = 0$ then $\sqrt{z} = 1$ and $z = 1$. Therefore besides the point $z = 0$ only the point $z = 1$ can be a branch point. The branch point can lie only in the pack (of two sheets) corresponding to the branch $f_0(z) - 1$ (because one must have $\sqrt{1} = 1$). We have

$$f_0(z) - 1 = \frac{(f_0(z) - 1)(f_0(z) + 1)}{f_0(z) + 1} = \frac{[f_0(z)]^2 - 1}{f_0(z) + 1} = \frac{z - 1}{f_0(z) + 1}.$$

By one turn around the point $z = 1$ the argument of the denominator does not change, because $f_0(1) + 1 = 2 \neq 0$. The argument of the numerator, by one turn around the point $z = 1$, varies by 2π. Thus $\arg(f_0(z) - 1)$ varies by 2π; $\arg \sqrt{f_0(z) - 1}$ varies by π, and thus the value of $\sqrt{f_0(z) - 1}$ changes. By one turn around the point $z = 0$ the value of the function $\sqrt{z}$ changes, therefore from the pack of two sheets, corresponding to $f_0(z) - 1$, one moves to the pack of two sheets corresponding to $f_1(z) - 1$, and vice versa. By a double turn around the point $z = 0$ the final value of the function $\sqrt{z}$ coincides with the initial value and $\arg(\sqrt{z} - 1)$ does not change (because $\sqrt{z} - 1 \neq 0$). By a double turn around the point $z = 0$ one comes back onto the first sheet. Combining together the results obtained we are able to build the scheme of the Riemann surface of the function $\sqrt{\sqrt{z} - 1}$, shown in Figure 106.

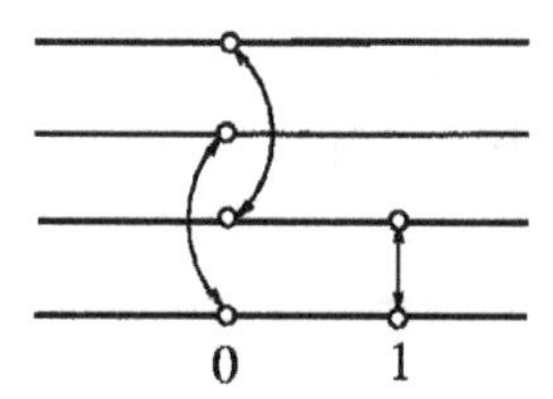

FIGURE 106

326. This problem is solved in the same way as Problem **325**: a) cf., Figure 107; b) cf., Figure 108. *Hint.* If $f_0(z)$ is a single-valued continuous branch of the function $\sqrt[3]{z}$ and $f_0(1) = 1$, then

$$f_0(z) - 1 = \frac{f_0^3 - 1}{f_0^2 + f_0(z) + 1} = \frac{z - 1}{f_0^2(z) + f_0(z) + 1},$$

where $f_0^2(1) + f_0(1) + 1 = 3 \neq 0$.

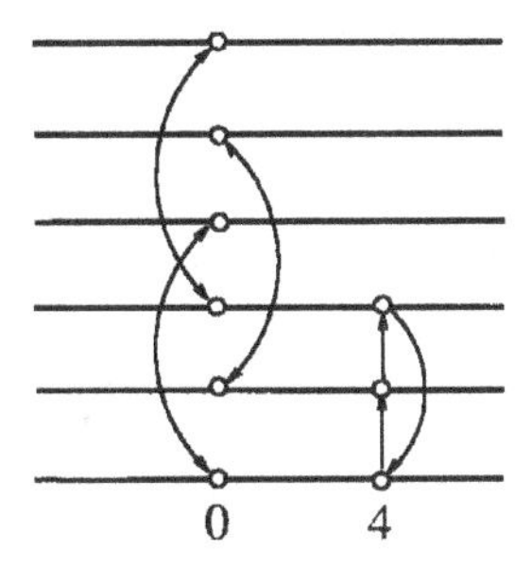

FIGURE 107

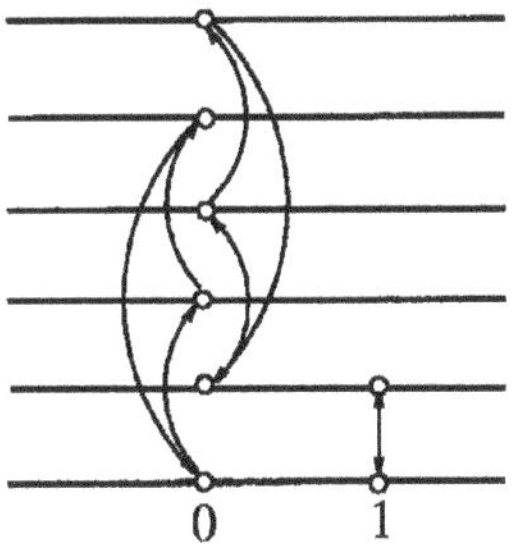

FIGURE 108

327. Since $z^2 + 1 = (z - i)(z + i)$, the value of $\sqrt{z^2+1}$ changes as a consequence of one turn around the point $z = i$ and of one turn round $z = -i$, i.e., these points are the branch points of the function $\sqrt{z^2+1} - 2$. The scheme in both cases (a) and (b) is shown in Figure 75 (Solution **303**).

If $\sqrt{z^2+1} - 2 = 0$ then $z^2 + 1 = 4$ and $z = \pm\sqrt{3}$ ($\sqrt{3}$ is the positive value of the root). Let $f_0(\sqrt{3}) = 0$, i.e., in this case we had chosen $\sqrt{4} = 2$. We are looking for the value of $f_0(-\sqrt{3})$. Join the point $z = \sqrt{3}$ to the point $z = -\sqrt{3}$ by a continuous curve not crossing the cuts.

In the case (a) one can take, for example, the segment joining the points $z = \sqrt{3}$ and $z = \sqrt{-3}$. It is easy to see that on moving along this segment $\arg(z + i) = \arg(z - (-i))$ increases by $2\pi/3$, whereas $\arg(z - i)$ decreases by $2\pi/3$. Therefore $\arg(z^2 - 1)$ does not change and consequently the value of $\sqrt{z^2+1} - 2$ does not change. In this way in the case (a) $f_0(-\sqrt{3}) = f_0(\sqrt{3}) = 0$.

In the case (b) on moving along an arbitrary curve joining the points $z = \sqrt{3}$ and $z = \sqrt{-3}$ and not crossing the cuts, $\arg(z + i)$ increases by $2\pi/3$, and $\arg(z - i)$ increases by $4\pi/3$. So $\arg(z^2 + 1)$ increases by 2π and $\arg\sqrt{z^2+1}$ increases by π: the value of $\sqrt{z^2+1}$ changes into the opposite. Therefore in the case(b) $f_0(-\sqrt{3}) = -2 - 2 = -4 \neq 0$, whereas $f_1(-\sqrt{3}) = 2 - 2 = 0$.

328. Let $g_0(z)$ and $g_1(z)$ be the single-valued continuous branches of the function $g(z) = \sqrt{z^2+1}$, such that $g_0(z) - 2 = f_0(z)$ (cf., solution **327**) and $g_1(z) - 2 = f_1(z)$. In the case(a) one has $g_0(\sqrt{3}) = g_0(-\sqrt{3}) = 2$ (cf., solution **327**) and

$$g_0(z) - 2 = \frac{g_0^2(z) - 4}{g_0(z) + 2} = \frac{z^2 - 3}{g_0(z) + 2} = \frac{(z - \sqrt{3})(z + \sqrt{3})}{g_0(z) + 2}.$$

By one turn around the points $z = \sqrt{3}$ and $z = -\sqrt{3}$ the argument of the denominator does not change because $g_0(\sqrt{3}) + 2 = g_0(-\sqrt{3}) + 2 = 4 \neq 0$, whereas the argument of the denominator increases by 2π. Therefore $\arg(g_0(z) - 2)$ increases by 2π and $\arg\sqrt{g_0(z) - 2}$ increases by π: then the value of $\sqrt{f_0(z)}$ changes. Consequently the branch points at $z = \sqrt{3}$ and $z = -\sqrt{3}$ lie in the same pack of sheets. In the same way one proves that in the case (b) these branch points lie on different packs. It remain to calculate how the passages amongst the sheets by turning around the points $z = i$ and $z = -i$ match each other.

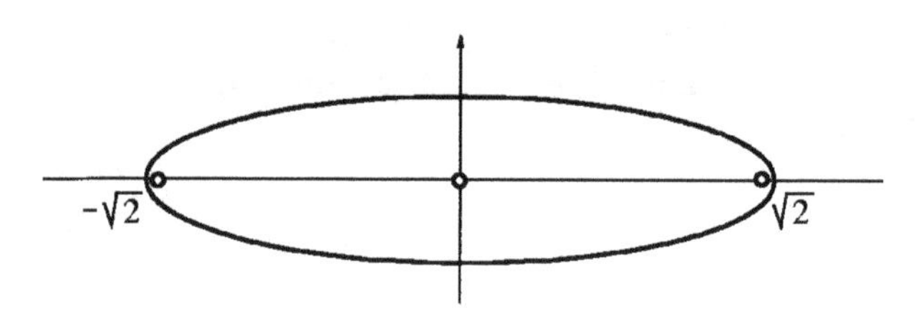

FIGURE 109

The continuous image under the mapping $w = \sqrt{z^2+1}$ of the circle C_R of radius $R = 1.1$, with centre at the point $z = 0$, is the curve shown in Figure 109 (consider the mappings $w = z^2$, $w = z^2 + 1$, $w = \sqrt{z^2+1}$). This curve does not turn around the point $z = 2$. Therefore by turning along the circle C_R, neither $\arg(\sqrt{z^2+1} - 2)$ nor the value of the function $h(z) = \sqrt{\sqrt{z^2+1} - 2}$ changes. Consequently turning around the point $z = i$, and later around the point $z = -i$, we must come back onto the same sheet (cf., Remark 1 §2.10). The required schemes are shown in Figure 110 and in Figure 111.

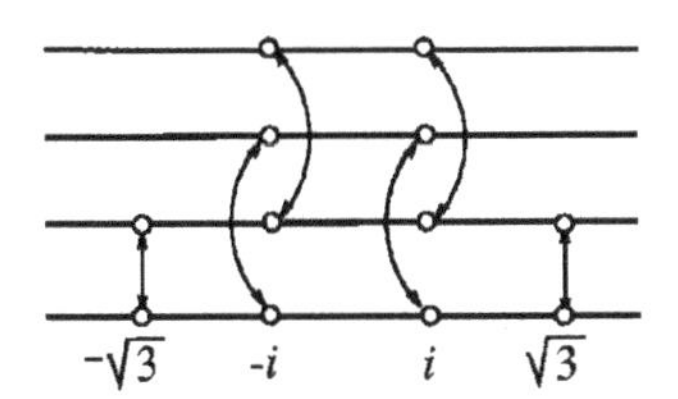

FIGURE 110

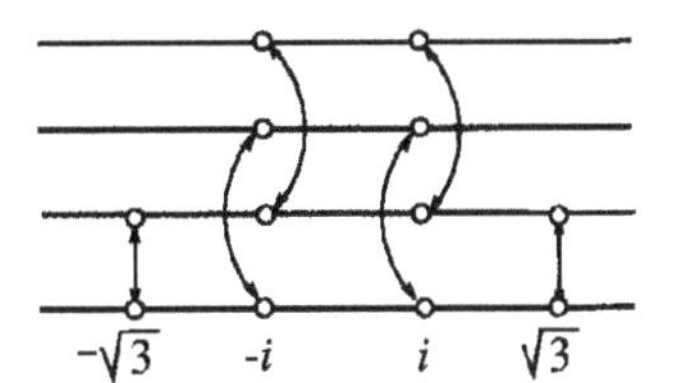

FIGURE 111

329. *Hint.* In the opposite case, covering the inverse curve C^{-1}, one should lose the uniqueness.

330. *Answer.* $z = 0: \begin{pmatrix} 1 & 2 & 3 & 4 \\ 3 & 4 & 1 & 2 \end{pmatrix}$, $z = 1: \begin{pmatrix} 1 & 2 & 3 & 4 \\ 2 & 1 & 4 & 3 \end{pmatrix}$;

b) $z = 0: \begin{pmatrix} 1 & 2 & 3 & 4 & 5 & 6 \\ 2 & 1 & 4 & 3 & 6 & 5 \end{pmatrix}$, $z = 1, -1: \begin{pmatrix} 1 & 2 & 3 & 4 & 5 & 6 \\ 3 & 4 & 5 & 6 & 1 & 2 \end{pmatrix}$;

c) $z = 0: \begin{pmatrix} 1 & 2 & 3 & 4 & 5 & 6 \\ 5 & 6 & 4 & 2 & 3 & 1 \end{pmatrix}$;

d) $z = 1: \begin{pmatrix} 1 & 2 & 3 & 4 & 5 & 6 & 7 & 8 \\ 6 & 7 & 8 & 5 & 2 & 3 & 4 & 1 \end{pmatrix}$,

$z = -1: \begin{pmatrix} 1 & 2 & 3 & 4 & 5 & 6 & 7 & 8 \\ 5 & 6 & 7 & 8 & 1 & 2 & 3 & 4 \end{pmatrix}$;

331. *Hint.* Use the result of Problem **57**: $e = g \cdot g^{-1}$; conditions (1) and (3) are obviously satisfied.

332. *Answer.* a) The cyclic group $\mathbb{Z}_2$, b) the cyclic group $\mathbb{Z}_3$, c) the cyclic group $\mathbb{Z}_n$, d) $\mathbb{Z}_3$, e) $\mathbb{Z}_4$.

333. *Hint.* For the function $\sqrt{z} + \sqrt{z-1}$ given in Problem **314**(a) we obtain that a turn around the point $z = 0$ involves a permutation of the first indices in the branches $h_{i,j}(z)$, and that a turn around the point $z = 1$ involves a permutation of the second indices. Hence the group we seek is the direct product of groups $\mathbb{Z}_2 \times \mathbb{Z}_2$ (cf., §1.7).

For the function $\sqrt{z^2-1} + \sqrt[4]{z-1}$ given in Problem **314**(d) let g_1 be the permutation of the sheets which corresponds to one turn around the point $z = 1$, and g_2 the permutation of the sheets which corresponds to one turn around the point $z = -1$. Thus g_2 permutes cyclically the first indices of the branches $h_{i,j}(z)$, whereas $g_1 g_2^{-1}$ permutes cyclically the second indices. Since $g_1 = (g_1 g_2^{-2}) g_2$ the subgroup generated by the permutations g_1 and g_2 coincides with the subgroup generated by the permutations g_2 and $g_1 g_2^{-1}$. Therefore the required group is the direct product $\mathbb{Z}_2 \times \mathbb{Z}_4$.

In the other cases Problem **333** is solved in a similar way.

Answer. 1. a) the direct product (§1.7) $\mathbb{Z}_2 \times \mathbb{Z}_2$; b) $\mathbb{Z}_3 \times \mathbb{Z}_2 \cong \mathbb{Z}_6$ (cf., **77**); c) $\mathbb{Z}_2 \times \mathbb{Z}_3 \cong \mathbb{Z}_6$; d) $\mathbb{Z}_2 \times \mathbb{Z}_4$. 2. a) $\mathbb{Z}_2$, b) $\mathbb{Z}_4$, c) $\mathbb{Z}_4$, d) $\mathbb{Z}_2 \times \mathbb{Z}_3 \cong \mathbb{Z}_6$. 3. a) $\mathbb{Z}_2$, b) $\{e\}$, c) $\mathbb{Z}_2$.

334. If the permutation g_1 exchanges the two packs of sheets, and the permutation g_2 changes the positions of the sheets in the same pack, then it is easy to see that the permutation $g_1 g_2 g_1^{-1}$ changes the positions of the sheets of the other pack. So the permutation group of both schemes contains a permutation which exchanges the packs, a permutation which permutes the sheets in one pack and a permutation which permutes the sheets in the other pack.

Our group, generated by these permutations, contains only the permutations such that every pack is sent to itself or to the other pack, whereas the sheets in each pack are arbitrarily permuted. Numbering the branches of one bunch (or, equivalently, the sheets of one pack) by the numbers 1 and 3, and those of the other bunch by 2 and 4, one obtains that every permutation of the group so defined corresponds to a symmetry of the square with vertices 1, 2, 3, 4 and, conversely, to every symmetry of this square there corresponds a permutation of the permutation group of the scheme just defined. Therefore our group is isomorphic in both cases to the group of symmetries of the square.

335. Let $w_1, w_2, \ldots, w_n$ be all values of (z_0) and let $z_1, \ldots, z_s$ be the branch points of the function $w(z)$. Numbering the sheets of the scheme of the Riemann surface of the function $w(z)$ in such a way that for every $i = 1, \ldots, n$ the value $w_i = w(z_0)$ corresponds to the ith sheet, we obtain that to every permutation of the values w_i there naturally corresponds a permutation of the sheets. We prove that under this correspondence the groups G_1 and G_2 coincide. Let a permutation g of the group G_1 produced by a turn along a continuous curve C, starting and ending at the point z_0. Suppose that the curve C crosses the cuts (according to which the Riemann surface has been built), drawn from the points z_{i_1}, $z_{i_2}, \ldots, z_{i_m}$. If at the point z_j there corresponds a permutation g_j, it is easy to see that to the curve C there corresponds a permutation of the sheets (together with a permutation of the values w_i) equal to $g_{i_m}^{\sigma_m}, \cdot \ldots \cdot g_{i_2}^{\sigma_2} \cdot g_{i_1}^{\sigma_1}$, where $\sigma_i = 1$ if the cut is traversed counterclockwise, and $\sigma_i = -1$ if the cut is traversed clockwise (cf., Remark 1 in §2.10).

It follows that $g = g_{i_m}^{\sigma_m}, \cdot \ldots \cdot g_{i_1}^{\sigma_1}$ and g belong to G_2. Inversely, if the element $g = g_{i_m}^{\sigma_m}, \cdot \ldots \cdot g_{i_1}^{\sigma_1}$ is given in the group G_2 (here $\sigma_i = \pm 1$)[3], then one easily defines a curve C which produces the same permutation of the values w_i in the group G_1. For example, in Figure 112 one sees a curve which corresponds to the permutation $g_2^{-1} g_3 g_1^{-1}$.

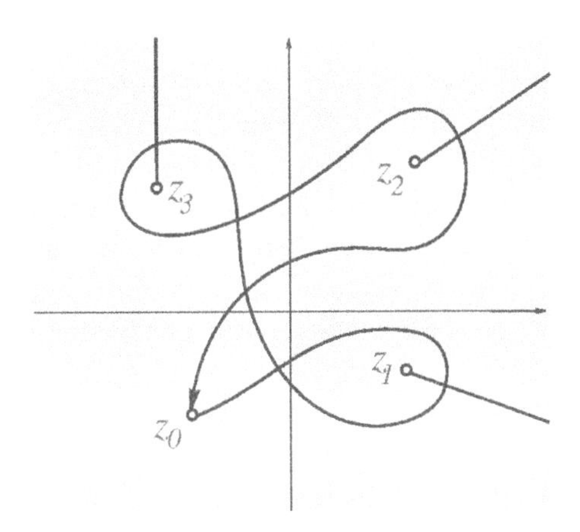

FIGURE 112

336. Let z_0 be a branch point of the function $h(z)$ and suppose that to one turn around this point there correspond the permutations d_1 and d_2 of the schemes of the functions $f(z)$ and $g(z)$. (If z_0 is not a branch point for one of the functions $f(z)$ or $g(z)$, then the permutation corresponding to d_1 or to d_2 is the identity permutation.) If the branches $h_{i,j}(z)$ of the function $h(z)$ are numbered by two indices i, j as we described in Proposition (a) of Theorem 8 (§2.11), then after a turn around the point z_0 the first and second indices turn out to be independently permuted (Theorem 8, Proposition (b)). Moreover, the permutation of the first indices is equal to d_1 and that of the second to d_2. So to one turn round the branch point there corresponds a permutation of the sheets of the scheme of the Riemann surface of the function $h(z)$, which can be considered as a pair of permutations (d_1, d_2). Since d_1 and d_2 are elements respectively of the groups F and G, then the pair (d_1, d_2) is an element of the direct product $F \times G$. These pairs, corresponding to all branch points of the function $h(z)$, generate a certain subgroup of the group $F \times G$.

[3]It follows from the definition of the permutation group of a given scheme in §2.12 that every element of this group can be put into the indicated form.

337. The scheme built by the formal method may contain some sets of sheets on which the branches coincide. (We have seen (Theorem 8 (c)) that in this case we have to identify the sheets in everyone of these sets in order to obtain the correct scheme of the Riemann surface of the function $h(z)$.) By virtue of the uniqueness, by one turn round an arbitrary branch point we move from the sheets of one set to the sheets of a different set of the scheme built by the formal method. Consequently any permutations d of the sheets of the scheme built by the formal method, corresponding to one turn round a branch point, permutes the sets without destroying them. If the permutations d_1 and d_2 permute the sets without destroying them, then also the permutation $d_1 d_2$ obviously permutes the sets without destroying them. Therefore all the permutations d_i of the sheets of a scheme built by the formal method, belonging to group H_1, permute the sets without destroying them. Let us put into correspondence with every permutation d_i a permutation d_i' of the sets. If to the permutation of sheets d_i there corresponds the sets permutation d_i' and, to the permutation d_j, the sets permutation d_j', then it easy to see that to the permutation $d_i d_j$ there corresponds the sets permutation $d_i' d_j'$. This means that the mapping we have defined of the group H_1 on the permutation group of the sets of sheets is a homomorphism.

Since to every set (considering the sets containing only one sheet as well) there corresponds a single sheet of the correct scheme of the Riemann surface of the function $h(z)$ (Theorem 8 (c)), and the passages amongst the sheets of the original scheme are transformed exactly into passages amongst the sets of sheets, the homomorphism we have defined is a surjective homomorphism of the group H_1 onto the group H_2.

338. Cf., **336** and **337**. By hypothesis the groups F and G (cf., **336**) are soluble. But thus the group $F \times G$ is also soluble (cf., **167**). Since the group H_1 (the permutation group of the sheets of the scheme built by the formal method) can be considered as a subgroup of the group $F \times G$ (cf., **336**), then the group H_1 is also soluble (cf., **162**). Since there exists a surjective homomorphism of the group H_1 onto the group H_2 (the permutation group of the correct scheme of the function $h(z)$ (cf., **337**)), it follows that the group H_2 is also soluble (cf., **163**).

339. *Hint.* Cf., Theorem 9, §2.11. If F and H are the monodromy groups for the schemes of the functions $f(z)$ and $h(z)$, then as in Problem **337** one proves the existence of a surjective homomorphism of the group F onto the group H. Afterwards one uses the result of Problem **163**.

340. To every sheet of the scheme of the Riemann surface of the function $f(z)$ there corresponds a pack of n sheets in the scheme of the Riemann surface of the function $h(z) = \sqrt[n]{f(z)}$ (Proposition (a) of Theorem 10, §2.11). The permutations of the scheme of the function $h(z)$, which correspond to the turns around the branch points of the function $h(z)$, permute the packs without destroying them (Proposition (b) of Theorem 10). But thus all permutations of the group H also permute the packs, without destroying them. Therefore to every permutation d of the group H there corresponds a permutation d' of the packs. Moreover, if to the permutation d_1 there corresponds the packs permutation d'_1 and to the permutation d_2 the packs permutation d'_2, then to the permutation $d_1 d_2$ there corresponds the packs permutation $d'_1 d'_2$. We obtain a homomorphism of the group H into the permutation group of the packs. The permutation of the packs, obtained by a turn around an arbitrary point z_0, corresponds to the permutation of the sheets by a turn around the point z_0 in the scheme of the Riemann surface of the function $f(z)$ (Proposition (c) of Theorem 10). Consequently the group of the permutations of the packs, generated by the group H, coincides with the group F (more precisely, it is isomorphic to F).

The abovedefined homomorphism is thus a surjective homomorphism of the group H onto the group F.

341. The kernel of the homomorphism, built in the solution of Problem **340**, consists in those permutations of the group H which transform each pack into itself. Let d_1 and d_2 be two permutations of such a type. If the sheets of the packs are numbered in this way: $f_{i,k}(z) = f_{i,0}(z)\epsilon_n^k$, then both permutations d_1 and d_2 permute cyclically the sheets of every pack (cf., Proposition (d) of Theorem 10). Consider an arbitrary pack. If d_1 cyclically displaces the sheets of this pack by l sheets, and d_2 displaces them by k sheets, then both permutations $d_1 d_2$ and $d_2 d_1$ displace the sheets of the given pack by $k+l$ sheets. In this way the permutations $d_1 d_2$ and $d_2 d_1$ permute identically the sheets in every pack, i.e., $d_1 d_2 = d_2 d_1$.

342. If ϕ is the homomorphism defined in the solution of Problem **340** and $\ker\phi$ is its kernel, then the quotient group $H/\ker\phi$ is isomorphic to the group F (Theorem 3, §1.13). Since the group $\ker\phi$ is commutative (cf., **341**) and the group F is soluble by hypothesis, the group H is soluble as well (cf., **166**).

343. Let $P_z(w) = 3w^5 - 25w^5 + 60w - z$. If w_0 is a multiple root of the equation $P_z(w) = 0$ then w_0 is a root of the equation $P'_z(w) = 0$, where $P_z'(w)$ is the derivative of polynomial $P_z(w)$ with respect to w (cf., **276**). We have $P_z'(w) = 15w^4 - 75w^2 + 60 = 15(w^4 - 5w^2 + 4) = 15(w-2)(w-1)(w+1)(w+2)$. Since the equation $P_z'(w) = 0$ has four roots of order 1, $w_0 = -2, -1, 1, 2$, only the values $w_0 = -2, -1, 1, 2$ can be multiple roots of the equation $P_z(w) = 0$ (of order 2). Putting these values in the equation

$$3w^5 - 25w^5 + 60w - z = 0,$$

one obtains that they can be roots of order two if z takes the values, $-16, -38, 38, 16$, respectively.

Answer. The roots of order two are the values: $w_0 = -2$ for $z = -16$, $w_0 = -1$ for $z = -38$, $w_0 = 1$ for $z = 38$, $w_0 = 2$ for $z = 16$.

344. Let $P_z(w) = 3w^5 - 25w^5 + 60w - z$. Set $z = z_0$ and consider the single-valued mapping of the w plane onto the complex τ plane defined by $\tau = P_{z_0}(w)$. In the w plane let C be a circle of radius r with centre at the point w_0 (Figure 113) and C' the image of the circle C under the mapping $\tau = P_{z_0}(w)$. Decompose the polynomial $P_{z_0}(w) = 3w^5 - 25w^5 + 60w - z_0$ into monomials of first degree (cf., **269**). We obtain $P_{z_0}(w) = 3(w-w_1)(w-w_2)(w-w_3)(w-w_4)(w-w_5)$, where all values w_i are roots of the equation $P_{z_0}(w) = 0$. By a counterclockwise turn along the circle C, the argument of the factor $(w - w_i)$ does not change if w_i lies outside the disc D, bounded by C, and increases by 2π if w_i lies inside this disc. Therefore, going counterclockwise along the circle C, the argument of the function $P_{z_0}(w)$ increases by $2\pi m$, where m is the number of roots (taking into account their multiplicities) of the equation $P_{z_0}(w) = 0$ which lie inside D. Consequently the curve C', the image of the circle C under the mapping $\tau = P_{z_0}(w)$, turns around the point $\tau = 0$ m times (Figure 114).

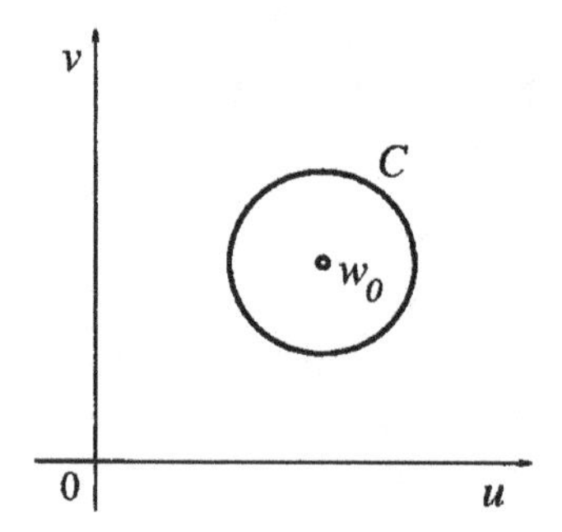

FIGURE 113

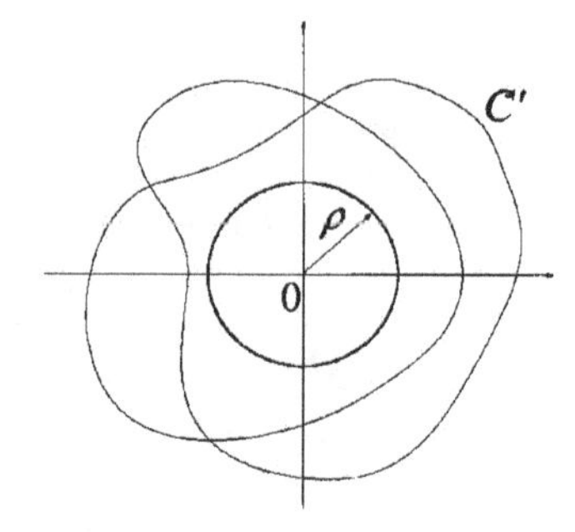

FIGURE 114

Since by hypothesis the point w_0, the centre of the circle C, is a root of the equation $P_{z_0}(w) = 0$, then $m \geq 1$. There thus exists a value $\rho > 0$ such that the disc of radius equal to ρ with centre at the point $\tau = 0$ has no intersection with the curve C' (Figure 114). Consider now a different complex number z_0' and consider another mapping $\tau = P_{z_0'}(w)$. Let C'' be the image of the circle C under the mapping $\tau = P_{z_0'}(w)$. Since $P_{z_0'}(w) = 3w^5 - 25w^5 + 60w - z_0' = 3w^5 - 25w^5 + 60w - z_0 + (z_0 - z_0') = P_{z_0}(w) + (z_0 - z_0')$, the curve C'' is obtained from the curve C' displacing it by the vector $z_0 - z_0'$ (cf., **246**). If the length of the vector $z_0 - z_0'$ is smaller than ρ, then the curve C'' is displaced with respect to C' by so small an amount that along C'' one turns around the point $\tau = 0$ as many times as along C'. (Equivalently, one may imagine, conversely, that the point $\tau = 0$ be displaced instead of the curve, see Figure 114). Since the curve C' turns m times around the point $\tau = 0$, the curve C'' will turn m times around the point $\tau = 0$ as well. Following the same reasoning as before, we obtain that inside the disc D there are $m \geq 1$ roots of the equation $P_{z_0'}(w) = 0$ (taking into account their multiplicities).

345. Let z_0 be an arbitrary point, different from $z = \pm 38$ and $z = \pm 16$. We thus have 5 different images of the point z under the mapping $w(z)$. Let these images be w_1, w_2, w_3, w_4, w_5. If a continuous curve C starts from the point z_0, then at every point w_i $(i = 1, \ldots, 5)$ at least a continuous image of the curve C under the mapping $w(z)$ starts. If two continuous images of the curve C were starting from the point w_i, then the curve C should have at least six continuous images. This is not possible because an equation of degree 5 cannot have more than five roots. Consequently the point z_0 is not a point of non-uniqueness of the function $w(z)$.

Consider now 5 discs D_1, D_2, D_3, D_4, D_5 with a certain radius r with centres at the points w_is. Choose r sufficiently small so that these discs be disjoint. By virtue of the result of Problem **344** there exists a disc D_0, with centre at the point z_0, such that for every point z_0' inside this disc there exists at least (and consequently only) one image in each one of the discs D_1, D_2, D_3, D_4, D_5 on the w plane. If C is a continuous curve which lies entirely in the disc D_0 all images of its points lie in the discs D_i $(i = 1, \ldots, 5)$. But thus a continuous image of the curve C under the mapping $w(z)$ cannot jump from one disc to another, and each one of the images of C lies entirely in one of the discs D_i $(i = 1, \ldots, 5)$. If the curve C, which lies entirely in the disc D_0, begins and ends at the point z_0' then the end points of its continuous image C' are images of the point z_0' under the mapping $w(z)$. Since the curve C' lies entirely in one of the discs D_i ($i = 1, \dots, 5$), and in this disc there is only one image of the point z_0', the curve C' begins and ends at a unique point.

So if C is a closed curve lying entirely inside the disc D_0 then the value of the function $w(z)$ at the final point of the curve C, defined by continuity, coincides with the value at the initial point. In particular, this holds for all circles with centre at z_0 and radius smaller than ρ. Consequently the point z_0 is not a branch point of the function $w(z)$.

346. From the result of Problem **343** it follows that for $z = z_0 = 38$ equation (2.8) has four roots: w_1, w_2, w_3, w_4, of which one (for example, w_1) has order 2, and the others are simple. Suppose that the point z_0' is close to the point z_0. Thus from the solution of Problem **344**, we obtain that near the point w_1 there are two images of the point z_0' under the mapping $w(z)$ and near the point w_2, w_3 and w_4 there is a sole image of the point z_0'. Let C be a circular curve of small radius, with centre z_0, starting and ending at z_0'. As in the solution of Problem **345** we obtain that the continuous images of the circle C under the mapping $w(z)$, which start from the points w_2, w_3, and w_4, end at the initial point, whereas the continuous images which start from one of the images of the point z_0', near the point w_1, may end on the other image of the point z_0', which lies near the point w_1 as well. Consequently at the point z_0 only two sheets can meet, whilst there are no passages between the other three sheets.

347. Let us draw a continuous curve C' from the point w_0 to the point w_1, not crossing the images under $w(z)$ of the points $z = \pm 38$ and $z = \pm 16$. It is possible to draw it, because the points $z = \pm 38$ and $z = \pm 16$ have a finite number of images. Now let C be the image of the curve C' under the mapping $z(w) = 3w^5 - 25w^3 + 60w$. Since $z(w)$ is a continuous function and C' a continuous curve, C is a continuous curve as well. Since z and w are related, under the mapping $z(w)$, by the relation $3w^5 - 25w^3 + 60w - z = 0$, which coincides with that given by the mapping $w(z)$, the curve C' is itself a continuous image of the curve C under the mapping $w(z)$. Since the curve C' does not pass through the images of the points $z = \pm 38$ and $z = \pm 16$, the curve C does not pass through the points $z = \pm 38$ and $z = \pm 16$. The initial and final points of the curve C are $z(w_0) = z_0$ and $z(w_1) = z_1$. Consequently C is the curve we sought.

348. By virtue of the result of Problem **347** one can move from an arbitrary sheet of the Riemann surface of the function $w(z)$ to any other, moving along a curve which does not pass through the points $z = \pm 38$ and $z = \pm 16$. Moreover, every passage from one sheet to another, whilst crossing a cut, coincides with the passage indicated on the scheme at that branch point from which this cut starts (cf., note 1 §2.10). Consequently the connection of the sheets at the branch points must be made so as to obtain a connected scheme. Since points different from $z = \pm 38$ and $z = \pm 16$ are not branch points (cf., **345**), and at each one of the points $z = \pm 38$ and $z = \pm 16$ only two sheets can join (cf., **346**), in order to obtain a connected scheme we must put one arrow in correspondence with each one of the points $z = \pm 38$ and $z = \pm 16$, i.e., all of these four points are branch points. All the distinct connected schemes are shown in Figure 115. Any connected scheme of the Riemann surface of the function $w(z)$ matches one of these three schemes after a permutation of the sheets and of the branch points (here we do not claim that all these three schemes can be realized).

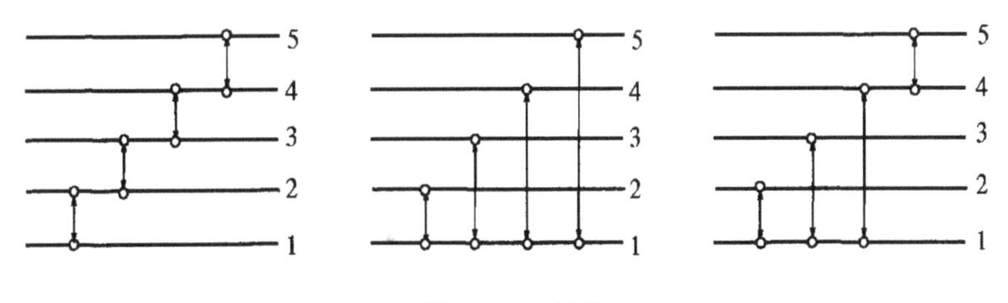

FIGURE 115

349. We prove that the permutation group of the scheme for all the three schemes shown in Figure 115 contains all the elementary transpositions (cf., §1.15), i.e., the transposition $(1,2)$, $(2,3)$, $(3,4)$, $(4,5)$. For the first scheme this is evident because these transpositions correspond to the branch points. In the second and in the third scheme one of the branch points corresponds to the transposition $(1,2)$. The transpositions $(2,3)$ and $(3,4)$ are obtained in both cases as the products $(2,3) = (1,2) \cdot (1,3) \cdot (1,2)$ and $(3,4) = (1,3) \cdot (1,4) \cdot (1,3)$. The transposition $(4,5)$ is obtained for the second scheme as the product $(4,5) = (1,4) \cdot (1,5) \cdot (1,4)$ and for the third scheme it simply coincides with one of the branch points.

Consequently the required group contains, in all cases, all elementary transpositions, and consequently (Theorem 4, §1.15) it coincides with the entire group of the permutations of degree 5, S_5.

350. From the result of Problem **349** we obtain that the monodromy group of the function $w(z)$ is the group S_5 of all permutations of degree 5, which is not soluble (cf., Theorem 5, 1.15). On the other hand if the function $w(z)$ were representable by radicals, then the corresponding monodromy group should be soluble (cf., Theorem 11, 2.13). From the contradiction so obtained it follows that the function $w(z)$ is not representable by radicals.

351. *Hint.* If such a formula existed then, taking in it the values $a_0 = 3$, $a_2 = -25$, $a_4 = 60$, $a_1 = a_3 = 0$ and $a_5 = z$, we would obtain that the function $w(z)$ (cf., **350**) is representable by radicals.

352. The function $w_1(z)$ expressing the roots of equation (2.9) in terms of the parameter z, possesses a Riemann surface which consists of a separated sheet, on which $w_1(z) \equiv 0$, and of 5 sheets which represent the scheme of the function $w(z)$ expressing the roots of the equation

$$3w^5 - 25w^3 + 60w - z = 0$$

in terms of the parameter z. Hence the monodromy group corresponding to the function $w_1(z)$ coincides with the monodromy group of the scheme of the function $w(z)$, i.e., with group S_5 of all degree 5 permutations, which is not soluble (cf., **349**). On the other hand if the function $w_1(z)$ were represented by radicals then the corresponding monodromy group would be soluble (cf., Theorem 11, 2.13). From the contradiction so obtained it follows that the function $w_1(z)$ is not representable by radicals and that the general equation of degree n, for $n > 5$, is not solvable by radicals.

Drawings of Riemann surfaces

The Riemann surface of a complex function $w(z)$, as has been defined in this book, is a collection of different copies of the z plane (the sheets) suitably joined each other along some cuts, in such a way that the function w, defined on these sheets, becomes a single-valued continuous function.

However, the Riemann surface of a function $w = f(z)$ sometimes can be realized by a suitable projection of its graph, which lives in $\mathbb{C}^2$. For example, the Riemann surface of the function $\sqrt{z}$ in Figure 27 is homeomorphic to the surface shown in Figure 116, which is the graph of the real part of $\sqrt{z}$, i.e., the projection of the graph of this function onto the three-dimensional space with coordinates $(x = \Re(z), y = \Im(z), u = \Re(w))$.

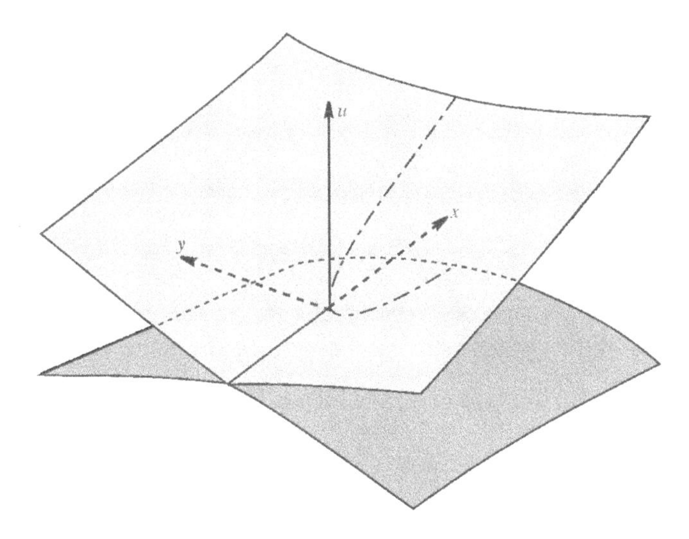

FIGURE 116

The drawings of Riemann surfaces in this Appendix are not obtained as projections of graphs, but they are 'artificial' surfaces constructed by the method just explained in §2.10.

Here I explain how to 'read' these drawings. The different sheets are joined in such a way that the passage from one sheet to another when one traverses a cut is realized by a smooth curve on the surface. The drawing notations are the following:

- Distinct grey colours indicate distinct sheets.

- Ramification points are indicated by small white discs.
- The cuts are black.
- A dashed vertical surface indicates a connection between two non-adjacent sheets (i.e., between which there are other sheets): it always corresponds to a cut.
- At every cut 2 sheets meet: going along a smooth curve on the surface one moves from one sheet to another (see Figure 117).

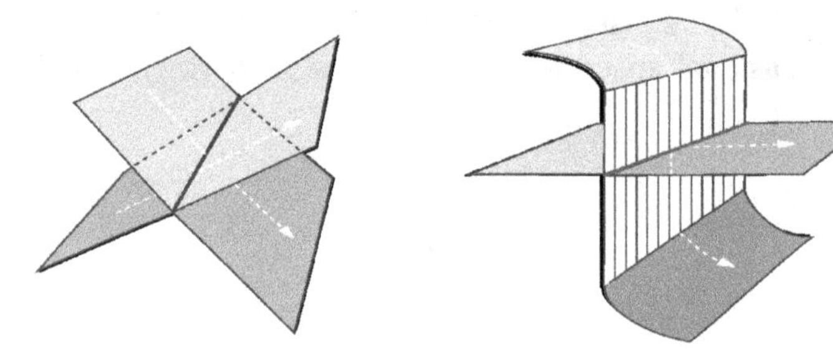

FIGURE 117

- The self-intersection lines of the surface which are not cuts are white. By a transversal crossing of a line of this type along a smooth curve lying on the surface one remains on the same sheet (see Figure 118).

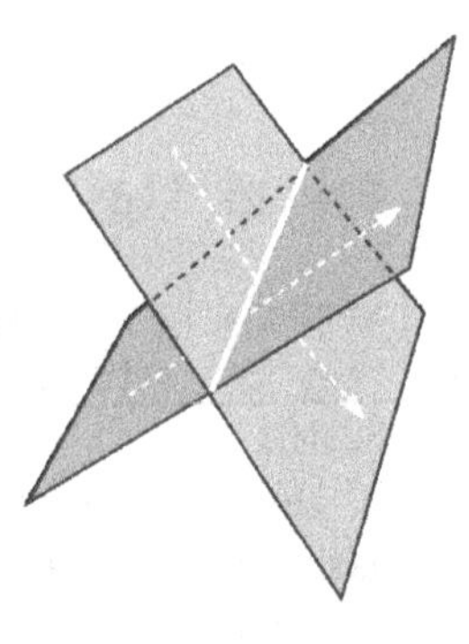

FIGURE 118

The following Riemann surfaces are considered in Problems **298**–**317**.

$\sqrt[n]{z}, \quad n = 6$

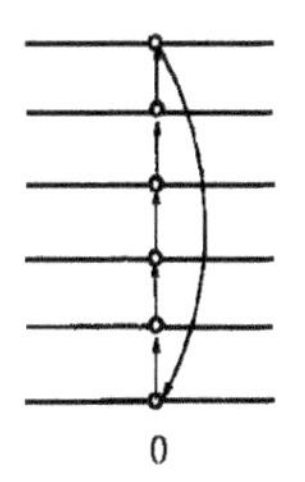

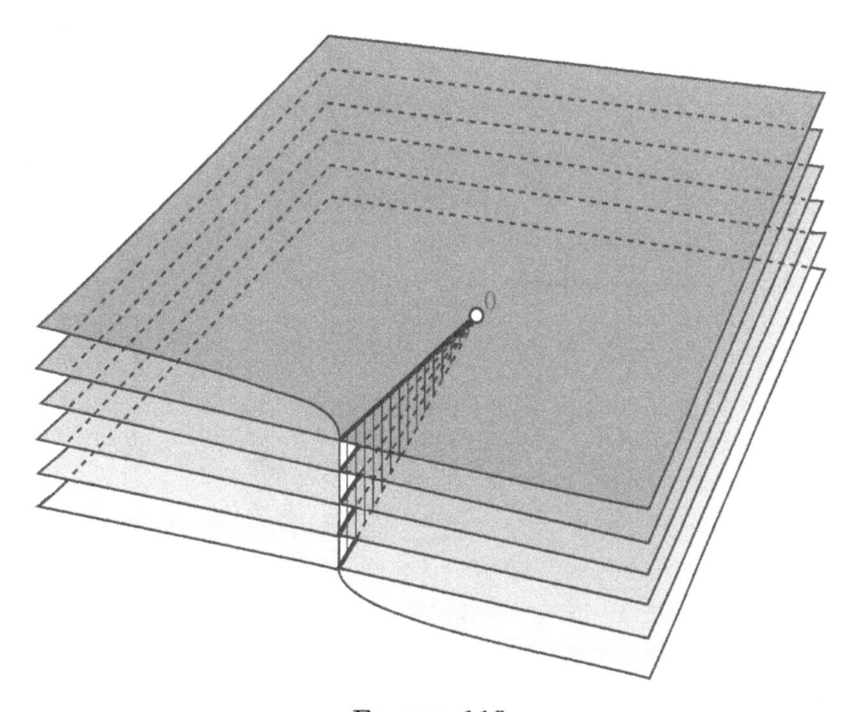

FIGURE 119

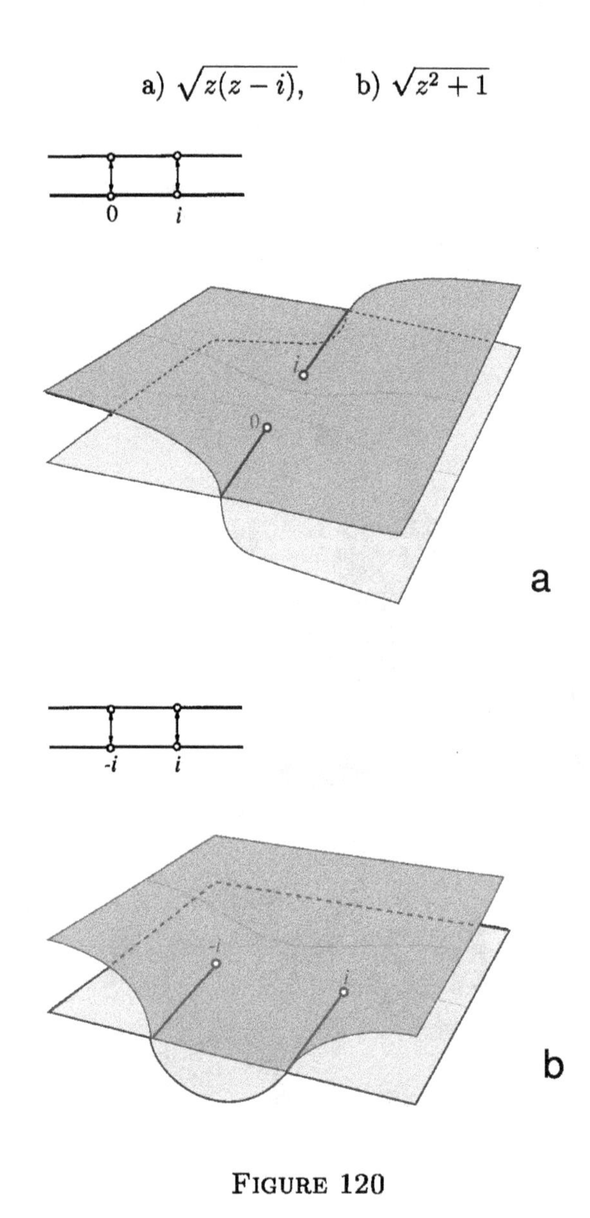

FIGURE 120

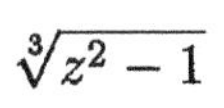

$\sqrt[3]{z^2 - 1}$

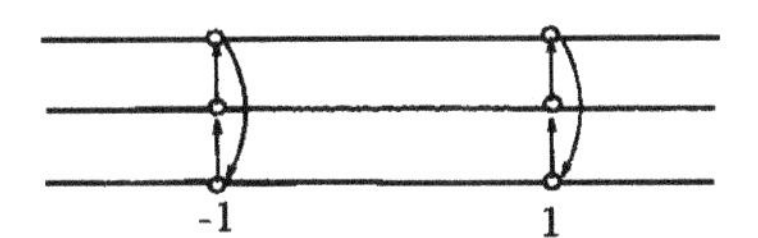

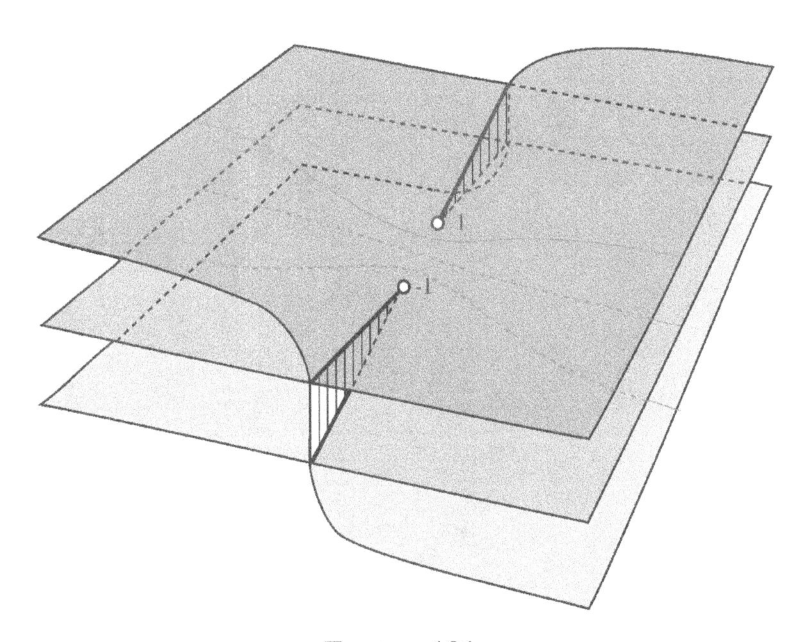

FIGURE 121

$$\sqrt[3]{z(z-1)^2}$$

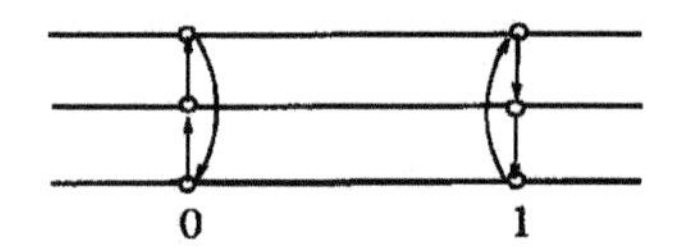

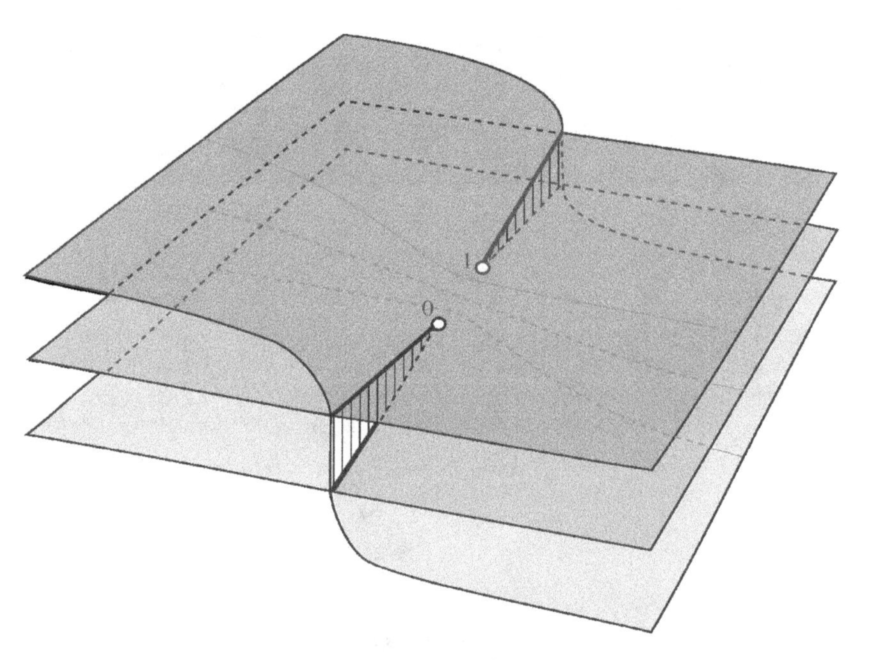

FIGURE 122

$\sqrt[4]{z^2}$

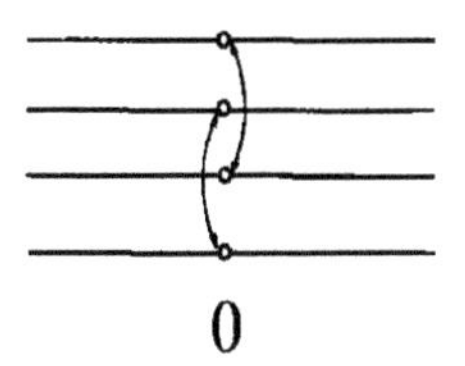

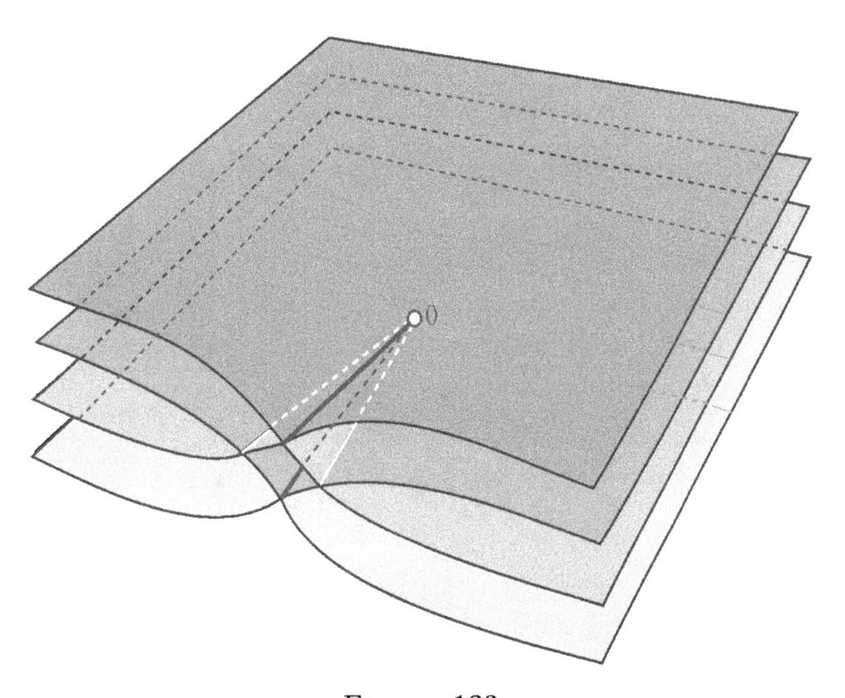

FIGURE 123

$$\sqrt[4]{(z^2-1)^3(z+1)^3}$$

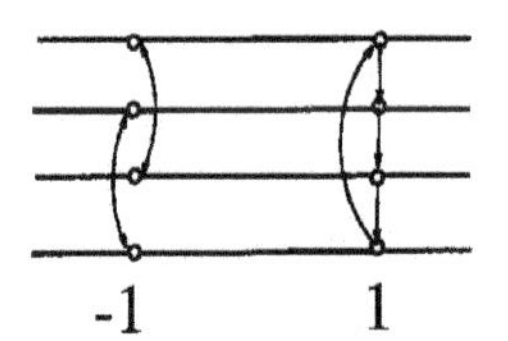

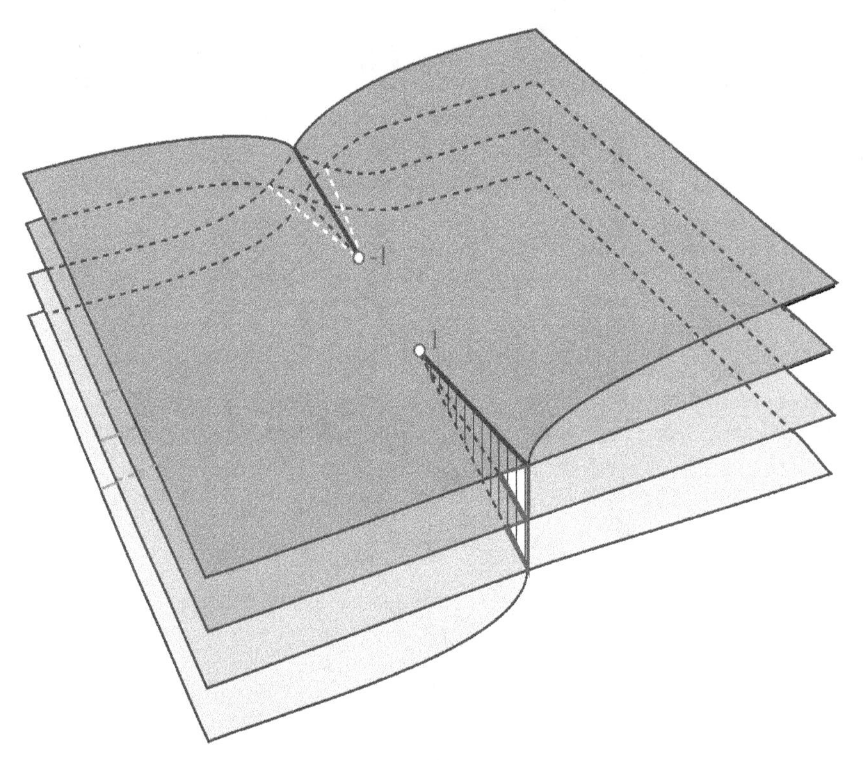

FIGURE 124

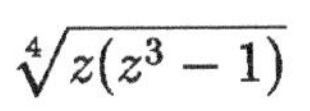

$\sqrt[4]{z(z^3 - 1)}$

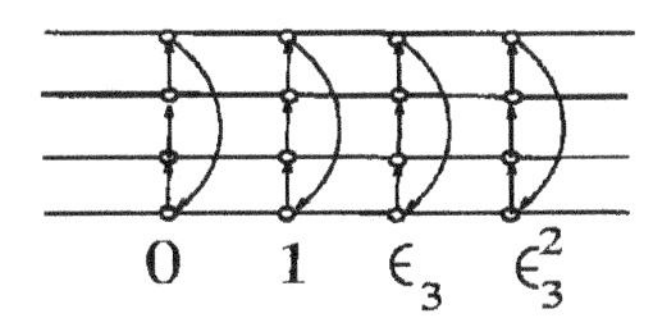

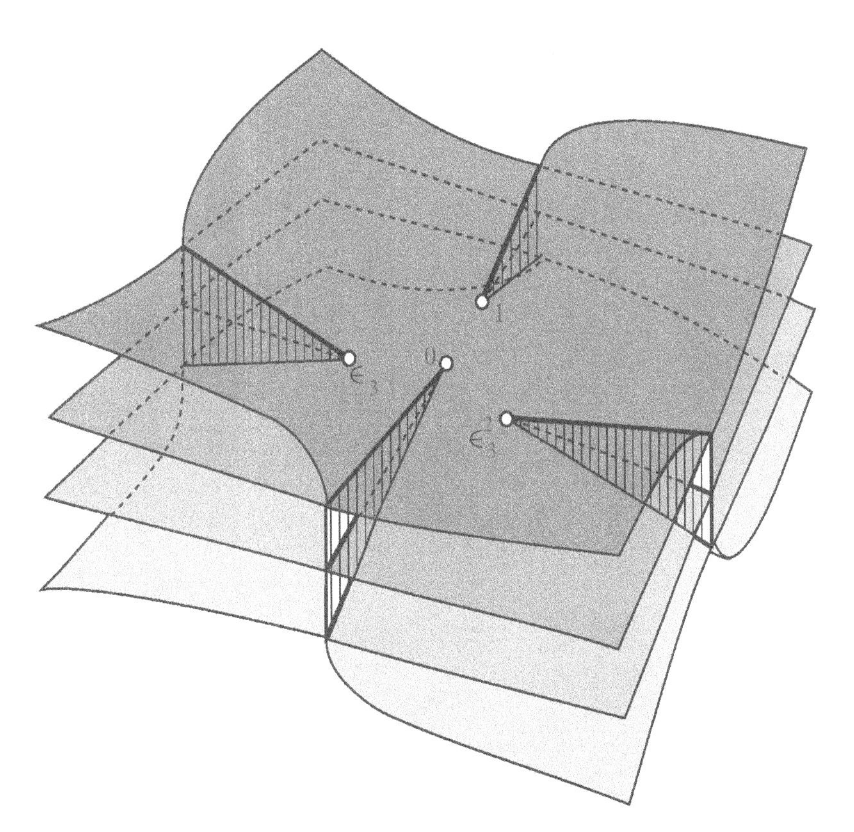

FIGURE 125

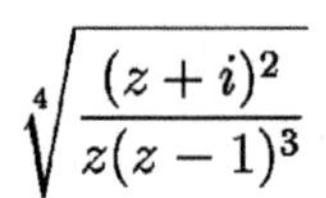

$$\sqrt[4]{\frac{(z+i)^2}{z(z-1)^3}}$$

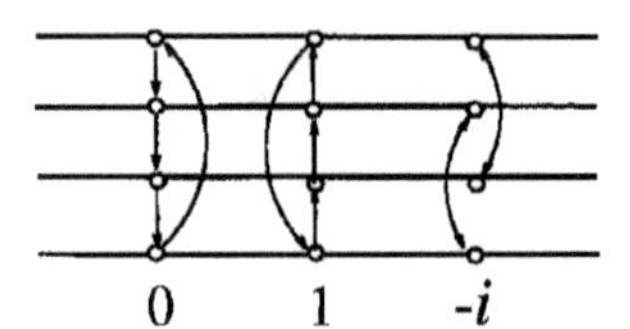

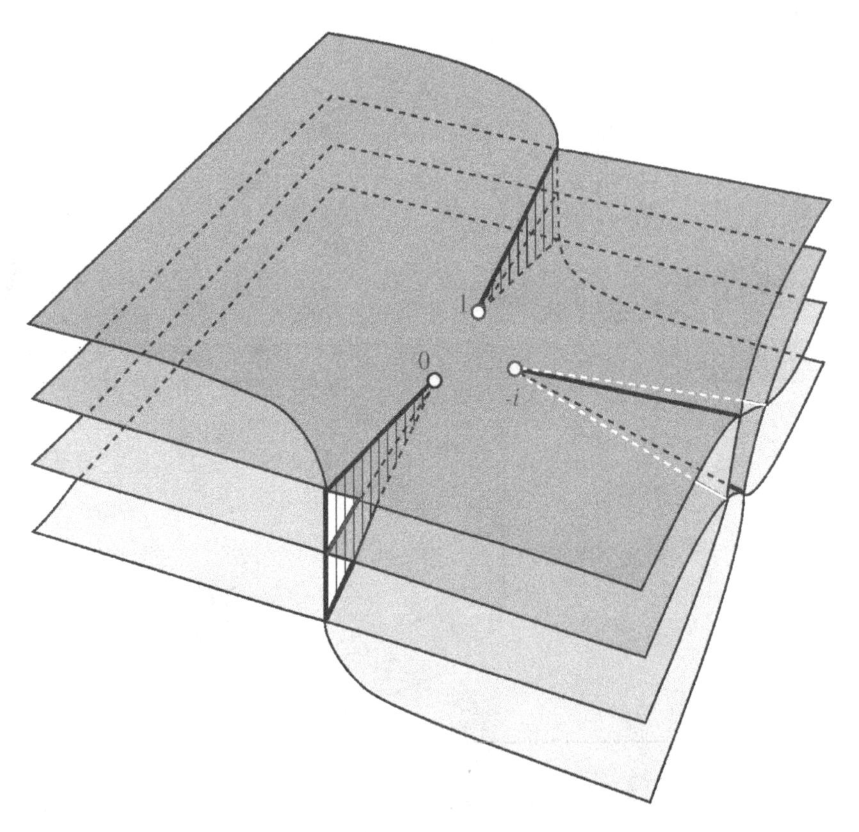

FIGURE 126

$$\sqrt{z^2-1}+\sqrt[4]{z-1}$$

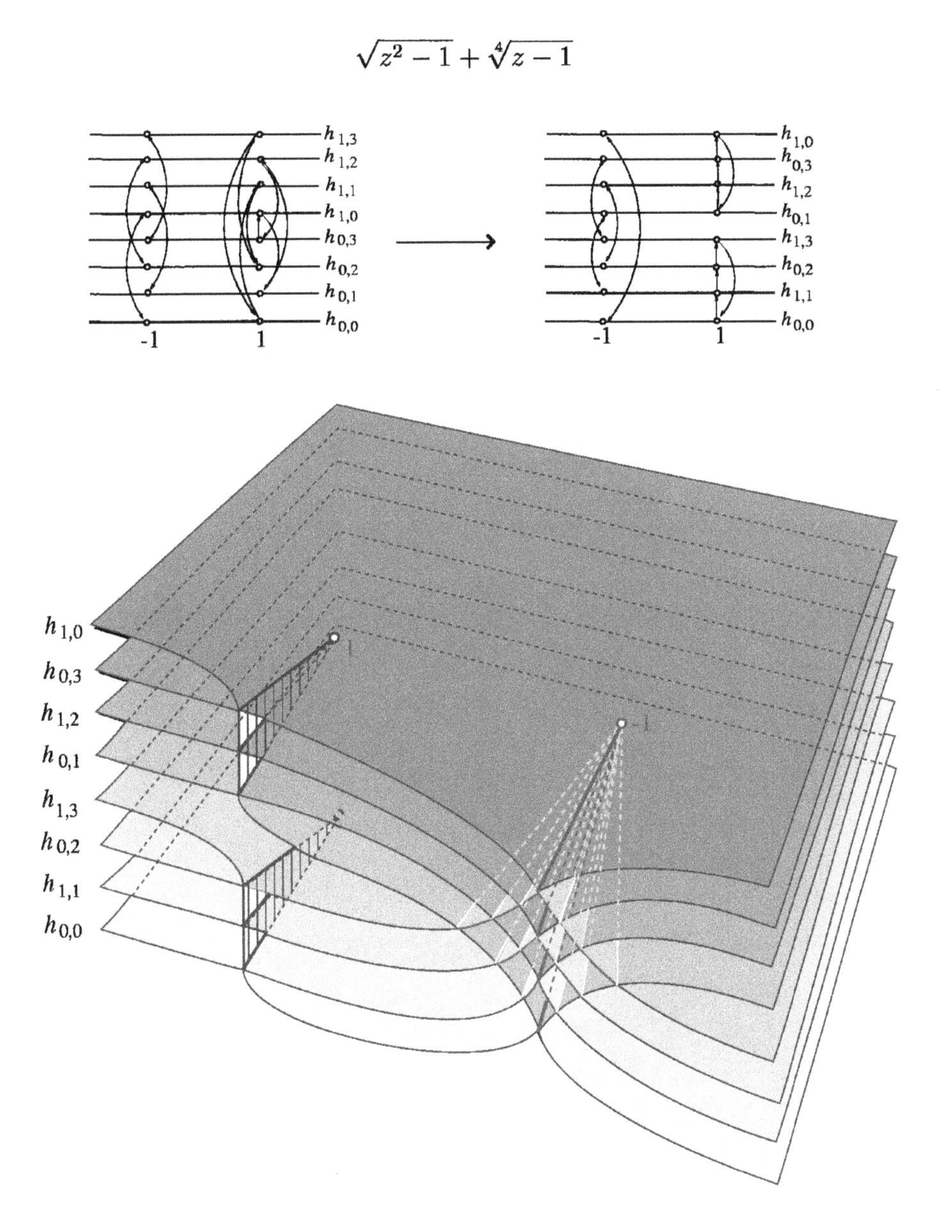

FIGURE 127

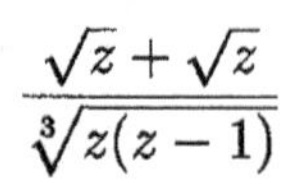

$$\frac{\sqrt{z}+\sqrt{z}}{\sqrt[3]{z(z-1)}}$$

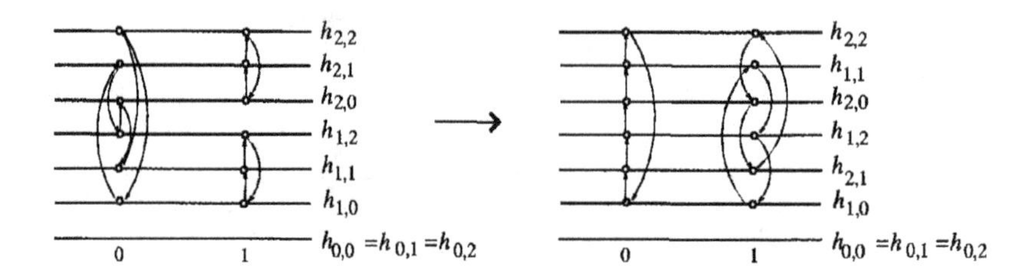

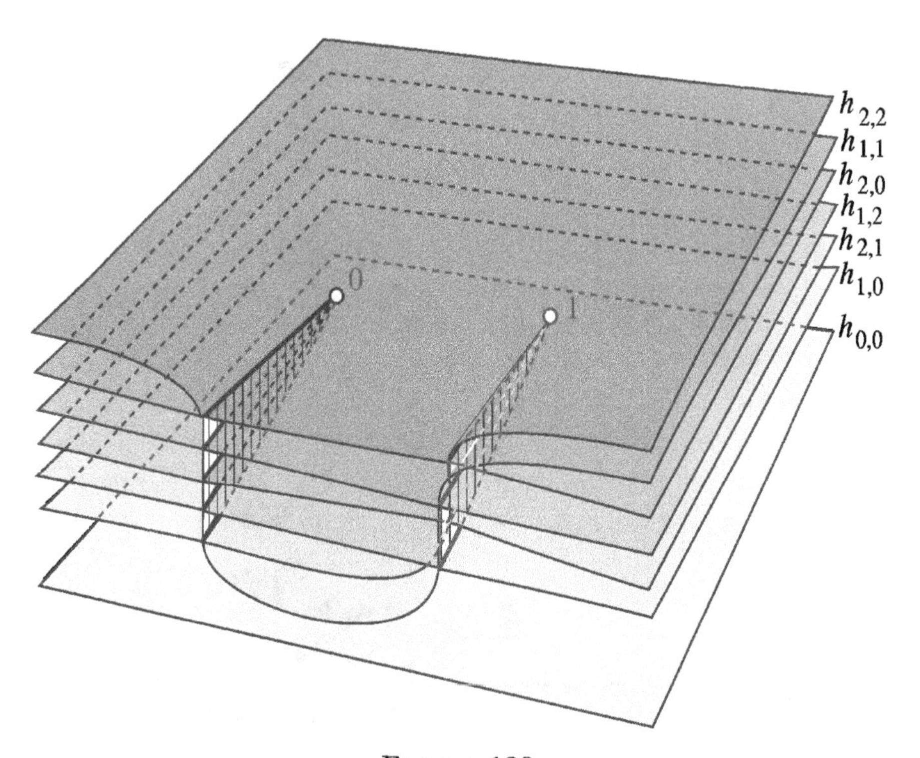

FIGURE 128

Appendix by A. Khovanskii: Solvability of equations by explicit formulae (Liouville's theory, differential Galois theory, and topological obstructions)

This Appendix is dedicated to the study of the solvability of differential equations by explicit formulae. This is a quite old problem: the first idea for solving it dates back to Abel. Today one knows three approaches to solving this problem. The first belongs to Liouville; the second approach considers the problem from the point of view of Galois theory: it is related to the names of Picard, Vessiot, Kolchin, and others; the third approach, topological, was first introduced in the case of functions of one variable in my thesis. I am infinitely grateful to my research director V.I. Arnold who aroused my interest in this subject.

I had always believed that the topological approach cannot be completely applied to the case of many variables. Only recently I discovered that this is not true and that in the multi-dimensional case one can obtain absolutely analogous results [25]–[27].

This Appendix contains the subject of my lectures to the Mathematical Society of Moscow and to the students of the École Normale Supérieure at the Independent University of Moscow (October 1994).

The section, concerning the functions of many variables, was added for this Appendix in autumn 2002.

I would like to thank T.V. Belokrinitska for her help during the editing of this Appendix and F. Aicardi for the translation into English.

A.1 Explicit solvability of equations

Some differential equations possess 'explicit solutions'. If it is the case the solution gives itself the answer to the problem of solvability. But in general all attempts to find explicit solutions of equations turn out to be in vain. One thus tries to prove that for some class of equations explicit solutions do not exist. We must now define correctly what this means (otherwise, it will not be clear what we really wish to prove). We choose the following way: we distinguish some classes of functions, and we say that an equation is explicitly solvable if its solution belongs to one of these classes. To different classes of functions there correspond different notions of solvability.

To define a class of functions we give a list of *basic functions* and a list of *allowed operations.*

The class of functions is thus defined as the set of all functions which are obtained from the basic functions by means of the allowed operations.

EXAMPLE 1. The class of the functions *representable by radicals.*

The list of basic functions: constants and the identity function (whose value is equal to that of the independent variable).

The list of the allowed operations: the arithmetic operations (addition, subtraction, multiplication, division) and the root extractions $\sqrt[n]{f}$, $n = 2, 3, \ldots$ of a given function f.

The function $g(x) = \sqrt[3]{5x + 2\sqrt[2]{x} + \sqrt[7]{x^3 + 3}}$ is an example of a function representable by radicals.

The famous problem of the solvability of the algebraic equations by radicals is related to this class. Consider the algebraic equation

$$y^n + r_1(x)y^{n-1} + \cdots + r_n(x) = 0 \tag{A.10}$$

in which $r_i(x)$ are rational functions of one variable. The complete answer to the problem of the solvability of the equation (A.10) by radicals consists in the Galois theory (see §A.8).

Note that already in the simplest class, that in Example 1, we encounter some difficulties: the functions we deal with are multi-valued.

Let us see exactly, for example, what is the sum of two multi-valued analytic functions $f(x)$ and $g(x)$. Consider an arbitrary point a, one of the germs f_a of the function $f(x)$ at the point a and one of the germs g_a of the function $g(x)$ at the same point a. We say that the function $\varphi(x)$, defined by the germ $f_a + g_a$, is representable as the sum of the functions $f(x)$ and $g(x)$. This sum is, however, not defined in a unique. For example, one easily sees that there are exactly two functions representable as the sum $\sqrt{x} + \sqrt{x}$, namely, $f_1 = 2\sqrt{x}$ and $f_2 \equiv 0$.

The closure of a class of multi-valued functions with respect to the addition is a class which contains, together with any two functions, all functions representable by their sum. One can say the same for all the operations on the multi-valued functions that we shall encounter in this chapter.

EXAMPLE 2. *Elementary functions.* Basic elementary functions are those functions which one learns at school and which are usually represented on the keyboard of calculators. Their list is the following: the constant function, the identity function (associating with every value x of the argument the value x itself), the nth roots $\sqrt[n]{x}$, the exponential $\exp x$, the logarithm $\ln x$, the trigonometrical functions: $\sin x$, $\cos x$, $\tan x$, $\arcsin x$, $\arccos x$, $\arctan x$. The allowed operations are: the arithmetic operations, the composition.

Elementary functions are expressed by formulae, for instance:

$$f(x) = \arctan(\exp(\sin x) + \cos x).$$

From the beginning of the study of analysis we learn that the integration of elementary functions is very far from being an easy task. Liouville proved, in fact, that the indefinite integrals of elementary functions are not, in general, elementary functions.

EXAMPLE 3. *Functions representable by quadratures.* The basic functions in this class are the basic elementary functions. The allowed operations are the arithmetic operations, the composition and the integration. A class is said to be *closed with respect to integration* if it also contains together with every function f a function g such that $g' = f$.

For example, the function

$$\exp\left(\int^x \frac{\mathrm{d}t}{\ln t}\right)$$

is representable by quadratures. But, as Liouville had proved, this function is not elementary.

Examples 2 and 3 can be modified. We shall say that a class of functions is closed with respect to the solutions of the algebraic equations if together with every set of functions $f_1, \dots, f_n$ it also contains a function y satisfying the equation

$$y^n + f_1 y^{n-1} + \cdots + f_n = 0.$$

EXAMPLE 4. If in the definition of the class of elementary functions we add the operation of solution of algebraic equations, we obtain the class of the *generalized elementary functions.*

EXAMPLE 5. The class of functions representable by *generalized quadratures* contains the functions obtained from the class of functions representable by quadratures by adding the operation of solution of algebraic equations.

A.2 Liouville's theory

The first exact proofs of the non-solvability of some equations neither by quadratures nor by elementary functions were obtained by Liouville in the middle of the XIX century. Here we briefly expound his results.

The reader can find a wider exposition of the Liouville method and of the works on analogous subjects by Chebychev, Mordukai-Boltovski, Ostrovski, and Ritt in book [1].

First of all Liouville showed that the classes of functions in Examples 2–5 can be constructed in a very simple way. Indeed, the set of basic elementary functions seems to be very large. Moreover, in the definition of this class one encounters some algebraic difficulties owed to the composition operation. Liouville at first proved that one can reduce a great deal the lists of basic functions, in one half of the cases leaving in it only the constants, and in the remaining cases leaving only the constants and the identity function. Secondly, he proved that in the list of the allowed operations the composition is superfluous. One can define all the necessary operations using only arithmetic operations and differentiation. This fact plays an essential role for the algebraization of the problem of the differential fields numerability.

Let us formulate the corresponding definitions in differential algebra.

A field of functions F is called a *differential field* if it is closed with respect to the differentiation, i.e., if $g \in F$ then $g' \in F$. One can also consider the abstract differential fields, i.e., the field in which one can define a supplementary differentiation operation, satisfying the Leibniz identity $(a \cdot b)' = a' \cdot b + a \cdot b'$.

Suppose that a differential field F contains another smaller differential field F_0, $F_0 \subseteq F$. An element $y \in F$ is said to be *algebraic* over the field F_0 if y satisfies an algebraic equation of the type

$$y^n + a_1 y^{n-1} + \cdots + a_n = 0,$$

where the coefficients a_is belong to the field F_0. In particular, an element y is said to be *radical* over the field F_0 if $y^k \in F_0$. An element y is said to be *integral* over the field F_0 if $y' \in F_0$. An element y is said to be *logarithmic* over the field F_0 if $y' = a'/a$, where $a \in F_0$. An element y is said to be *exponential integral* over the field F_0 if $y' = ay$, $a \in F_0$. An element y is said to *exponential* over the field F_0 if $y' = a'y$.

The extension of a field F_0 by means of an element y, denoted by $F_0\{y\}$, is called the *minimal differential field containing F_0 and y*. The field $F_0\{y\}$ consists of the rational functions in $y, y', \ldots, y^{(k)}, \ldots$ with coefficients in F_0.

- 1) An element y is said to be *representable by radicals over the field* F_0 if there exists a sequence $F_0 \subseteq F_1 \subseteq \cdots \subseteq F_k$, such that every extension $F_i \subseteq F_{i+1}$ is obtained by adding one radical to the field F_i, and the field $F_0\{y\}$ is contained in F_k. A sequence of this type is called a *tower*.

  By this method one also defines other types of representability of an element y over a field F_0. The towers in these definitions are built by means of the corresponding types of extensions $F_i \subseteq F_{i+1}$:

- 2) An element y is said *elementary over the field F_0* when one can add logarithmic and exponential elements.

- 3) An element y is said to be *representable by quadratures over the field* F_0 when adding integrals and exponential integrals is allowed.

- 4) An element y is called a *generalized elementary element over the field* F_0 when one can add algebraic, exponential, and logarithmic elements.

- 5) An element y is said to be *representable by generalized quadratures over the field* F_0 when one can add algebraic, integral and exponential integral elements.

THEOREM 1. (Liouville) *A function is elementary (a generalized elementary function) if and only if it is an elementary (generalized elementary) element over the field of rational functions $\mathcal{R}$. A function is representable by quadratures (representable by generalized quadratures) if and only if it is representable by quadratures (representable by generalized quadratures) over the field of complex numbers $\mathbb{C}$.*

For example, it follows from Theorem 1 that the basic elementary function $f(x) = \arctan x$ is representable by quadratures over the field $F_0 = \mathbb{C}$. Indeed, this becomes clear from the equation $f' \equiv \frac{1}{1+x^2}$, $x' \equiv 1$.

To prove, for example, the part of Theorem 1 which concerns functions representable by quadratures it suffices to verify first that there exist analogous representations for all the basic elementary functions, and, furthermore, that the class of functions representable by quadratures over the field $\mathbb{C}$ is closed with respect to the composition.

Liouville constructed a nice theory about the solvability of equations. Let us show two examples of his results.

THEOREM 2 (Liouville). *The indefinite integral $y(x)$ of the algebraic function $A(x)$ of one complex variable is representable by generalized elementary functions if and only if it is representable in the form*

$$y(x) = \int^x A(t)dt = A_0(x) + \sum_{i=1}^{k} \lambda_i \ln A_i(x),$$

where the $A_i(x)$, for $i = 0, 1 \dots, k$, are algebraic functions.

A priori the integral of an algebraic function could be given by a very complicated formula. It could have the form

$$y = \exp(\exp(\exp(\exp(\exp(x))))).$$

Theorem 2 says that this does not happen. Either the integral of an algebraic function can be written in a simple way, or in general it is not a generalized elementary function.

THEOREM 3 (LIOUVILLE). *The differential linear equation*

$$y'' + p(x)y' + q(x)y = 0, \tag{A.11}$$

where $p(x)$ and $q(x)$ are rational functions, is solvable by generalized quadratures if and only if its solution can be written in the form

$$y = \exp\left(\int^x R(t)dt\right),$$

where $R(x)$ is an algebraic function.

A priori the solution of equation (A.11) could be expressed by very complicated formulae. Theorem 3 says that this is nowhere the case. Either the equation has sufficiently simple roots, or in general it cannot be solved by generalized quadratures.

Liouville found a series of results of this type. The common idea is the following: simple equations have either simple solutions, or in general have no solutions in a given class (by quadratures, by elementary functions, etc.).

The strategy of the proof in Liouville's theory is the following: prove that if a simple equation has a solution which is represented by a complicated formula then this formula can be always simplified.

Liouville, undoubtedly, was inspired by the results by Lagrange, Abel, and Galois on the non-solvability by radicals of algebraic equations. Differently from the Galois theory, Liuoville's theory does not involve the notion of the group of automorphisms. Liouville, however, uses, in order to simplify his formulae, 'infinitely small automorphisms'.

Let us return to Theorem 2 on the integrability of algebraic functions. The following corollary follows from this theorem.

COROLLARY. *If the integral of an algebraic function A is a generalized elementary function then the differential form $A(x)dx$ has some unavoidable singularities on the Riemann surface of the algebraic function A.*

It is well known that on every algebraic curve with positive genus there exist non-singular differential forms (the so called abelian differentials of first type). It follows that algebraic functions whose Riemann surfaces have positive genus are not, in general, integrable by generalized elementary functions.

This was already known by Abel, who discovered it as he was proving the non-solvability by radicals of a fifth-degree generic equation. Observe also that the Abel proof of the non-solvability by radicals is based on topological arguments. I do not know whether the topological properties of the Riemann surfaces of functions representable by generalized quadratures are different from those of the Riemann surfaces of generalized elementary functions. Indeed, I am unable to prove through topological arguments that the integral of an algebraic function is not an elementary function: each one of such integrals is by definition a function representable by generalized quadratures. However, if an algebraic function depends on a parameter its integral may depend on the parameter in an arbitrarily complicated manner. One can prove that *the integral of an algebraic function, as function of one parameter, can be not representable by generalized quadratures*, and consequently *can be not a generalized elementary function of the parameter* (cf., example in §A.9).

A.3 Picard–Vessiot's theory

Consider the linear differential equation

$$y^{(n)} + r_1(x)y^{(n-1)} + \cdots + r_n(x)y = 0, \tag{A.12}$$

in which the $r_i(x)$s are rational functions of complex argument.

Near a non-singular point x_0 there exist n linearly independent solutions $y_1, \ldots, y_n$ of equation (A.12). In this neighbourhood one can consider the functions field $\mathcal{R}\{y_1, \ldots, y_n\}$, obtained by adding to the field of rational functions $\mathcal{R}$ all solutions y_i and all their derivatives $y_i^{(p)}$ until order $(n-1)$. (The derivatives of higher order are obtained from equation (A.12).

The field of functions $\mathcal{R}\{y_1, \ldots, y_n\}$ is a differential field, i.e., it is closed with respect to the differentiation, as well as the field $\mathcal{R}$ of rational functions. One calls *automorphism of the differential field* F an automorphism σ, of the field F, which also preserves the differentiation, i.e., $\sigma(g') = [\sigma(g)]'$. Consider an automorphism σ of the differential field $\mathcal{R}\{y_1, \ldots, y_n\}$ which fixes all elements of the field $\mathcal{R}$. The set of all automorphisms of this type forms a group which is called the *Galois group* of equation (A.12). Every automorphism σ of the Galois group sends a solution of the equation to a solution of the equation. Hence to each one of such automorphisms there corresponds a linear transform M_σ of the space V^n of solutions. The automorphism σ is completely defined by the transform M_σ, because the field $\mathcal{R}\{y_1, \ldots, y_n\}$ is generated by the functions y_is. In general, not every linear transform of the space V^n is an automorphism σ of the Galois group. The reason is that the automorphism σ preserves all differential relations holding among the solutions. The Galois group can be considered as a special group of linear transforms of the solutions. It turns out that this group is algebraic.

So the Galois group of an equation is the algebraic group of linear transforms of the space of solutions that preserves all differential relations betweeen the solutions.

Picard began to translate systematically the Galois theory in the case of linear differential equations. As in the original Galois theory, one also finds here an one to one correspondence (the *Galois correspondence*) between the intermediate differential field and the algebraic subgroups of the Galois group.

Picard and Vessiot proved in 1910 that the solvability of an equation by quadratures and by generalized quadratures depends exclusively on its Galois group.

PICARD–VESSIOT'S THEOREM. *A differential equation is solvable by quadratures if and only if its Galois group is soluble. A differential equation is solvable by generalized quadratures if and only if the connected component of unity in its Galois group is soluble.*

The reader can find the basic results of the differential Galois theory in the book [2]. In [3] he will find a brief exposition of the actual state of this theory together with a rich bibliography.

Observe that from the Picard–Vessiot theorem it is not difficult to deduce that if equation (A.12) is solvable by generalized quadratures then it has a solution of the form $y_1 = \exp(\int^x A_1(t), dt)$, where $A_1(x)$ is an algebraic function. If the equation has an explicit solution y_1 then one can decrease its order, taking as the new unknown function $z = (y/y_1)'$. The function z satisfies a differential equation having an explicit form and a lower order. If the initial equation was solvable, the new equation for function z is also solvable. By the Picard–Vessiot theorem it must therefore have a solution of the type $z_1 = \exp \int A_2(x)dx$, where A_2 is an algebraic function, etc.. We see in this way that if a linear equation is solvable by generalized quadratures, the formulae expressing the solutions are not exceedingly complicated. Here the Picard–Vessiot approach coincides with the Liouville approach. Moreover, the criterion of solvability by generalized quadratures can be formulated without mentioning the Galois group. Indeed, equation (A.12) of order n is solvable by generalized quadratures if and only if has a solution of the form $y_1 = \exp \int^x A(x)dx$, and the equation of order $(n-1)$ for the function z is solvable by generalized quadratures.

This theorem was enunciated and proved by Murdakai–Boltovskii exactly in this form. Murdakai and Boltovskii obtained at the same time this result in 1910 using the Liouville method, independently of the works of Picard and Vessiot. The Mordukai–Boltovskii theorem is a generalization of the Liouville theorem (cf., Theorem 3 in the preceding section) to linear differential equations of any order.

A.4 Topological obstructions for the representation of functions by quadratures

There exists a third approach to the problem of the representability of a function by quadratures. (cf., [4]–[10]). Consider the functions representable by quadratures as multi-valued analytic functions of a complex variable. It turns out that there are some topological restrictions on the kind of disposition on the complex plane of the Riemann surface of a function representable by quadratures. If the function does not satisfy these conditions, it cannot be represented by quadratures.

This approach, besides the geometrical evidence, possesses the following advantage. The topological obstructions are related to the character of the multi-valued function. They hold not only for functions representable by quadratures, but also for a wider class of functions. This class is obtained by adding to the functions representable by quadratures all the meromorphic functions and allowing the presence of such functions in all formulae. Hence the topological results on the non-representability by quadratures are stronger that those of algebraic nature. The reason of this is that the composition of two functions is not an algebraic operation. In differential algebra, instead of the composition of two functions one considers the differential equation that they satisfy. But, for instance, the Euler function Γ does not satisfy any algebraic differential equation; therefore it is useless to seek an equation satisfied, for example, from the function $\Gamma(\exp x)$. The unique known results on the non-representability of functions by quadratures and, for instance, by the Euler functions Γ are those obtained by our method.

On the other hand, by this method one cannot prove the non-representability by quadratures of an arbitrary meromorphic single-valued function.

Using the Galois differential theory (and, to be precise, its linear–algebraic part, related to the matrix algebraic groups and their differential invariants) one can prove that the sole reason for the non-solvability by quadratures of the linear differential equations of Fuchs type (cf., §A.11) is of topological nature. In other words, when there are no topological obstructions for the solvability by quadratures for a differential equation of Fuchs type this equation is solvable by quadratures.

The topological obstructions for the representation of a function by quadratures and by generalized quadratures are the following:

First, functions representable by generalized quadratures and, as a special case, by quadratures can have at most a countable set of singular points on the complex plane. (cf., §A.5) (even though for the simplest functions representable by quadratures the set of singular points may be everywhere dense!).

Second, the monodromy group of a function representable by quadratures is necessarily soluble (cf., §A.7) (whereas for the simplest functions representable by quadratures the monodromy group may already contain a continuum of elements!).

There also exist analogous topological restrictions on the disposition of the Riemann surface for functions representable by generalized quadratures. However, these restrictions cannot be simply formulated: in this case the monodromy group is not considered as an abstract group, but as the group of permutations of the sheets of the Riemann surface. In other words, in the formulation of such restrictions not only the monodromy group intervenes, but also the *monodromy pair* of the function. The monodromy pair of a function consists of its monodromy group and of a stationary subgroup for some germ (cf., §A.9). We shall see this geometrical approach to the problem of solvability more precisely.

A.5 $\mathcal{S}$-functions

We define a class of functions which will be the object of this section.

DEFINITION. One calls $\mathcal{S}$-*function* an analytic multi-valued function of a complex variable if the set of its singular points is at most countable.

Let us make this definition more precise. Two regular germs f_a and g_b, defined at points a and b on the Riemann sphere S^2, are said to be equivalent if the germ g_b is obtained from the germ f_a by a regular continuation along some curve. Every germ g_b equivalent to the germ f_a is called a regular germ of the analytic multi-valued function f generated by the germ f_a.

A point $b \in S^2$ is said to be *singular for the germ* f_a if there exists a curve $\gamma\,[0,1] \to S^2$, $\gamma(0) = a$, $\gamma(1) = b$, such that the germ f_a cannot be regularly continued along this curve, but for every t, $0 \leq t < 1$, this germ can be continued along the shortened curve $\gamma\,[0,t] \to S^2$. It is easy to see that the sets of singular points for equivalent germs coincide.

A regular germ is called an $\mathcal{S}$-*germ* if the set of its singular points is at most countable. An analytic multi-valued function is called an $\mathcal{S}$-function if each one of its regular germs is an $\mathcal{S}$-germ.

We proved the following theorem.

THEOREM ON THE CLOSURE OF THE CLASS OF $\mathcal{S}$-FUNCTIONS (see [6],[8],[10]). *The class $\mathcal{S}$ of all $\mathcal{S}$-functions is closed with respect to the following operations:*

- *1) differentiation, i.e., if $f \in \mathcal{S}$, then $f' \in \mathcal{S}$;*
- *2) integration, i.e., if $f \in \mathcal{S}$, then $\int f(x)dx \in \mathcal{S}$;*
- *3) composition, i.e., if $g,\ f \in \mathcal{S}$, then $g \circ f \in \mathcal{S}$;*
- *4) meromorphic operation, i.e., if $f_i \in \mathcal{S}$, $i = 1, \dots, n$, $F(x_1, \dots, x_n)$ is a meromorphic function of n variables and $f = F(f_1, \dots, f_n)$, then $f \in \mathcal{S}$[4];*
- *5) solution of algebraic equations, i.e., if $f_i \in \mathcal{S}$, $i = 1, \dots, n$, and $f^n + f_1 f^{n-1} + \dots + f_n = 0$, then $f \in \mathcal{S}$;*
- *6) solution of linear differential equations, i.e., if $f_i \in \mathcal{S}$, $i = 1, \dots, n$, and $f^{(n)} + f_1 f^{(n-1)} + \dots + f_n = 0$, then $f \in \mathcal{S}$.*

COROLLARY. *If the multi-valued function f can be obtained from single-valued $\mathcal{S}$-functions by the operations of integration, differentiation, meromorphic operations, compositions, solutions of algebraic and linear differential equations, then the function f has at most a countable set of singular points. In particular, a function having a non countable set of singular points is not representable by generalized quadratures.*

A.6 Monodromy group

The *monodromy group* of an $\mathcal{S}$-function f with a set A of singular points is the group of all permutations of the sheets of the Riemann surface of f which are visited when one moves around the points of set A.

[4]More precisely, *the meromorphic operation* defined by the meromorphic function $F(x_1, \dots, x_n)$ puts into correspondence with the functions $f_1, \dots, f_n$ a new function $F(f_1, \dots, f_n)$. The arithmetic operations and the exponential are examples of meromorphic operations, corresponding to the functions $F_1(x, y) = x + y$, $F_2(x, y) = x \cdot y$, $F_3(x, y) = x/y$ and $F_4(x) = \exp x$.

More precisely, let F_a be the set of all germs of the $\mathcal{S}$-function f at the point a, not belonging to the set A of singular points. Consider a closed curve γ in $S^2 \setminus A$ beginning at the point a. The continuation of every germ of the set F_a along the curve γ leads to a germ of the set F_a.

Consequently to every curve γ there corresponds a mapping of set F_a into itself, and to homotopic curves in $S^2 \setminus A$ there corresponds the same mapping. To the composition of curves there corresponds the mapping composition. One has thus defined an homomorphism τ of the fundamental group of the set $S^2 \setminus A$ in the group $S(F_a)$ of the bijective mappings of the set F_a into itself. One calls *monodromy group* of the $\mathcal{S}$-function f the image of the fundamental group $\pi_1(S^2 \setminus A, a)$ in the group $S(F_a)$ under the homomorphism τ.

We show some results which are useful in the study of functions representable by quadratures as functions of one complex variable.

EXAMPLE. Consider the function $w(z) = \ln(1 - z^\alpha)$, where $\alpha > 0$ is an irrational number. The function w is an elementary function given by a very simple formula. However, its Riemann surface is very complicated. The set A of its singular points consists of the points $0, \infty$ and of the points $a_k = e^{2\pi ki/\alpha}$, where k is any integer. Since α is irrational the points a_k are densely distributed on the unitary circle. It is not difficult to prove that the fundamental group $\pi_1(S^2 \setminus A)$ and the monodromy group of the function w are continuous. One can also prove that the image under the homomorphism τ of the fundamental group $\pi_1(S^2 \setminus \{A \cup b\})$ of the complement of $A \cup b$, where $b \neq a_k$ is an arbitrary point on the unit circle, is a proper subgroup of the monodromy group of the function w. (That the elimination of a single point can produce a radical change in the monodromy group makes all proofs essentially difficult).

A.7 Obstructions for the representability of functions by quadratures

We have proved the following theorem.

THEOREM ([6],[8],[10]). *The class of all $\mathcal{S}$-functions having a soluble monodromy group is closed with respect to the composition, the meromorphic operations, the integration and the differentiation.*

We thus obtain the following corollary.

RESULT ON QUADRATURES. *The monodromy group of a function f representable by quadratures is soluble. Moreover, also the monodromy group of every function f which is obtained from single-valued $\mathcal{S}$-functions by means of compositions, meromorphic operations, integration and differentiation is soluble.*

We see now the application of this result to algebraic equations.

A.8 Solvability of algebraic equations

Consider the algebraic equation

$$y^n + r_1 y^{n-1} + \cdots + r_n = 0, \tag{A.13}$$

where the r_is are rational functions of complex variable x.

Near to a non-singular point x_0 there are all solutions $y_1, \ldots, y_n$ of the equation (A.13). In this neighbourhood one can consider the field of all functions $\mathcal{R}\{y_1, \ldots, y_n\}$ that is obtained by adding to the field $\mathcal{R}$ all solutions y_i.

Consider the automorphisms σ of the field $\mathcal{R}\{y_1, \ldots, y_n\}$ which fix every element of the field $\mathcal{R}$. The totality of these automorphisms forms a group which is called the *Galois group of the equation* (A.13). Every automorphism σ of the Galois group transforms a solution of the equation into a solution of the equation; consequently to every automorphism σ there corresponds a permutation S_σ of the solutions. The automorphism σ is completely defined by the permutation S_σ because the field $\mathcal{R}\{y_1, \ldots, y_n\}$ is generated by the functions y_i. In general, not all permutations of the solutions can be continued to an automorphism σ of the Galois group: the reason is that the automorphisms σ preserve all relations existing among the solutions.

The Galois group of an equation is thus the permutation group of the solutions that preserves all relations among the solutions.

Every permutation S_γ of the set of solutions can be continued, as an automorphism of the monodromy group, to an automorphism of the entire field $\mathcal{R}\{y_1, \ldots, y_n\}$. Indeed, with functions $y_1, \ldots, y_n$, along the curve γ, every element of the field $\mathcal{R}\{y_1, \ldots, y_n\}$ is continued meromorphically. This continuation gives the required automorphism, because during the continuation the arithmetic operations are preserved and every rational function returns to its preceding value because of the uniqueness.

In this way the monodromy group of the equation is contained in the Galois group: in fact, the Galois group coincides with the monodromy group. Indeed, the functions of the field $\mathcal{R}\{y_1, \ldots, y_n\}$ that are fixed under the action of the monodromy group are the single-valued functions. These functions are algebraic, but every algebraic single-valued function is a rational function. Therefore the monodromy group and the Galois group have the same field of invariants, and thus by the Galois theory they coincide.

According to Galois theory, the equation (A.13) is solvable by radicals over the field of rational functions if and only if its Galois group is soluble over this field. In other words, the Galois theory proves the following theorems:

1) *An algebraic function y whose monodromy group is soluble is representable by radicals.*

2) *An algebraic function y whose monodromy group is not soluble is not representable by radicals.*

Our theorem makes the result (2) stronger:

An algebraic function y whose monodromy group is not soluble cannot be represented through single-valued $\mathcal{S}$-functions by means of meromorphic operations, compositions, integrations, and differentiations.

If an algebraic equation is not solvable by radicals then it remains non solvable using the logarithms, the exponentials, and the other meromorphic functions on the complex plane. A stronger version of this statement in given in §A.15.

A.9 The monodromy pair

The monodromy group of a function is not only an abstract group but is the group of transitive permutations of the sheets of its Riemann surface. Algebraically this object is given by a pair of groups: the permutation group and a subgroup of it, the stationary group of a certain element.

One calls the *monodromy pair of an $\mathcal{S}$-function* a pair of groups consisting of the monodromy group of this function and the stationary subgroup of a sheet of the Riemann surface. The monodromy pair is defined correctly, i.e., this pair of groups up to isomorphism does not depend on the choice of the sheet.

DEFINITION. The pair of groups $[\Gamma, \Gamma_0]$ is called an *almost soluble pair of groups* if there exists a sequence of subgroups

$$\Gamma = \Gamma_1 \supseteq \cdots \supseteq \Gamma_m, \quad \Gamma_m \subset \Gamma_0,$$

such that for every i, $1 \leq i \leq m-1$ the group Γ_{i+1} is a normal divisor of the group Γ_i, and the quotient group Γ_i/Γ_{i+1} is either commutative or finite.

Any group Γ can be considered as the pair of groups $[\Gamma, e]$, where e is the unit subgroup (the group containing only the unit element). We say that the group Γ is *almost soluble* if the pair $[\Gamma, e]$ is almost soluble.

THEOREM ([6],[8],[10]). *The class of all $\mathcal{S}$-functions having a monodromy pair almost soluble is closed with respect to the composition, the meromorphic operations, the integration, the differentiation, and the solutions of algebraic equations.*

We thus obtain the following corollary.

RESULT ON GENERALIZED QUADRATURES. *The monodromy pair of a function f representable by generalized quadratures is almost soluble. Moreover, also the monodromy pair of every function f, which is obtained from single-valued $\mathcal{S}$-functions by means of the composition, the meromorphic operations, the integration, the differentiation and the solutions of algebraic equations is almost soluble.*

Let us now consider some examples of functions not representable by generalized quadratures. Suppose the Riemann surface of a function f be a universal covering of $S^2 \setminus A$, where S^2 is the Riemann sphere and A is a finite set, containing at least three points. Thus *the function f cannot be expressed in terms of $\mathcal{S}$-functions by means of generalized quadratures, compositions, and meromorphic operations.* Indeed, the monodromy pair of this function consists of a free non commutative group and its unit subgroup. One easily sees that such a pair of groups is not almost soluble.

EXAMPLE. Consider the function z which realizes the conformal transformation of the upper semi-plane into the triangle with vanishing angles, bounded by three arcs of circle. The function z is the inverse of the modular Picard function. The Riemann surface of the function z is a universal covering of the sphere without three points; consequently the function z cannot be expressed in terms of single-valued $\mathcal{S}$-functions by means of generalized quadratures, compositions, and meromorphic operations.

Observe that the function z is strictly related to the elliptic integrals

$$K(k) = \int_0^1 \frac{dx}{\sqrt{(1-x^2)(1-k^2x^2)}}; \quad K'(k) = \int_0^{\frac{1}{k}} \frac{dx}{\sqrt{(1-x^2)(1-k^2x^2)}}.$$

Every one of the functions $K(k)$, $K'(k)$, and $z(w)$ can be obtained from the others by quadratures. It follows that none of the integrals $K(k)$ and $K'(k)$ can be expressed in terms of single-valued S-functions by means of generalized quadratures, compositions, and meromorphic operations.

In the following section we will generalize the example above, finding all polygons, bounded by arcs of circles, to which the upper semi-plane can be sent by functions representable by generalized quadratures.

A.10 Mapping of the semi-plane to a polygon bounded by arcs of circles

A.10.1 Application of the symmetry principle

In the complex plane consider a polygon G bounded by arcs of circles. By Riemann's theorem, there exists a function f_G sending the upper semi-plane to polygon the G. This mapping was studied by Riemann, Schwarz, Christoffel, Klein, and others (cf., for example, [11]). Let us recall some classical results which will be useful.

Denote by $B = \{b_j\}$ the pre-image of the set of the vertices of the polygon G under the mapping f_G, by $H(G)$ the group of conformal transformations of the sphere generated by the inversions with respect to the sides of the polygon, and by $L(G)$ the subgroup of homographic mappings (the quotient of two linear functions). $L(G)$ is a subgroup of index 2 of the group $H(G)$. From the Riemann–Schwarz symmetry principle one obtains the following results.

PROPOSITION.

- *1) The function f_G can be meromorphically continued along any curve avoiding the set B.*

- *2) All germs of the multi-valued functions f_G in a non-singular point $a \notin B$ are obtained by applying to a given germ f_a the group $L(G)$ of homographic mappings.*

- *3) The monodromy group of the function f_G is isomorphic to the group $L(G)$.*

- *4) The singularities of the function f_G are of the following types at the points b_j. If at the vertex a_j of the polygon G that corresponds* *to the point b_j the angle is equal to $\alpha_j \neq 0$, then the function f_G, through a homographic transformation, is put into the form $f_G(z) = (z-b_j)^{\beta_j}\varphi(z)$, where $\beta_j = \alpha_j/2\pi$, and the function φ is holomorphic in a neighbourhood of the point b_j. If the angle α_j is equal to zero then the function f_G is put by an homographic transformation into the form $f_G(z) = \ln(z) + \varphi(z)$, where the function φ is holomorphic in a neighbourhood of b_j.*

From our results it follows that if the function f_G is representable by generalized quadratures then the group $L(G))$ and the group $H(G)$ are almost soluble.

A.10.2 Almost soluble groups of homographic and conformal mappings

Let π be the epimorphism of the group $SL(2)$ of matrices of order 2 with unit determinant onto the group of the homographic mappings L,

$$\pi : \begin{pmatrix} a & b \\ c & d \end{pmatrix} \to \frac{az+b}{cz+d}.$$

Since $\ker \pi = \mathbb{Z}_2$, the group $\widetilde{L} \subseteq L$ and the group $\pi^{-1}(\widetilde{L}) = \Gamma \subseteq SL(2)$ are both almost soluble. The group Γ is a group of matrices: therefore Γ *is almost soluble if and only if it has a normal subgroup Γ_0 of finite index which admits a triangular form.* (This version of Lie's theorem is true also in higher dimensions and plays an important role in differential Galois theory). Since the group Γ_0 consists of matrices of order 2, the group Γ_0 can be put into triangular form in one of the three following cases:

- 1) group Γ_0 has only one one-dimensional eigenspace;
- 2) group Γ_0 has two one-dimensional eigenspaces;
- 3) group Γ_0 has a two-dimensional eigenspace.

Consider now the group of homographic mappings $\widetilde{L} = \pi(\Gamma)$. The group $\widetilde{L}$ of homographic mappings is almost soluble if and only it has a normal subgroup $L_0 = \pi(\Gamma_0)$ of finite index, and the set of invariant points consists of either a unique point, or two points, or of the whole Riemann sphere.

The group of conformal mappings $\widetilde{H}$ contains the group $\widetilde{L}$ of index 2 (or of index 1) consisting of the homographic mappings. Hence for the almost soluble group $\widetilde{H}$ of conformal mappings an analogous proposition holds.

LEMMA. *A group of conformal mappings of the sphere is almost soluble if and only if it satisfies at least one of these conditions:*

- 1) *the group has only one invariant point;*
- *2) the group has an invariant set consisting of two points;*
- *3) the group is finite.*

This lemma follows from the preceding propositions because the set of invariant points for a normal divisor is invariant under the action of the group. It is well known that *a finite group $\widetilde{L}$ of homographic mappings of the sphere is sent by a homographic transformation of coordinates to a group of rotations.*

It is not difficult to prove that if the product of two inversions with respect to two different circles corresponds under the stereographic projection to a rotation of the sphere, then these circles correspond to great circles. Hence *every finite group $\widetilde{H}$ of conformal mappings generated by the inversions with respect to some circles is sent by a homographic transformation of coordinates to a group of motions of the sphere, generated by reflections.*

All the finite groups of the motions of the sphere generated by reflections are well known. They are exactly the symmetry groups of the following objects:

- 1) the regular pyramid with a regular n-gon as basis;
- 2) the n-dihedron, i.e., the solid made from two regular pyramids joining their bases;
- 3) the tetrahedron;
- 4) the cube or the octahedron;
- 5) the dodecahedron or the icosahedron.

All these groups of symmetries, except the group of the dodecahedron–icosahedron, are soluble. The sphere whose centre coincides with the centre of gravity of the solid is cut by the symmetry planes of the solid along a net of great circles. Lattices corresponding to the stated solids are called the finite nets of great circles. The stereographic projections of these finite nets are shown in Figures 129–133.

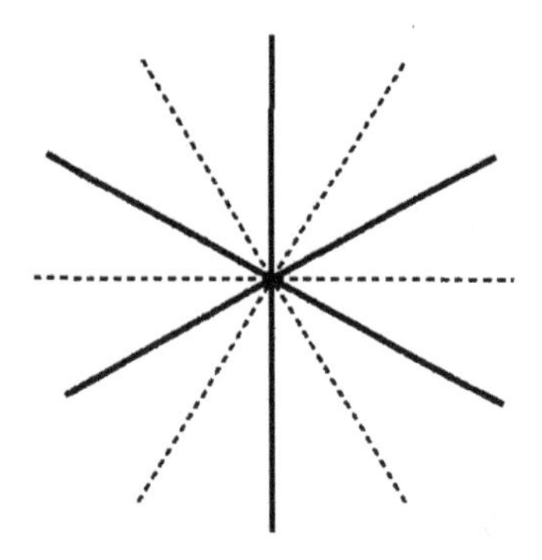

FIG. 129: PYRAMID

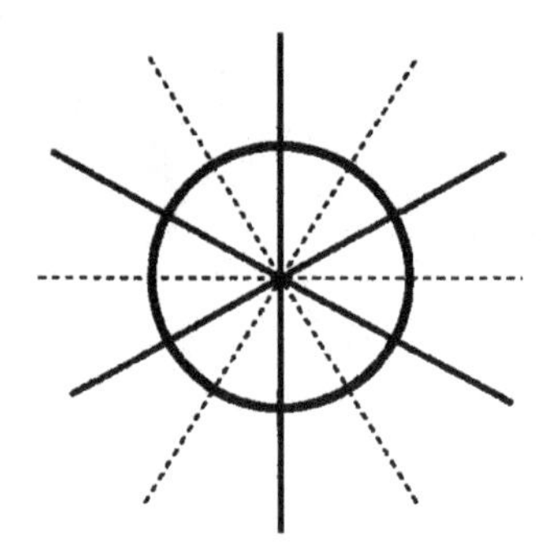

FIG. 130: 6-DIHEDRON

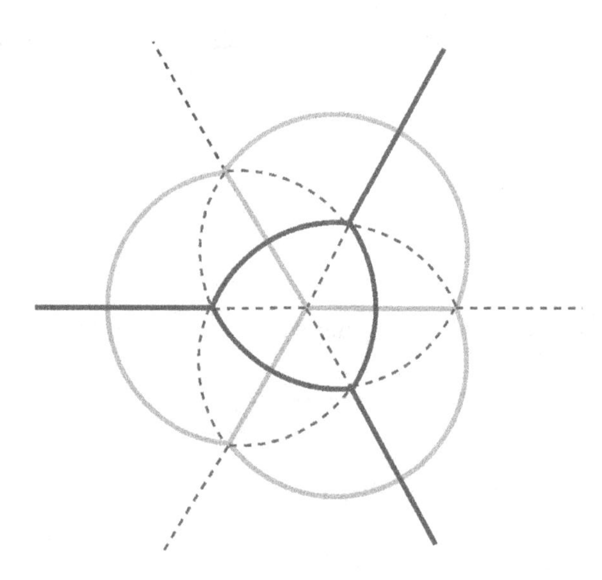

FIG. 131: TETRAHEDRON–TETRAHEDRON

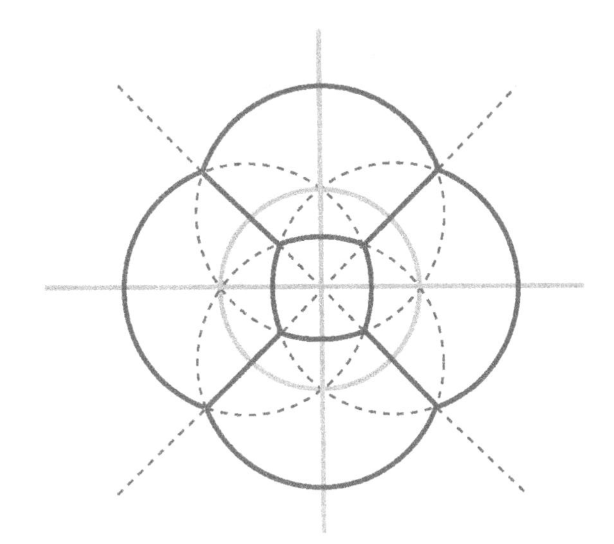

FIG. 132: CUBE–OCTAHEDRON

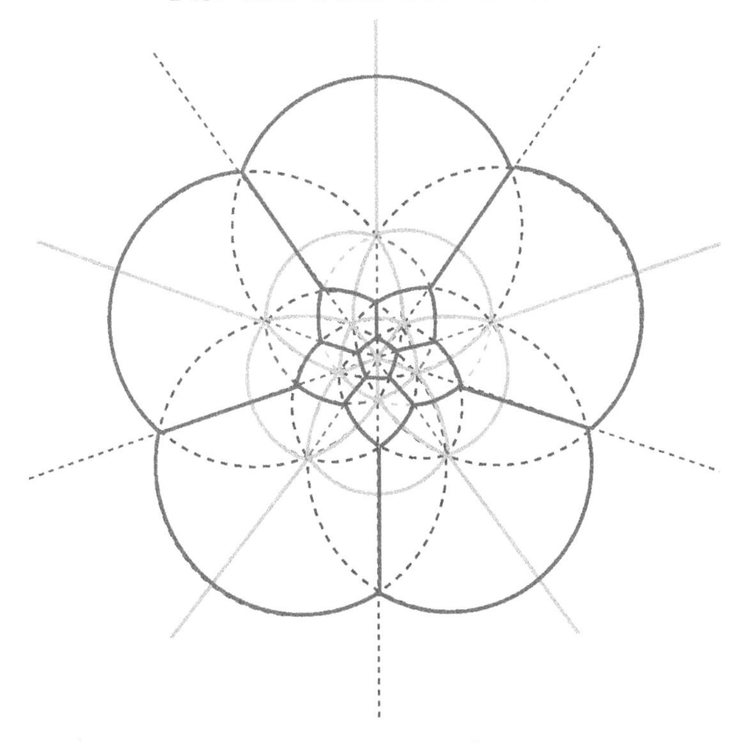

FIG. 133: DODECAHEDRON–ICOSAHEDRON

A.10.3 The integrable case

Let us come back to the problem of the representability of the function f_G by generalized quadratures.

We consider now the different possible cases and we prove that the condition we have found is not only necessary but also sufficient for the representability of the function f_G by generalized quadratures.

FIRST INTEGRABILITY CASE. The group $H(G)$ has an invariant point. This means that the continuations of the edges of the polygon G intersect in a point. Sending this point to infinity by a homographic transformation, we obtain the polygon $\overline{G}$ bounded by segments of straight lines (cf., Figure 134).

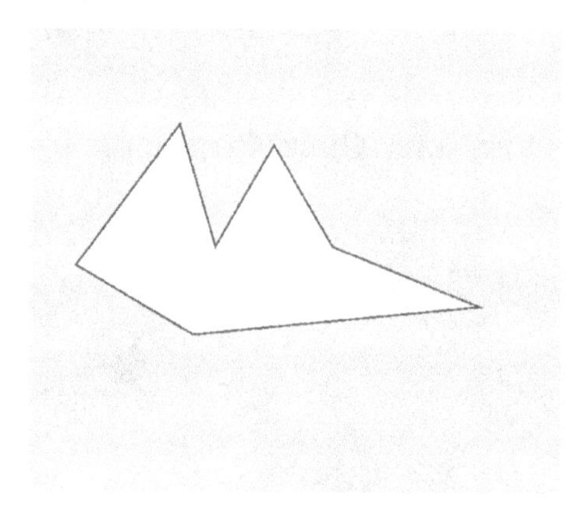

FIGURE 134

All mappings in $L(\overline{G})$ have the form $z \to az + b$. All germs of the function $\overline{f} = f_{\overline{G}}$ at a non-singular point c are obtained by applying to a given germ $\overline{f}_c$ the group $L(\overline{G})$, $\overline{f}_c \to a\overline{f}_c + b$. The germ $R_c = \overline{f}''_c/\overline{f}'_c$ is invariant under the action of the group $L(\overline{G})$. This means that the germ R_c is the germ of a single-valued function. A singular point b_j of the function R_c can be only a pole (cf., the Proposition in §A.10.1). Thus the function R_c is rational. The equation $\overline{f}''/\overline{f}' = R$ is integrable by quadratures. This case of integrability is well known. The function $\overline{f}$ in this case is called the *Christoffel–Schwarz integral.*

SECOND INTEGRABILITY CASE. The invariant set of the group $H(G)$ consists of two points. This means that there are two points with the following properties: for every side of the polygon G these points either are obtained by an inversion with respect to this side or belong to the continuation of this side. Sending one of these points to the origin and the other one to infinity by a homographic transformation, we obtain the polygon $\overline{G}$ bounded by arcs of circles with centre at the point 0 and by segments of rays coming from the point 0 (cf., Figure 135).

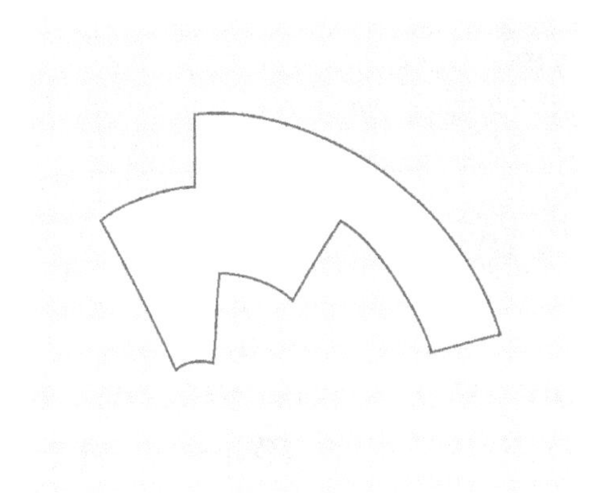

FIGURE 135

All transformations of the group $L(\overline{G})$ are of the form $z \to az$, $z \to b/z$. All germs of the function $\overline{f} = f_{\overline{G}}$ at a non-singular point c are obtained by applying to a given germ $\overline{f}_c$ the transformations of the group $L(\overline{G})$

$$\overline{f}_c \to a\overline{f}_c, \ \overline{f}_c \to b/\overline{f}_c.$$

The germ $R_c = (\overline{f}'_c/\overline{f}_c)^2$ is invariant under the action of the group $L(\overline{G})$ and it is the germ of the single-valued function R. The only singularities of the function R are poles (cf., the Proposition in §A.10.1). Thus the function R_c is rational. The equation $R = (\overline{f}''/\overline{f}')^2$ is integrable by quadratures.

THIRD INTEGRABILITY CASE. The group $H(G)$ is finite. This means that polygon G is sent by a homographic transformation to a polygon $\overline{G}$ whose sides lie on a finite net of great circles (see Figures 129–133). The group $L(G)$ is finite, and as a consequence the function f_G has a finite number of values. Since all singularities of the function f_G are of 'jump' type ((cf., the Proposition in §A.10.1) the function f_G is an algebraic function.

Let us analyze the case in which the group $H(G)$ is finite and soluble. This happens if and only if the polygon G is sent by a homographic transformation to a polygon $\overline{G}$ whose sides lie on a net of great circles different from that of the dodecahedron–icosahedron. In this case the group $L(G)$ is soluble, and the function f_G in expressed in terms of rational functions by means of arithmetic operations and of radicals (cf., §A.8).

From our results a theorem follows:

THEOREM ON THE POLYGONS BOUNDED BY OF ARCS OF CIRCLES ([6],[8],[10]). *For an arbitrary polygon G not belonging to the three cases of integrability above, the function f_G not only is not representable by generalized quadratures, but it cannot be expressed in terms of single-valued $\mathcal{S}$-functions by means of generalized quadratures, compositions, and meromorphic operations.*

A.11 Topological obstructions for the solvability of differential equations

A.11.1 The monodromy group of a linear differential equation and its relation with the Galois group

Consider the linear differential equation

$$y^{(n)} + r_1 y^{(n-1)} + \cdots + r_n y = 0, \tag{A.14}$$

where the r_is are rational functions of the complex variable x. The poles of the functions r_i and ∞ are called the singular points of the equation (A.14).

Near a non-singular point x_0 the solutions of the equations form a space V^n of dimension n. Consider now an arbitrary curve $\gamma(t)$ on the complex plane, beginning at x_0 and ending at the point x_1 and avoiding the singular points a_i. The solutions of the equation can be analytically continued along the curve, remaining solutions of the equation. Hence to every curve γ there corresponds a linear mapping M_γ of the space $V^n_{x_0}$ of the solutions at the point x_0 in the space $V^n_{x_1}$ of the solutions at the point x_1.

If one changes the curve γ, avoiding the singular points and leaving its ends fixed, the mapping M_γ does not vary. Hence to a closed curve there corresponds a linear transform of the space V^n into itself. The totality of these linear transforms of the space V^n forms a group which is called the *monodromy group of the equation* (A.14). So the monodromy group of an equation is the group of the linear transforms of the solutions which correspond to different turns around the singular points. The monodromy group of an equation characterizes the property of its solutions being multi-valued.

Near a non-singular point x_0 there are n linearly independent solutions, $y_1, \dots, y_n$, of the equation (A.14). In this neighbourhood one can consider the field of functions $\mathcal{R}\{y_1, \dots, y_n\}$ that is obtained by adding to the field of rational functions $\mathcal{R}$ all solutions y_i and all their derivatives.

Every transformation M_γ of the monodromy group of the space of solutions can be continued to an automorphism of the entire field $\mathcal{R}\{y_1, \dots, y_n\}$. Indeed, with functions $y_1, \dots, y_n$, along the curve γ every element of the field $\mathcal{R}\{y_1, \dots, y_n\}$ can be analytically continued. This continuation gives the required automorphism, because during the continuation the arithmetic operations and the differentiation are preserved, and the rational functions come back to their initial values because of their uniqueness.

In this way *the monodromy group of an equation is contained in its Galois group.*

The field of the invariants of the monodromy group is a subfield of $\mathcal{R}\{y_1, \dots, y_n\}$, consisting of the single-valued functions. Differently from the algebraic case, for differential equations the field of invariants under the action of the monodromy group can be bigger than the field of rational functions.

For example, for the differential equation (A.14), in which all the coefficients $r_i(x)$s are polynomials, all solutions are single-valued. But of course the solutions of such equations are not always polynomials. The reason is that here the solutions of differential equations may grow exponentially in approaching the singular points. One knows an extension of the class of linear differential equations for which there are no similar complications, i.e., for which the solutions, whilst approaching the singular points, grow at most as some power. Differential equations which possess this property are called *equations of Fuchs' type.*

For differential equations of Fuchs' type the Frobenius theorem holds.

THEOREM 1. *For the differential equations of Fuchs' type the subfield of the differential field $\mathcal{R}\{y_1, \dots, y_n\}$ that consists of single-valued functions coincides with the field of rational functions.*

According to the differential Galois theory, from the Frobenius theorem it follows that the algebraic closure of the monodromy group M (i.e., the smallest algebraic group containing M) coincides with the Galois group.

The differential Galois theory thus gives the following criterion of solvability of differential equations of Fuchs' type.

THEOREM 2. *A differential equation of Fuchs' type is solvable by quadratures or by generalized quadratures if its monodromy group is, respectively, soluble or almost soluble.*

The differential Galois theory provides at the same time two results:

- *1) if the monodromy group of a differential equation of Fuchs' type is soluble (almost soluble) then this equation is solvable by quadratures (by generalized quadratures).*
- *2) if the monodromy group of a differential equation of Fuchs' type is not soluble (almost soluble) then this equation is not solvable by quadratures (by generalized quadratures).*

Our theorem makes the result (2) stronger. Indeed, it is easy to see that for almost every solution of the differential equation (A.14) the monodromy pair is $[M, e]$, where M is the monodromy group of the equation, and e its trivial subgroup. We thus have the following:

THEOREM 3 ([6],[8]). *If the monodromy group of the differential equation (A.14) is not soluble (almost soluble) then almost every solution of this equation is not representable in terms of single-valued* $\mathcal{S}$*-functions by means of compositions, meromorphic operations, integrations, differentiations, and solutions of algebraic equations.*

Is the monodromy group of a given linear differential equation soluble (almost soluble)? This question turns out to be quite difficult. However, there exists an interesting example in which the answer to this question is very simple.

A.11.2 Systems of differential equations of Fuchs' type with small coefficients

Consider a system of linear differential equations of Fuchs type, i.e., a system of the type

$$y' = Ax, \tag{A.15}$$

where $y = y_1, \ldots, y_n$ is the unknown vectorial function and A is an $n \times n$ matrix consisting of rational functions of the complex variable x, having the following form:

$$A(x) = \sum_{i=1}^{k} \frac{A_i}{x - a_i},$$

where the A_is are constant matrices.

If the matrices A_is are at the same time put into triangular form, then the system (A.15), as every triangular system, is solvable by quadratures. There are undoubtedly non-triangular systems which are solvable. However, if the matrices A_is are sufficiently small, then such systems do not exist. More precisely, we have obtained the following results:

THEOREM 4([9]). *A non-triangular system (A.15) with matrices A_is sufficiently small, $\|A_i\| < \varepsilon(a_1, \ldots, a_k, n)$, is strictly non-solvable, i.e., it is not solvable even using all single-valued $\mathcal{S}$-functions, compositions, meromorphic operations, integrations, differentiations, and solutions of algebraic equations.*

The proof of this theorem uses the Lappo-Danilevskij theory [12].

A.12 Algebraic functions of several variables

Up to now we have considered only single-valued functions. We are ready to make two observations concerning functions of several variables, the proofs of which do not require new notions and are obtained by the same method we used for single-valued functions.

Consider the algebraic equation

$$y^n + r_1 y^{n-1} + \cdots + r_n = 0, \tag{A.16}$$

where the r_is are rational functions of k complex variables $x_1, \ldots, x_k$.

1) According to the Galois theory the equation (A.16), having a soluble monodromy group, is solvable by radicals. *But if the monodromy group of the equation (A.16) is not soluble then not only is the equation not solvable by radicals, but it cannot be solved even by using radicals of entire functions of several variables, arithmetic operations and compositions.* This statement can be considered as a variation of the Abel theorem about the non-solvability of algebraic equations of degree higher than four. (A stronger result is presented in §A.15).

2) The equation (A.16) defines an algebraic function of k variables. What are the conditions for representing a function of k variables by algebraic functions of a smaller number of variables, using compositions and arithmetic operations? The 13th Hilbert problem consists of this question[5]. If one excludes the remarkable results [13],[14] on this subject[6],

---

[5]The problem of the composition was formulated by Hilbert for classes of continuous functions, not for algebraic functions. A.G. Vitushkin considered this problem for smooth functions and proved the non-representability of functions of n variables with continuous derivatives up to order p by functions of k variables with continuous derivatives up to order q having a lower 'complexity', i.e., for $k/q < n/p$ [15]. Afterwards he applied his method to the study of complexity in the problem of tabularizations [16].

Vitushkin's results was also proved by Kolmogorov, developing his own theory of the ϵ-entropy for classes of functions, measuring their complexity as well: this entropy, expressed by the logarithm of the number of ϵ-different functions, grows, for decreasing ϵ, as $(1/\epsilon)^{n/p}$ [17].

The solution of the problem in the Hilbert initial formulation turned out to be the opposite of that conjectured by Hilbert himself: Kolmogorov [18] was able to represent continuous functions of n variables by means of continuous functions of 3 variables, Arnold [19] represented the functions of three variables by means of functions of two, and finally Kolmogorov [20] represented functions of two variables as the composition of functions of a single variable with the help of the sole addition. (*Translator's note.*)

[6]V. I. Arnold [13] invented a completely new approach to the proof of the non-representability of an algebraic entire function of several variables as a composition of algebraic entire functions of fewer variables. This approach is based on the study of the cohomology of the complement of the set of the branches of the function, which leads to the study of the cohomology of the braid groups.

We must remark that here the definition of representability of an algebraic function differs from the classical definition. Classical formulae for the solutions by radicals of equations of degree 3 and 4 cannot be completely considered mere compositions: these multi-valued expressions by radicals contain, with the required roots, 'parasite' values also. The new methods show that these parasite values are unavoidable: even equations of degree 3 and 4 are not ***strictly*** (i.e., without parasite values) solvable by radicals.

In particular, Arnold proved [13] that if $n = 2^r$ $(r \geq 2)$ the algebraic function $\lambda(z)$ of n complex variables $z = (z_1, \ldots, z_n)$ defined by the equation

$$\lambda^n + z_1\lambda^{n-1} + \cdots + z_n = 0$$

is not strictly representable in any neighbourhood of the origin as a composition of algebraic entire functions (division is not allowed) with fewer than $n-1$ variables and of single-valued holomorphic functions of any number of variables. V.Ya. Lin [14] proved the same proposition for any $n \geq 3$.

The work of Arnold had a great resonance: the successive calculations of the cohomologies with other coefficients of the generalized braid groups allowed more and more extended results to be found.

The methods of the theory of the cohomologies of the braid groups, elaborated in

up to now there had been no proof that there exist algebraic functions of several variables which are not representable by algebraic functions of a single variable.

We know, however, the following result:

THEOREM ([4],[5]). *An entire function y of two variables (a, b), defined by the equation*

$$y^5 + ay + b = 0,$$

cannot be expressed in terms of entire functions of a single variable by means of compositions, additions, and subtractions.

The reason is the following. To every singular point p of an algebraic function one can associate a *local monodromy group*, i.e., the group of the permutations of the sheets of the Riemann surface which is obtained by going around the singularities of the function along curves lying in an arbitrarily small neighbourhood of the point p. For algebraic functions of one variable this local group is commutative; as a consequence the local monodromy group of an algebraic function which is expressed by means of sums and differences of integer functions *must be soluble*. But the local monodromy group of the function

$$y^5 + ay + b = 0,$$

near the point $(0, 0)$ is the group $S(5)$ of all permutations of five elements, which is not soluble. This explains the statement of the theorem.

Observe that if the operation of division is allowed, then the above argument no longer holds. Indeed, the division is an operation killing the continuity and its application destroys the locality. In fact, the function y satisfying

$$y^5 + ay + b = 0$$

can be expressed by means of division in terms of a function $g(x)$ of one variable, defined by the equation

$$g^5 + g + x = 0,$$

and of the function of one variable $f(a) = \sqrt[4]{a}$. It is not difficult to see that

$$y(a, b) = g(b/\sqrt[4]{a^5})\sqrt[4]{a}.$$

the study of compositions of algebraic functions, were afterwards applied by Vassiliev and Smale [21],[22],[23] to the problem of finding the topologically necessary number of ramifications in the numeric algorithms for the approximate calculation of roots of polynomials. (The number of ramifications is of the order of n for a polynomial of degree n).(*Translator's note.*)

A.13 Functions of several complex variables representable by quadratures and generalized quadratures

The multi-dimensional case is more complicated than the one-dimensional case. We have to reformulate the basic definitions and, in particular, to change slightly the definition of representability of functions by quadratures and by generalized quadratures. In this section we give a new formulation of the problem.

Suppose there have been fixed a class of basic functions and a set of allowed operations. Is a given function (being, for instance, the solution of a given algebraic or differential equation, or the result of one of the other allowed operations) representable in terms of the basic functions by means of the allowed operations? First of all, we are interested in exactly this problem but we give to it a slightly different meaning. We consider the distinct *single-valued branches* of a multi-valued function as single-valued functions on different domains: we consider also every multi-valued function as the set of its single-valued branches. We apply the allowed operations (such as the arithmetic operations or the composition) only to the single-valued branches on different domains. Since our functions are analytic, it suffices to consider as domains only small neighbourhoods of points. The problem now is the following: *is it possible to express a given germ of a function at a given point in terms of the germs of the basic functions by means of the allowed operations?* Of course, here the answer depends on the choice of the single-valued germ of the multi-valued function at that point. However, it happens that (for the class of basic functions we are interested in) either the representation sought does not exist for any germ of the single-valued function at any point, or, on the contrary, all germs of the given multi-valued function are expressed by the same representation at almost all points. In the former case we say that *no branches of the given multi-valued function can be expressed in terms of the branches of the basic functions by means of the allowed operations*; in the latter case we say that this representation *exists.*

First of all, observe the difference between this formulation of the problem and that of the problem expounded in §A.1. *For analytic functions of a single variable there exists amongst the allowed operations, in fact, the operation of analytic continuation.*

Consider the following example. Let f_1 be an analytic function, defined in a domain U of the plane $\mathbb{C}^1$, which cannot be continued beyond the boundaries of a domain U, and let f_2 be the analytic function in the domain U defined by the equation $f_2 = -f_1$. According to the definition in §A.1, the zero function is representable in the form $f_1 + f_2$ *for all the values of the argument.* From this new point of view the equation $f_1 + f_2 = 0$ is satisfied only inside the domain U, not outside it. Previously we were not interested in the existence of a *unique domain* in which all required properties should hold on the single-valued branches of the multi-valued function: a result of an operation could hold on a domain, another result in another domain on the analytic continuations of the functions obtained. For the $\mathbb{S}$-functions of a single variable one can obtain the needed topological limitations even with this extended notion of the operation on analytic multi-valued functions. For functions of several variables we can no longer use this extended notion and we must adopt a new formulation with some restrictions, which can seem less (but which is perhaps more) natural.

Let us start by giving precise definitions. Fix the standard space $\mathbb{C}^n$ with coordinates system $x_1, \ldots, x_n$.

DEFINITION. 1) *The germ of a function φ at a point $a \in \mathbb{C}^n$ can be expressed in terms of the germs of the functions $f_1, \ldots, f_n$ at the point a by means of the integration* if it satisfies the equation $d\varphi = \alpha$, where $\alpha = f_1 dx_1 + \cdots + f_n dx_n$. For the germs of the given functions $f_1, \ldots, f_n$ the germ φ exists if and only if the 1-form α is closed. The germ φ is thus defined up to additive constants.

2) *The germ of a function φ at the point $a \in \mathbb{C}^n$ can be expressed in terms of the germs of the functions $f_1, \ldots, f_n$ at the point a by means of the exponential and of integrations* if it satisfies the equation $d\varphi = \alpha\phi$, where $\alpha = f_1 dx_1 + \cdots + f_n dx_n$). For the germs of the given functions $f_1, \ldots, f_n$ the germ φ exists if and only if the 1-form α is closed. The germ φ is thus defined up to multiplicative constants.

3) *The germ of a function y at a point $a \in \mathbb{C}^n$ can be expressed in terms of the germs of the functions $f_0, \ldots, f_k$ at the point a by means of a solution of an algebraic equation* if the germ f_0 does not vanish and satisfies the equation

$$f_0 y^k + f_1 y^{k-1} + \cdots + f_k = 0.$$

DEFINITION. 1) *The class of germs of functions in $\mathbb{C}^n$ representable by quadratures* (over the field of the constants) is defined by the following choice: the germs of the basic functions are the germs of the constant functions (at every point of the space $\mathbb{C}^n$); the allowed operations are the arithmetic operations, the integration, and raising to the power of the integral.

2) *The class of germs of functions in $\mathbb{C}^n$ representable by generalized quadratures* (over the field of the constants) is defined by the following choice: the germs of the basic functions are the germs of the constant functions (at every point of the space $\mathbb{C}^n$); the allowed operations are the arithmetic operations, the integration, the raising to the power of the integral, and the solution of algebraic equations.

Notice that the above definitions can be translated almost literally in the case of abstract differential fields, provided with n commutative differentiation operations $\partial/\partial x_1, \ldots, \partial/\partial x_n$. In such a generalized form these definitions are owed to Kolchin.

Now consider the class of the germs of functions representable by quadratures and by generalized quadratures in the spaces $\mathbb{C}^n$ of any dimension $n \geq 1$. Repeating the Liouville argument (cf., Theorem 1 in §A.2), it is not difficult to prove that *the class of the germs of functions of several variables representable by quadratures and by generalized quadratures contains the germs of the rational functions of several variables and the germs of all elementary basic functions; these classes of germs are closed with respect to the composition.* (The closure with respect to the composition of a class of germs of functions representable by quadratures means the following: if $f_1, \ldots, f_m$ are germs of functions representable by quadratures at a point $a \in \mathbb{C}^n$ and g is a germ of a function representable by quadratures at the point $b \in \mathbb{C}^m$, where $b = (f_1(a), \ldots, f_m(a))$, then the germ $g(f_1, \ldots, f_m)$ at the point $a \in \mathbb{C}^n$ is the germ of a function representable by quadratures).

A.14 $\mathcal{SC}$-germs

Does there exist a class of germs of functions of several variables sufficiently wide (containing the germs of functions representable by generalized quadratures, the germs of entire functions of several variables and closed with respect to the natural operations such as the composition) for which the monodromy group is defined? In this section we define the class of $\mathcal{SC}$-germs and we state the theorem about the closure of this class with respect to the natural operations: this gives an affirmative answer to the question posed. I discovered the class of $\mathcal{SC}$-germs relatively recently: up to that time I believed the answer were negative.

In the case of functions of a single variable it was useful to introduce the class of the $\mathcal{S}$-functions. Let us start with a direct generalization of the class of $\mathcal{S}$-functions to the multi-dimensional case.

A subspace $A \subset M$ in a connected k-dimensional analytic manifold M is said to be *thin* if there exists a countable set of open subsets $U_i \subset M$ and a countable set of analytic subspaces $A_i \subset U_i$ in these open subsets such that $A \subseteq \bigcup A_i$. An analytic multi-valued function on the manifold M is called an $\mathcal{S}$-*function* if the set of its singular points is thin. Let us make this definition more precise.

Two regular germs f_a and g_b, given at the points a and b of the manifold M, are said to be *equivalent* if the germ g_b is obtained by a regular continuation of the germ f_a along some curve. Every germ g_b, equivalent to the germ f_a, is also called a *regular* germ of the analytic multi-valued function f generated by the germ f_a.

A point $b \in M$ is said to be *singular for the germ* f_a if there exists a curve $\gamma\,[0,1] \to M$, $\gamma(0) = a$, $\gamma(1) = b$, such that the germ f_a cannot be regularly continued along this curve, but for every t, $0 \leq t < 1$, this germ can be continued along the shortened curve $\gamma\,[0,t] \to M$. It is easy to see that the sets of singular points for equivalent germs coincide.

A regular germ is said to be an $\mathcal{S}$-*germ* if the set of its singular points is thin. An analytic multi-valued function is called an $\mathcal{S}$-function if every one of its regular germs is an $\mathcal{S}$-germ.

REMARK. For functions of one complex variable we have already given two definitions of $\mathcal{S}$-functions. The first one is the above definition, the second one is given by the theorem in §A.5. These definitions obviously coincide.

For $\mathcal{S}$-functions of several variables the notions of *monodromy group* and of *monodromy pair* are automatically translated.

Let us clarify way the multi-dimensional case is more complicated than the one-dimensional case.

Imagine the following situation. Let $f(x,y)$ be a multi-valued analytic function of two variables with a set A of branch points, where $A \subset \mathbb{C}^2$ is an analytic curve on the complex plane. It can happen that at one of the points $a \in A$ there exists an analytic germ f_a of the multi-valued analytic function f (by the definition of the set A of branch points, at the point a not every germ of the function f exists; yet some of them can exist). Now let $g_1(t)$ and $g_2(t)$ be two analytic functions of the complex variable

t given by the mapping of the complex line $\mathbb{C}$ on the complex plane $\mathbb{C}^2$, such that the image of the line $\mathbb{C}$ is contained in A, i.e., $(g_1(t), g_2(t)) \in A$ for every $t \in \mathbb{C}$. Let b be the pre-image of the point a under this mapping, i.e., $a = (g_1(b), g_2(b))$. What can we say about the multi-valued analytic function, on the complex line, generated by the germ of $f(g_1, g_2)$ at the point b, obtained as the result of the composition of the germs of the rational function g_1, g_2 at the point b and of the germ of the function f at the point a? It is clear that the analytic properties of this function depend essentially on the continuation of the germ f_a along the singular curve A.

Nothing like this may happen under the composition of functions of a single variable. Indeed, the set of singularities of an $\mathcal{S}$-function of one variable consists of isolated points. If the image of the complex space under an analytic mapping g is entirely contained in the set of the singular points of a function f, then the function g is a constant. It is evident that if the function g is constant, after having defined f on its set of singular points, the function $f(g)$ turns out to to be constant, too.

In the one-dimensional case, for our purpose it suffices to study the values of an analytic multi-valued function only in the complement of its singular points. In the multi-dimensional case we have to study the possibility of continuing those germs of functions which meet along their set of singularities (if, of course, the germ of the function is defined in an arbitrary point of the set of singularities). It happens that the germs of multi-valued functions sometimes are automatically continued along their set of singularities [24]: this thus allows us to pass all difficulties.

An important role is played by the following definition:

DEFINITION. A germ f_a of an analytic function at the point a of the space $\mathbb{C}^n$ is called an $\mathcal{SC}$-*germ* if the following condition is satisfied. For every connected complex analytic manifold M, every analytic mapping $G M \to \mathbb{C}^n$, and every pre-image c of the point a, $G(c) = a$, there exists a thin subset $A \subset M$ such that for every curve $\gamma\,[0,1] \to M$, beginning at the point c, $\gamma(0) = c$, and having no intersection with the set A, except, at most, at the initial point, i.e., $\gamma(t) \notin A$ for $t > 0$, the germ f_a can be analytically continued along the curve $G \circ \gamma\,[0,1] \to \mathbb{C}^n$.

PROPOSITION. *If the set of singular points of an $\mathcal{S}$-function is an analytic set then every germ of this function is an $\mathcal{SC}$-germ.*

This proposition follows directly from the results set out in [24].

It is evident that every $\mathcal{SC}$-germ is the germ of an $\mathcal{S}$-function. For the

$\mathcal{SC}$-germs the notions of monodromy group and of monodromy pair are thus well defined.

In the sequel we will need the notion of a holonomic system of linear differential equations. A system of N linear differential equations $L_j(y) = 0$, $j = 1, \ldots, N$,

$$L_j(y) = \sum a^j_{i_1,\ldots,i_n} \frac{\partial^{i_1+\cdots+i_n} y}{\partial x_1^{i_1} \ldots \partial x_n^{i_n}} = 0$$

for the unknown function y, whose coefficients $a^j_{i_1,\ldots,i_n}$ are analytic functions of n complex variables $x_1, \ldots, x_n$, is said to be *holonomic* if the space of its solutions has a finite dimension.

THEOREM ON THE CLOSURE OF THE CLASS OF $\mathcal{SC}$-GERMS. *The class of the $\mathcal{SC}$-germs in $\mathbb{C}^n$ is closed with respect to the following operations:*

- *1) differentiation, i.e., if f is an $\mathcal{SC}$-germ at a point $a \in \mathbb{C}^n$, then for every $i = 1, \ldots, n$ the germs of the partial derivatives $\partial f/\partial x_i$ are also $\mathcal{SC}$-germs at the point a;*

- *2) integration, i.e., if $df = f_1 dx_1 + \cdots + f_n dx_n$, where $f_1, \ldots, f_n$ are $\mathcal{SC}$-germs at a point $a \in \mathbb{C}^n$, then also f is an $\mathcal{SC}$-germ at the point a;*

- *3) composition with the $\mathcal{SC}$-germs of m variables, i.e., if $f_1, \ldots, f_m$ are $\mathcal{SC}$-germs at a point $a \in \mathbb{C}^n$ and g is an $\mathcal{SC}$-germ at the point $(f_1(a), \ldots, f_m(a))$ in the space $\mathbb{C}^m$, then $g(f_1, \ldots, f_m)$ is an $\mathcal{SC}$-germ at the point a as well;*

- *4) solutions of algebraic equations, i.e., if $f_0, \ldots, f_k$ are $\mathcal{SC}$-germs at a point $a \in \mathbb{C}^n$, the germ f_0 is not zero and the germ y satisfies the equation $f_0 y^k + f_1 y^{k-1} + \cdots + f_k = 0$, then the germ y is also an $\mathcal{SC}$-germ at the point a;*

- *5) solutions of holonomic systems of linear differential equations, i.e., if the germ of a function y at a point $a \in \mathbb{C}^n$ satisfies the holonomic system of N linear differential equations*

$$L_j(y) = \sum a^j_{i_1,\ldots,i_n} \frac{\partial^{i_1+\cdots+i_n} y}{\partial x_1^{i_1} \ldots \partial x_n^{i_n}} = 0,$$

all of whose coefficients $a^j_{i_1,\ldots,i_n}$ are $\mathcal{SC}$-germs at the point a, then y is also an $\mathcal{SC}$-germ at the point a.

COROLLARY. *If a germ of a function f can be obtained from the germs of single-valued $\mathcal{S}$-functions having an analytic set of singular points by means of integrations, of differentiations, meromorphic operations, compositions, solutions of algebraic equations, and solutions of holonomic systems of linear differential equations, then this germ of f is an $\mathcal{SC}$-germ. In particular, a germ which is not an $\mathcal{SC}$-germ cannot be represented by generalized quadratures.*

A.15 Topological obstructions for the representability by quadratures of functions of several variables

This section is dedicated to the topological obstructions for the representability by quadratures and by generalized quadratures of functions of several complex variables. These obstructions are analogous to those holding for functions of one variable considered in §§A.7–A.9.

THEOREM 1. *The class of all $\mathcal{SC}$-germs in $\mathbb{C}^n$ having a soluble monodromy group, is closed with respect to the operations of integration and of differentiation. Moreover, this class is closed with respect to the composition with the $\mathcal{SC}$-germs of m variables ($m \geq 1$) having soluble monodromy groups.*

RESULT ON QUADRATURES. *The monodromy group of any germ of a function f representable by quadratures is soluble. Moreover, every germ of a function, representable by the germs of single-valued $\mathcal{S}$-functions having an analytic set of singular points is also soluble by means of integrations, of differentiations, and compositions.*

COROLLARY. *If the monodromy group of the algebraic equation*

$$y^k + r_1 y^{k-1} + \cdots + r_k = 0$$

in which the r_is are rational functions of n variables is not soluble, then any germ of its solutions not only is not representable by radicals, but cannot be represented in terms of the germs of single-valued $\mathcal{S}$-functions having an analytic set of singular points by means of integrations, of differentiations, and compositions.

This corollary represents the strongest version of the Abel theorem.

THEOREM 2. *The class of all* $\mathcal{SC}$*-germs in* $\mathbb{C}^n$ *having an almost soluble monodromy pair is closed with respect to the operations of integration, differentiation, and solution of algebraic equations. Moreover, this class is closed with respect to the composition with the* $\mathcal{SC}$*-germs of* m *variables* $(m \geq 1)$ *having an almost soluble monodromy pair.*

RESULT ON GENERALIZED QUADRATURES. *The monodromy pair of a germ of a function* f*, representable by generalized quadratures, is almost soluble. Moreover, the monodromy pair of every germ of a function* f *representable in terms of the germs of single-valued* $\mathcal{S}$*-functions having an analytic set of singular points by means of integrations, differentiations, compositions, and solutions of algebraic equations is also almost soluble.*

A.16 Topological obstruction for the solvability of the holonomic systems of linear differential equations

A.16.1 The monodromy group of a holonomic system of linear differential equations

Consider a holonomic system of N differential equations $L_j(y) = 0$, $j = 1, \ldots, N$,

$$L_j(y) = \sum a^j_{i_1,\ldots,i_n} \frac{\partial^{i_1+\cdots+i_n} y}{\partial x_1^{i_1} \ldots \partial x_n^{i_n}} = 0,$$

where y is the unknown function, and the coefficients $a^j_{i_1,\ldots,i_n}$ are rational functions of the n complex variables $x_1, \ldots, x_n$.

One knows that for any holonomic system there exists a singular algebraic surface Σ in the space $\mathbb{C}^n$ that have the following properties. Every solution of the system can be analytically continued along an arbitrary curve avoiding the hypersurface Σ. Let V be the finite-dimensional space of the solutions of a holonomic system near a point x_0 which lies outside the hypersurface Σ. Consider an arbitrary curve $\gamma(t)$ in the space $\mathbb{C}^n$ with the initial point x_0, not crossing the hypersurface Σ. The solutions of the system can be analytically continued along the curve γ, remaining solutions of the system. Consequently to every curve γ of this type there corresponds a linear transformation M_γ of the space of solutions V in itself. The totality of the linear transformations M_γ corresponding to all curves γ forms a group, which is called the *monodromy group of the holonomic system.*

Kolchin generalized the Picard–Vessiot theory to the case of holonomic systems of differential equations. From the Kolchin theory we obtain two corollaries concerning the solvability by quadratures of the holonomic systems of differential equations. As in the one-dimensional case, a holonomic system is said to be *regular* if approaching the singular set Σ and infinity its solutions grow at most as some power.

THEOREM 1. *A regular holonomic system of linear differential equations is soluble by quadratures and by generalized quadrature if its monodromy group is, respectively, soluble and almost soluble.*

Kolchin's theory proves at the same time two results.

- 1) *If the monodromy group of a regular holonomic system of linear differential equations is soluble (almost soluble) then this system is solvable by quadratures (by generalized quadratures).*

- 2) *If the monodromy group of a regular holonomic system of linear differential equations is not soluble (is not almost soluble) then this system is not solvable by quadratures (by generalized quadratures).*

Our theorem makes the result (2) stronger.

THEOREM 2. *If the monodromy group of a holonomic system of equations of linear differential equations is not soluble (is not almost soluble), then every germ of almost all solutions of this system cannot be expressed in terms of the germs of single-valued* $\mathcal{S}$*-functions having an analytic set of singular points by means of compositions, meromorphic operations, integrations and differentiations (by means of compositions, meromorphic operations, integrations, differentiations and solutions of algebraic equations).*

A.16.2 Holonomic systems of equations of linear differential equations with small coefficients

Consider a system of linear differential equations completely integrable of the following form

$$dy = Ay, \tag{A.17}$$

where $y = y_1, \ldots, y_n$ is the unknown vector function and A is an $(n \times n)$ matrix consisting of differential 1-forms with rational coefficients in the space $\mathbb{C}^n$, satisfying the condition of complete integrability $dA + A \wedge A = 0$ and having the following form:

$$A = \sum_{i=1}^{k} A_i \frac{dl_i}{l_i},$$

where the A_is are constant matrices and the l_is are linear non-homogeneous functions in $\mathbb{C}^n$.

If the matrices A_i can be put at the same time into triangular form, then the system (A.17), as every completely integrable triangular system, is solvable by quadratures. There undoubtedly exist integrable non-triangular systems. However, when the matrices A_i are sufficiently small such systems do not exist. More precisely, we have proved the following theorem.

THEOREM 3. *A completely integrable non-triangular system (A.17), with the moduli of the matrices A_i sufficiently small, is strictly not solvable, i.e., its solution cannot be represented even through the germs of all single-valued $\mathcal{S}$-functions, having an analytic set of singular points, by means of compositions, meromorphic operations, integrations, differentiations, and solutions of algebraic equations.*

The proof of this theorem uses a multi-dimensional variation of the Lappo-Danilevskij theorem [28].

Bibliography

[1] Ritt J., *Integration in Finite Terms. Liouville's Theory of Elementary Methods*, N. Y. Columbia Univ. Press, 1948.

[2] Kaplanskij I., *Vvedenie v differentsial'nuyu algebru.* MIR, 1959.

[3] Singer M.F., *Formal Solutions of Differential Equations.* J. Symbolic computation, **10**, 1990, 59–94.

[4] Khovanskii A., *The Representability of Algebroidal Functions as Compositions of Analytic Functions and One-variable Algebroidal Functions.* Funct. Anal. and Appl. **4**, 2, 1970, 74–79.

[5] Khovanskii A., *On Superpositions of Holomorphic Functions with Radicals.* (in Russian) Uspehi Matem. Nauk, **26**, 2, 1971, 213–214.

[6] Khovanskii A., *On the Representability of Functions by Quadratures.* (in Russian) Uspehi Matem. Nauk, **26**, 4, 1971, 251–252.

[7] Khovanskii A., *Riemann surfaces of functions representable by quadratures.* Reports of VI All-Union Topological Conference, Tiblisi, 1972.

[8] Khovanskii A., *The representability of functions by quadratures.* PhD thesis, Moscow, 1973.

[9] Khovanskii A., Ilyashenko Yu., *Galois Theory of Systems of Fuchs-type Differential Equations with Small Coefficients.* Preprint IPM, **117**, 1974.

[10] Khovanskii A., *Topological Obstructions for Representability of Functions by Quadratures.* Journal of dynamical and control systems, **1**, 1, 1995, 99–132.

[11] Klein F., *Vorlesugen uber die hypergeometriche funktion.* Berlin, 1933.

[12] Lappo-Danilevskij I.A., *Primenenie funktsij ot matrits k teorii lineinyh sistem obyknovennyh differentsial'nyh uravnenij.* GITTL, 1957.

[13] Arnold V.I., *Cohomology classes of algebraic functions invariant under Tschirnausen transformations.* Funct. Anal. and Appl., **4**, 1, 1970, 74–75.

[14] Lin V. Ya., *Superpositions of Algebraic Functions.* Funct. Anal. and Appl., **10**, 1, 1976, 32–38.

[15] Vitushkin A.G., *K tridnadtsatoj problem Gilberta.* DAN SSSR, 1954. **95**, 4, 1954, 701–74 (also: Nauka, Moskow, 1969, 163–170).

[16] Vitushkin A.G., *Nekotorye otsenki iz teorii tabulirovaniya.* DAN SSSR, **114**, 1957, 923–926.

[17] Kolmogorov A.N., *Otsenki chisla elementov ϵ-setej v razlichnyh funktsionalnykh klassah i ih primenenie k voprosu o predstavimosti funktsij neskolkih peremennyh superpositsyami funktsij men'shego chisla peremennyh.* UMN, **10**, 1, 1955, 192 -194.

[18] Kolmogorov A.N., *O predstavlenii neprerivnyh funkcij neskol'kikh peremennyh superpositsyami neprerivnyh funktsij men'shego chisla peremennyh.* DAN SSSR, **108**, 2, 1956, 179–182.

[19] Arnol'd V.I., *O funktsiyah treh peremennyh.* DAN SSSR, **4**, 1957, 679–681.

[20] Kolmogorov A.N., *O predstavlenii neprerivnyh funktsij neskol'kih peremennyh v vide superpositsij neprerivnyh funkcij odnogo peremennogo.* DAN SSSR, **114**, 5, 1957, 953–956.

[21] Vassiliev V.A., *Braid Group Cohomology and Algorithm Complexity.* Funct. Anal. and Appl., **22**, 3, 1988, 182–190.

[22] Smale S., *On the Topology of Algorithms. I.* J. Complexity, **4**, 4, 1987, 81–89.

[23] Vassiliev V.A., *Complements of Discriminants of Smooth Maps, Topology and Applications.* Transl. of Math. Monographs, **98**, AMS Providence, 1994.

[24] Khovanskii A., *On the Continuability of Multivalued Analytic Functions to an Analytic Subset.* Funct. Anal. and Appl. **35**, 1, 2001, 52–60.

[25] Khovanskii A., *A multidimensional Topological Version of Galois Theory.* Proceeding of International Conference "Monodromy in Geometry and Differential Equations ", 25–30 June, Moscow, 2001.

[26] Khovanskii A., *On the Monodromy of a Multivalued Function Along Its Ramification Locus.* Funct. Anal. and Appl. **37**, 2, 2003, 134–141.

[27] Khovanskii A., *Multidimensional Results on Nonrepersentability of Functions by Quadratures.* Funct. Anal. and Appl. **37**, 4, 2003, 141–152.

[28] Leksin V.P., *O zadache Rimana–Gilberta dlya analiticheskih semeistv predstavlenii.* Matematicheskie zametki **50**, 2, 1991, 89–97.

Appendix (V.I. Arnold)

The topological arguments for the different types of non-solvability (of equations by radicals, of integrals by elementary functions, of differential equations by quadratures etc.) can be expressed in terms of very precise questions.

Consider, for example, the problem of the integration of algebraic functions (i.e, the search for Abelian integrals). The question in this example consists in knowing *whether these integrals and their inverse functions* (for example, the elliptic sinus) *are topologically equivalent to elementary functions.*

The topological equivalence of two mappings f and g of M to N means the existence of a homeomorphisms h of M into M and a homeomorphism k of N into N which transform f into g, i.e, such that $f(h(x)) \equiv k(g(x))$. The absence amongst the objects of a class B of an object topologically equivalent to the objects of a class A means the topological non-reducibility of A to B (of the Abelian integrals and of the elliptic functions to the elementary functions, etc.).

In my lectures in the years 1963–1964 I expounded the topological proof of all three aforementioned versions of the Abel problems (cf., [6], [7]), but the book extracted from my lectures contains only the topological proof of the non-solvability by radicals of the algebraic equations of degree 5.

Since I am unable to give references of the unpublished proofs of the two remaining enunciations of the topological non-solvability, here it is convenient to call them 'problems'. I underline only that, although the non-solvability of every problem follows from the non-solvability in the topological sense explained above, the assertion about the topological non-solvability is stronger and it is not proved by means of calculations, showing the non-existence of the formulae sought.

This topological point of view of the non-solvability is also applied to many other problems; for example, to the results by Newton [4], [5], to the problem of the Lyapounov stability of the equilibrium states of a dynamical system [1], to the problem of the topological classification of the singular points of differential equations [1], to the question of the 16th Hilbert problem about the limit cycles (cf., [2]), to the topological formulation of the 13th Hilbert problem about the composition of complex algebraic functions [3], and to the problem of the non-existence of first integrals in Hamiltonian systems (as a consequence of the presence of many closed isolated curves) [8].

These applications of the idea of the topological non-solvability, coming out of the range of the Abel theory, can be found in the papers [1]–[8].

1. Arnold V.I., *Algebraic Unsolvability of the Problem of Lyapounov Stability and the Problem of Topological Classification of Singular Points of an Analytic System of Differential Equations.*, Funct. Anal. and Appl., **4**, 3, 1970, 173–180.

2. Arnol'd V.I., Olejnik O.A., *Topologiya deistvitel'nyh algebraicheskih mnogoobrazij.* Vestnik MGU, ser. 1, matem - mekhan., **6**, 1979, 7–17

3. Arnol'd V.I., *Superpozitsii.* In the book: A.N. Kolmogorov, Izbrannye trudy, matematika i mehanika, Nauka, 1985, 444–451.

4. Arnol'd V.I., *Topologicheskoe dokazatel'stvo transtsendentnosti abelevykh integralov v "Matematicheskih nachalah natural'noj filosofii" N'yutona.* Istoriko-Matematicheskie Issledovaniya, **31**, 1989, 7–17.

5. Arnold V.I., Vassiliev V.A., *Newton's "Principia" read 300 years later.* Notices Amer. Math. Soc. **36**, 9, 1989, 1148–1154 [Appendix: **37**, 2, 144].

6. Arnold V.I., *Problèmes solubles et problèmes irrésolubles analytiques et géométriques.* In: Passion des Formes. Dynamique Qualitative, Sémiophysique et Intelligibilité. Dédié à R. Thom, ENS Éditions: Fontenay-St. Cloud, 1994, 411–417.

7. Petrovskij I.G., *Hilbert's Topological Problems, and Modern Mathematics.* Russ. Math. Surv. **57**, 4, 2002, 833-845.

8. Arnol'd V.I., *O nekotoryh zadachah teorii dinamicheskih sistem.* In: V.I. Arnol'd — Izbrannoe 60, Fazis, 1997, 533–551 (also: Topol. Methods Nonlinear Anal., **4**, 2, 1994, 209–225).

Index

Printed in the USA
CPSIA information can be obtained
at www.ICGtesting.com
CBHW080816140924
14513CB00005B/51